Fundamentals of Machine Learning for Life Sciences

Machine learning touches our lives in quiet and remarkable ways. It helps doctors detect illnesses sooner by recognizing subtle patterns in scans and helping them make sense of medical data with speed, judgment, and care. It helps us care for our fields and forests by tracking changes that unfold over time. It helps us study the weather with a memory far longer than our own, and notices small signs that equipment may soon need attention. And when we find ourselves in unfamiliar places, it helps us translate words, find our bearings, and discover new corners to explore. And yet, even as it becomes part of daily life, its inner workings can still feel distant when you first encounter them. *Fundamentals of Machine Learning* aims to bring it within reach.

This book offers a clear and steady introduction to how machines learn from data. It explains how models begin to understand, decide, improve, and sometimes falter. Ideas build gradually, one upon another, supported by real examples and datasets in R. The focus is insight over jargon, clarity over complexity. As these ideas become familiar, they also hold the promise of supporting the works of scientists, engineers, and students—by opening new pathways of exploration.

Inside the Book

- How learning algorithms discover patterns
- Supervised, unsupervised, and other ways machines learn
- Regression, decision trees, neural networks, and more
- Working with data and understanding results
- Ethics, fairness, and responsible use

Warm, practical, and approachable, *Fundamentals of Machine Learning* encourages readers to build confidence step by step, to make sense of new ideas in their own time, and to discover how understanding machine learning can enrich the way we work, learn, and see the world.

Machine Learning for Life Sciences

Series editor Ankur Awadhiya

Fundamentals of Machine Learning for Life Sciences: A Simple, Step-by-Step Approach with Examples in R

Authored by Ankur Awadhiya

Fundamentals of Machine Learning for Life Sciences

A Simple, Step-by-Step Approach with Examples in R

Ankur Awadhiya

CRC Press
Taylor & Francis Group
Boca Raton London New York

CRC Press is an imprint of the
Taylor & Francis Group, an **informa** business

Designed cover image: Ankur Awadhiya
First edition published 2026
by CRC Press
2385 NW Executive Center Drive, Suite 320, Boca Raton, FL 33431

and by CRC Press
4 Park Square, Milton Park, Abingdon, Oxon, OX14 4RN

CRC Press is an imprint of Taylor & Francis Group, LLC

ISBN: 978-1-041-27556-5 (hbk)
ISBN: 978-1-041-27559-6 (pbk)
ISBN: 978-1-003-75498-5 (ebk)

DOI: 10.1201/9781003754985

Typeset in Alegreya-Regular font
by KnowledgeWorks Global Ltd.

To Ushi

Contents

List of figures xiii
List of tables xv
Preface........ xvii
About the author........ xix

PART I Introduction

Chapter 1 What is machine learning? 3

1.1 Relationship between machine learning and artificial intelligence 5
1.2 Relationship between machine learning and deep learning . 6
1.3 A brief history of machine learning 8
1.4 What do we do in machine learning? 9
1.5 Approaches to machine learning........ 10
1.6 Applications of machine learning........ 23
1.7 Limitations of machine learning........ 24
1.8 Ethics of machine learning........ 26
1.9 How is this book organized? 28
1.10 Chapter summary 29

Chapter 2 Basic setup........ 31

2.1 R: An introduction........ 32
2.2 The RStudio integrated development environment........ 33
2.3 Packages in R 36
2.4 Basic input and output operations 39
2.5 Fundamentals of the tidyverse 42
2.6 Chapter summary 45

Chapter 3 Machine learning in practice....47

3.1 Prediction....47

3.2 Predictor and outcome variables....47

3.3 Model....48

3.4 Model parameters....49

3.5 Model fitting....49

3.6 Hyperparameter tuning....50

3.7 Validation and cross-validation....52

3.8 Data....53

3.9 Training and testing....54

3.10 Confusion matrix....55

3.11 A note on metrics....56

3.12 Tidymodels....62

3.13 Chapter summary....65

PART II Machine Learning Algorithms

Chapter 4 Linear regression....69

4.1 What is linear regression?....69

4.2 Galton: the father of linear regression....72

4.3 The Galton families dataset....73

4.4 Observing simple linear regression in Galton families dataset....79

4.5 Estimating regression parameters....83

4.6 The `lm()` function....93

4.7 Tidymodels example....99

4.8 When to use linear regression, and when not to....104

4.9 Chapter summary....105

Chapter 5 Polynomial regression 107

5.1 What is polynomial regression? 107

5.2 The Boston dataset 108

5.3 Polynomial regression 114

5.4 Tidymodels example 122

5.5 Hyperparameter tuning 127

5.6 When to use polynomial regression, and when not to 136

5.7 Chapter summary 136

Chapter 6 Logistic regression 139

6.1 What is logistic regression? 140

6.2 The Diabetes dataset 142

6.3 Logistic regression 148

6.4 Tidymodels example 150

6.5 Balancing dataset 157

6.6 Receiver operating characteristic 169

6.7 When to use logistic regression, and when not to 171

6.8 Chapter summary 172

Chapter 7 K-nearest neighbors 175

7.1 What is KNN? 175

7.2 The Iris dataset 177

7.3 Plotting the dataset 181

7.4 Tidymodels example 186

7.5 Hyperparameter tuning 192

7.6 Example: Identifying handwriting 201

7.7 When to use K-Nearest Neighbors, and when not to 217

7.8 Chapter summary 219

Chapter 8 Support vector machines 221

8.1 What is SVM? 221
8.2 The Palmer Penguin dataset........ 223
8.3 Tidymodels example 232
8.4 When to use support vector machines, and when not to .. 238
8.5 Chapter summary 240

Chapter 9 Decision trees and forests 241

9.1 What is a decision tree? 243
9.2 Constructing a decision tree 247
9.3 Pruning a decision tree........ 248
9.4 The breast cancer dataset 248
9.5 Tidymodels example 251
9.6 Instability of decision trees 266
9.7 Exemplifying instability in a decision tree 268
9.8 From trees to (random) forests 272
9.9 Boosted trees........ 278
9.10 Comparison: identifying handwriting 283
9.11 When to use tree-based models, and when not to........ 290
9.12 Chapter summary 292

Chapter 10 Neural networks and deep learning 295

10.1 Neurons: building blocks of the nervous system 295
10.2 From neurons to computer neural networks 298
10.3 Deep learning networks........ 300
10.4 Timeline of significant events 302
10.5 Training a neural network........ 304
10.6 The Boston dataset 308
10.7 Tidymodels example 311
10.8 Hyperparameter tuning........ 315

10.9 When to use neural networks, and when not to 325
10.10 Chapter summary 326

PART III Epilogue

Chapter 11 What to do next? 329

11.1 Exploring stacks 329
11.2 Exploring datasets in other fields 332
11.3 Exploring frontiers of machine learning 332
11.4 Gaining a knowledge of maths 335
11.5 Exploring tidymodels 335
11.6 Exploring other packages 335
11.7 Exploring other languages 336
11.8 Scrutinizing nuances like regularization 336
11.9 Learning interpretation of results 339
11.10 Writing your own machine learning code(s) 340
11.11 Combining strategies 342
11.12 Chapter summary 342

References 345

Index 349

List of Figures

1.1 Relationship between artificial intelligence and machine learning. 6
1.2 Relationship between machine learning and deep learning. 7
1.3 A schematic of supervised learning. 12
1.4 A schematic of unsupervised learning. 14
1.5 A schematic of reinforcement learning. 16
1.6 A schematic of self-supervised learning. 19
1.7 A schematic of transfer learning. 21

2.1 A view of the RStudio IDE. 34

4.1 A plot of number of hours studying versus test scores. 70
4.2 A schematic of linear regression. 71

5.1 A schematic of polynomial regression process. 122

6.1 A plot of number of study hours versus the probability of passing. 139
6.2 A plot of the logistic function. 141
6.3 A schematic of the logistic regression process. 148

7.1 A schematic of variations in iris flowers. 177
7.2 Pixel values for the digit 5, taken from a sample in the MNIST dataset. 204

8.1 A schematic representation of the workings of a support vector machine. 222

9.1 A schematic of a decision tree. 242
9.2 A schematic of bootstrapping. 273
9.3 Pixel values for the digit 5 in MNIST dataset. 284
9.4 A simple decision tree to decide whether or not to use a decision tree. 290

10.1 A schematic of the structure of a neuron.. 297

10.2 A schematic of a simple deep learning network with one input, two hidden, and one output layers.. 301

11.1 A schematic of the process of stacking. 330

List of Tables

1.1 Positive and negative reinforcement and punishment in operant conditioning. 17

3.1 An example of a binary confusion matrix. 55

3.2 Confusion matrix of a model designed to classify emails. 55

3.3 An example of a non-binary confusion matrix. 56

Preface

Machine learning is changing our world at an incredible pace. Everywhere we look—healthcare, conservation, business, art—its influence is growing. Yet for many beginners, the field can feel intimidating. The terms sound unfamiliar, the concepts seem abstract, and the math can appear overwhelming. If you've ever felt that way, you're not alone. And that's exactly why I wrote this book.

When I first began exploring machine learning, I kept running into two kinds of books. One type treated the subject like a mysterious black box: *"Don't worry about how it works—just use it."* The other type dove so deep into equations that it became hard to connect any of it to real-world applications. It often felt like trying to understand an entire forest by looking at a single leaf—or trying to grasp a leaf by staring at the whole forest.

What I really needed was a guide that spoke plainly, connected ideas clearly, and showed me exactly how to apply what I learned. When I couldn't find that resource, I cobbled together my understanding from many different books, courses, and programming languages.

Years later, when I began teaching machine learning, I went searching again for that perfect book—one I could confidently recommend to my students. But I found the same gap: highly theoretical texts on one side, purely practical ones on the other. So I decided to write the book I had once wished for.

This book is designed to make machine learning approachable and enjoyable. You'll start with a gentle introduction—no intimidating jargon—and gradually move toward working with real algorithms and real datasets. Every concept is paired with code you can run yourself, so you can experiment freely and watch the ideas come to life. As you progress, you'll notice familiar patterns across different algorithms, helping everything click into place and building your confidence chapter by chapter.

My goal is not just to teach you machine learning—it's to help you *enjoy* the learning process. To spark curiosity. To make you feel empowered. To show that machine learning isn't magic—it's a set of tools you can understand, use, and even master.

Whether you're an aspiring data scientist, a working professional expanding your skill set, or simply someone curious about this transformative field, I hope this book inspires you to explore boldly and learn with joy.

Happy learning!

Dr. Ankur Awadhiya, IFS

About the author

Dr. Ankur Awadhiya (b. 1987) is an officer of the Indian Forest Service and an engineer by training. He received his B.Tech and Ph.D. from IIT Kanpur, earned his AIGNFA (Master's in Forestry) from the Indira Gandhi National Forest Academy, and completed an Honours Post Graduate Diploma in Advanced Wildlife Management from the Wildlife Institute of India.

He has pursued advanced specializations in Artificial Intelligence from Oxford, Mind and Decision Making from Cambridge, and Applied Machine Learning from Johns Hopkins. He also holds professional certifications in Data Science, Computer Science, and Artificial Intelligence from Harvard, along with a MicroMasters in Data, Economics, and Development Policy from MIT.

As Deputy Director at the Forest Survey of India, Dr. Awadhiya works at the intersection of technology and ecology, integrating Artificial Intelligence and Machine Learning into natural resource monitoring and management. His interests span biodiversity conservation, forest management, and scientific research. A prolific writer, he has over fifty publications and holds two patents.

Dr. Awadhiya is the recipient of several prestigious honors, including the NTSE Scholarship, KVPY Fellowship, Vanyaprani Sanrakshan Puraskar, Shri P. Srinivas Memorial Prize, K. P. Sagriya Shreshta Vaniki Puraskar, and the S. K. Seth Prize. Beyond his professional life, he enjoys teaching, photography, painting, filmmaking, and creative writing.

Part I

Introduction

1 What is machine learning?

Machine learning is a scientific discipline that enables computers to acquire knowledge and skills through data analysis, without explicit programming. By identifying patterns within extensive datasets, today's computers can make predictions, categorize information, and execute tasks autonomously, just as humans learn from experience.

Envision teaching a computer to identify a cat from amongst dogs and cats. The classical approach is to explicitly define the criteria for categorizing a cat, often through the characteristics that define it—such as its weight, coat color, or fur density. But that requires us to identify these features, measure them out in the world, and set limits on what would define a cat—essentially too much human work. At the same time, it is also possible that despite our best efforts, we do a pretty shoddy job nevertheless, for hard thresholds often do not work well for most natural things. Nature does not usually fall into neat boxes, and whenever we try to force it into strict cutoffs, exceptions appear almost immediately. Suppose that, on the basis of a few examples of dogs and cats that we have seen around us, we decide to keep a simple rule: "if the weight is above 10 kilograms, it must be a dog, and if it is below 10 kilograms, it must be a cat". While this may work for a few individuals in the locality on the basis of which the threshold was decided, it will certainly fail when we consider individuals from other localities that are (quite) different from the ones we have around us. A Chihuahua, which is a very small dog, often weighs only two or three kilograms, so by this rule it would get classified as a cat. On the other hand, a Maine Coon, which is one of the largest breeds of domestic cats, can weigh 10 to 12 kilograms and would wrongly get classified as a dog.

If we instead tried to use size in centimeters as the deciding criterion, the problem would still remain. For example, we might say: "if the height at the shoulder is above 40 cm, it must be a dog, and if it is less than 40 cm, it must be a cat". But this too fails quickly. A Dachshund or a Chihuahua, both of which are dogs, can be much smaller than 40 cm. On the other hand, a large cat such as a Maine Coon, or even a tall domestic cat standing upright, can cross that threshold. At the same time, for the largest dog breeds such as Great Danes or Irish Wolfhounds, a 40 cm limit appears far too low and almost arbitrary, since these dogs can easily be twice that size. So even size measured in centimeters, though it might seem clearer than weight, will not provide a reliable dividing line.

If we tried to set a rule on the basis of tails, we might say: "dogs usually have curved, pointy tails, while cats have straight, cylindrical tails". This may look reasonable at first, but it will again quickly break down. Some dog breeds, like greyhounds, have straight tails that look very similar to those of cats. Other dogs, such as Afghan

Hounds, have long, hairy plume-like tails that do not fall into either of these categories at all. On the other hand, certain cats, like the Japanese Bobtail, can have short and curved tails. Many pet dogs also have their tails cropped, which means they will no longer fit on either side of the rule.

The same problem will also come up with ears. We might try to say: "dogs have floppy ears, while cats have pointed ears". But again, the rule will fail. German Shepherds and Dobermans are dogs with pointed ears, while many cats, like Scottish Folds, have ears that fold down and look floppy. Other breeds complicate the matter even further — Bulldogs and Pugs have small, button-like ears that are neither truly floppy nor truly pointy, so they do not fit into either side of the rule. And of course, some dogs have their ears cropped by humans, changing their natural shape altogether. So even when we try to fix rules for ears, they may work for a few animals in one area, but as soon as we look at more breeds, the rules break down.

Color too is misleading. We might assume dogs are mostly brown and cats are black and white. That may seem to hold if we only look at a few individuals around us, but it fails once we look more widely. There are white dogs, spotted dogs, and black dogs, just as there are brown cats, ginger cats, and cats with many other color combinations. So a color-based rule would also fall apart very quickly.

The same is also true with any hard combination of these characteristics. After sometime, the whole exercise will appear futile—we need to find another approach. But in real life, when we look at an animal, we can easily tell whether it is a dog or a cat. How do we humans do it? There is a crux—we do not rely on any one feature, or any hard threshold, or even a combination of hard thresholds for several features. To decide whether an individual is a dog or a cat, we don't make measurements. What we and our brains do is to notice many things together: the approximate size, the approximate weight, the approximate shape of the ears, the approximate type of tail, the face, the color, and even the behavior of the animal. Each of these gives us some evidence about whether the animal is a dog or a cat. None of them alone is perfect, and the "fuzzy" logic we employ is not about stacking them together as more hard rules. Instead, our fuzzy logic works differently: it accepts that each feature gives only partial information, and it allows us to express this information in degrees. For example, based on approximate weight, an animal may have a 70% chance of being a dog. But when we also look at its ears, the chance may go up to 80% if they match common dog traits, or it may drop to 40% if they look more like a cat's. Looking at the tail and color might adjust the likelihood further, either strengthening or weakening the case. In this way, our fuzzy logic—or soft evidence integration— does not make a yes-or-no decision at each step, but weighs all the evidence together to arrive at a balanced conclusion. And we "learn" these features by observing the plethora of cats and dogs throughout our lifetimes. Whenever we wrongly classify an animal, we take note of that exception and update our logic.

This captures the intuition behind machine learning. Rather than providing the computer with precise rules, we provide it with an abundance of images and data depicting various animals, including cats. The computer scrutinizes these images and endeavors to identify patterns, in an attempt to identify cats (and other animals). Just as a child exposed to few animals often makes mistakes, the computer also errs with little data. But just as our accuracy increases with exposure to more and more animals in our lifetime, the computer's accuracy also increases with more and more data. Through a series of analyses, the computer slowly develops the capacity to discern distinctive features of cats, such as ears, whiskers, and furry bodies. Over time, its proficiency in distinguishing cats from other animals progressively improves.

To sum up, machine learning enables computers to acquire the capacity to learn in a manner akin to humans (or other animals) by identifying patterns within data. This approach has facilitated the realization of numerous remarkable accomplishments by computers, including comprehension of spoken language, autonomous vehicle operation, and even weather forecasting, all without the need to specify the specifics of how to do these in the code. This also opens up huge avenues of applying similar logic to our own fields, without the need to explicitly learn lots of programming!

1.1 RELATIONSHIP BETWEEN MACHINE LEARNING AND ARTIFICIAL INTELLIGENCE

Machine learning and artificial intelligence (AI) are closely interconnected concepts within the domain of computer science. They are frequently discussed in tandem due to their mutually reinforcing nature in the development of intelligent systems. To comprehend their relationship, we may consider AI as the broader objective of designing machines capable of executing tasks that typically demand human cognition, such as comprehending speech or recognizing visual stimuli.

Within this framework, machine learning is a specific approach to achieving this objective. It is a method within AI—a subset of sorts—that focuses on developing algorithms that enable computers to learn from and make decisions based on data [Fig. 1.1]. Rather than being programmed with explicit instructions for every task, machine learning systems are trained on vast amounts of data. As they process this data, they identify patterns and make predictions or decisions based on their acquired knowledge.

In simpler terms, if AI represents the broad vision of creating intelligent machines, machine learning is one of the key techniques employed to realize that vision. Machine learning enables AI systems to enhance their performance over time by acquiring knowledge from novel data and experiences, akin to how humans learn through practice and experience.

Historically, AI preceded machine learning (ML) (Nilsson, 2009). Notably, there are several AI methods that are not based on machine learning. These non–machine learning AI approaches encompass **Symbolic AI** (also referred to as GOFAI, "Good old fashioned artificial intelligence") which employs symbols and logical reasoning to execute intelligent tasks based on explicit rules and relationships. GOFAI is frequently utilized in expert systems for decision-making purposes. **Rule-based sys-**

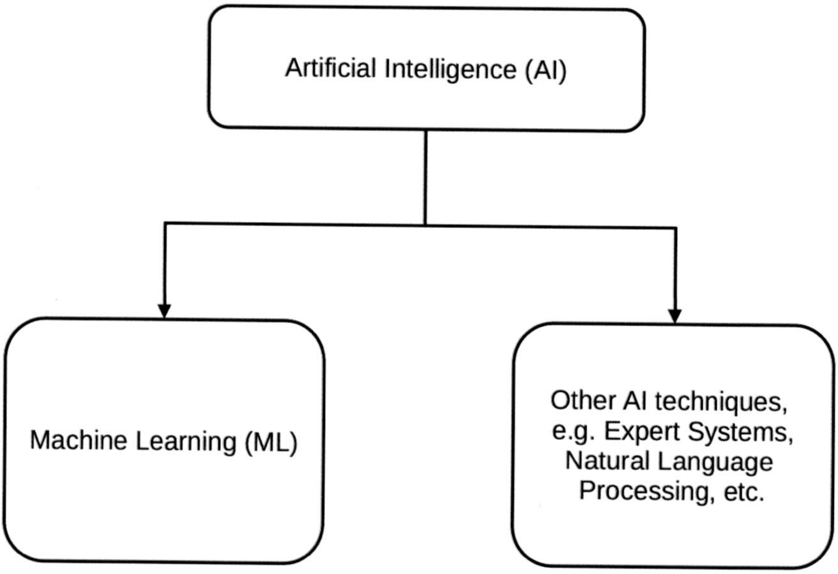

FIGURE 1.1: Relationship between artificial intelligence and machine learning.

tems, another non–machine learning approach, rely on if-then rules to derive actions based on specific conditions. These systems are commonly employed in automated decision systems and games. **Expert systems** replicate a human expert's decision-making by utilizing a knowledge base and predefined rules. Consequently, they are suitable for domains that necessitate expert interpretation. **Bayesian networks**, which are probabilistic graphical models, employ Bayes' theorem to draw inferences and predictions, particularly in uncertain environments. Finally, **fuzzy logic** extends traditional Boolean logic to accommodate partial truths, enabling degrees of truth between entirely true and entirely false; this application is often seen in control systems, such as those utilized in washing machines.

1.2 RELATIONSHIP BETWEEN MACHINE LEARNING AND DEEP LEARNING

Machine learning and deep learning are related fields within the broader domain of AI. Deep learning is a specialized technique or subset within machine learning [Fig. 1.2]. To comprehend their relationship, we may begin by considering machine learning as a general approach where computers are trained to identify patterns and make decisions based on data. This training entails algorithms that modify their behavior in response to the examples they process, enabling the system to enhance its performance over time.

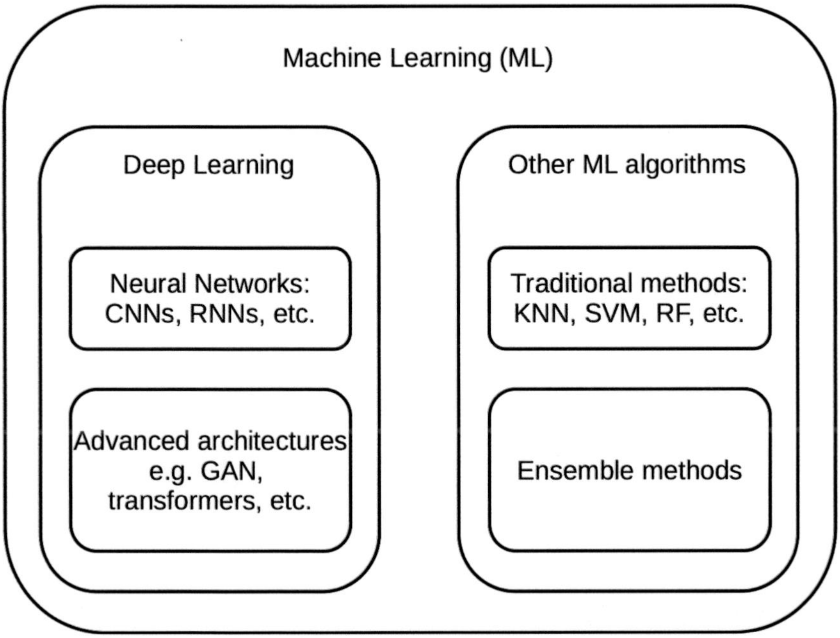

FIGURE 1.2: Relationship between machine learning and deep learning.

Deep learning is a more advanced subset of machine learning that employs neural networks inspired by the human brain's functional architecture. We will explore and design our own neural networks in a later chapter of this book. Briefly, neural networks consist of multiple layers of interconnected nodes, or "neurons"—with each layer responsible for extracting distinct features from the input data. For instance, in image recognition, the initial layer may discern rudimentary edges, while subsequent layers may identify more intricate shapes and patterns. The outputs of one layer serve as the inputs for the subsequent layer, enabling the execution of complex computations.

The primary distinction between deep learning and other conventional neural network methodologies lies in the depth of these neural networks. While conventional machine learning models may possess a limited number of layers and necessitate manual feature engineering,[1] deep learning models employ a significantly greater

[1]Feature engineering is the process of selecting, modifying, or creating variables (features) from raw data to enhance the performance of machine learning models. An illustration of this process is the development of a novel feature for a machine learning model by computing the interaction term between two existing features, such as multiplying "age" and "income" to capture the combined impact.

number of layers and possess the capability to automatically discern crucial features from raw data without the need for manual feature engineering. This inherent ability to learn intricate patterns directly from data without explicit human intervention renders deep learning particularly potent for tasks such as image and speech recognition.

1.3 A BRIEF HISTORY OF MACHINE LEARNING

The history of machine learning can be traced back to the mid-20th century, evolving from foundational theories in statistics and AI into a pivotal field in contemporary technology. The origins of machine learning can be found in early research on algorithms and statistical methods developed to emulate human learning processes and decision-making capabilities.

The advent of computers in the 1950s catalyzed the flourishing of the concept of machine learning. Pioneers such as Alan Turing proposed groundbreaking ideas about AI and learning machines (Turing, 1950). Turing's seminal work laid the groundwork for the notion that machines could be meticulously designed to acquire knowledge through experience. In 1956, the Dartmouth Conference (McCarthy et al., 1956) witnessed the formalization of the field's research agenda with the coining of the term "artificial intelligence". Initially, AI research focused on symbolic methods, which employed symbols and logical reasoning to execute intelligent tasks based on explicit rules and relationships. Additionally, rule-based systems, relying on if-then rules to derive actions based on specific conditions, emerged as pivotal attempts to construct machines capable of performing tasks that typically require human intelligence.

During the 1960s and 1970s, the field of machine learning experienced significant progress with the development of its initial learning algorithms. Notably, Arthur Samuel's pioneering work in game-playing algorithms, particularly in checkers, exemplified one of the earliest practical applications of machine learning (Samuel, 1959). Samuel's algorithms employed trial-and-error methodologies to enhance their performance, providing compelling evidence that machines possessed the capacity to learn from data. Additionally, Frank Rosenblatt's perceptron, an early form of neural network, laid the groundwork for future advancements in neural network technology (Rosenblatt, 1958).

Subsequently, the field experienced a period of stagnation characterized by the "AI winter" during the late 1970s and 1980s (Crevier, 1993). This was primarily attributed to the limitations of early computational resources and the overly optimistic projections regarding the capabilities of AI, which resulted in a hype followed by disillusionment and a decrease in funding. Despite these obstacles, substantial progress was made in the advancement of fundamental algorithms and theoretical frameworks.

The resurgence of machine learning in the late 1990s and early 2000s was propelled by several key factors: the exponential growth in computational power, the accessibility of vast datasets, and advancements in algorithms. A notable milestone during this era was the development of support vector machines (SVMs) by Vladimir Vapnik and collaborators (Vapnik, 1995). SVMs offered a robust classification method and became a cornerstone in the field of machine learning.

The 2010s witnessed a paradigm shift in the field of machine learning, with deep learning emerging as a dominant sub field. Deep learning, characterized by neural networks with multiple layers, enables automatic learning of hierarchical features from data. This era was propelled by the advent of vast datasets and high-performance graphics processing units (GPUs), which facilitated the training of intricate models. Prominent accomplishments during this period included the development of deep convolutional neural networks (CNN). AlexNet, a deep CNN, achieved victory in the ImageNet competition in 2012, demonstrating its effectiveness in image classification tasks (Krizhevsky et al., 2012).

Machine learning has since become an integral component of numerous applications, including natural language processing (NLP), computer vision, autonomous vehicles, and personalized recommendations. The advent of big data[2] and advancements in hardware have further propelled the field, making machine learning an essential component of modern technology.

Today, machine learning continues to evolve through ongoing research into more sophisticated algorithms, ethical considerations, and applications in diverse fields. As we advance, the integration of machine learning into various aspects of daily life underscores its profound impact on technology and society, marking its transformation from a theoretical concept to a transformative force in the modern world.

1.4 WHAT DO WE DO IN MACHINE LEARNING?

In the realm of machine learning, we develop systems capable of acquiring knowledge[3] from data[4] and subsequently making decisions or predictions based on that acquired knowledge. This process entails several distinct stages:

1. Data Collection: The initial phase involves gathering pertinent data that the system will utilize for learning purposes. This data can encompass a

[2] Big data refers to exceptionally large and complex datasets that cannot be efficiently processed using traditional data management tools. It is commonly characterized by the three Vs: high volume of data, high velocity of data generation and processing, and high variety of data types and sources.

[3] Knowledge encompasses the information, data, and rules acquired, deduced, or stored by a system to effectively address challenges and make informed decisions.

[4] Data, singular datum, comprises raw, unprocessed facts and figures from diverse sources. It can be quantitative or qualitative, manifesting as numerical values, textual content, or measurements. Unorganized data appears as an assemblage of numbers and letters without discernible patterns or inherent significance. In this way, data lacks inherent meaning or context, merely representing distinct elements. With meaning and context added, we move from data to information.

diverse range of formats, including images, textual data, numerical values, and user interactions.

2. Data Preparation: Following data collection, we prepare the data for further processing. This entails handling missing values, normalizing the data, and potentially transforming it into a format that is more amenable to the machine learning algorithm's functionality.
3. Model Selection: In this stage, we select a suitable machine learning model or algorithm that aligns with the specific task at hand. Different models excel at various types of problems—some are more adept at classification (sorting data into predefined categories), while others excel at regression (predicting numerical values).
4. Model Training: During this phase, the prepared data is fed into the selected model. The model processes the data and adjusts its internal parameters to identify patterns or relationships. This process is referred to as training or model fitting.
5. Evaluation: Post training, the model's performance is assessed using a separate dataset that has not been previously utilized. This evaluation facilitates the determination of the model's predictive accuracy on novel, unseen data.
6. Model Tuning: Based on the evaluation results, necessary adjustments may be made to the model by modifying its parameters or exploring alternative algorithms to enhance its performance.
7. Deployment: Once the model demonstrates satisfactory performance, it can be deployed for real-time prediction or decision-making based on incoming data. For instance, it can be integrated into a recommendation system, a spam-filtering mechanism, or a predictive maintenance tool.
8. Monitoring and Maintenance: Even after deployment, the model requires continuous monitoring to ensure its continued effectiveness. Occasionally, retraining or updates may be necessary as new data or patterns emerge.

We will further elaborate on these stages with practical examples in the later chapters of this book.

1.5 APPROACHES TO MACHINE LEARNING

As outlined in the preceding section's third step, various machine learning models excel in disparate domains and have specific data requirements. Consequently, the field of machine learning encompasses a diverse range of approaches, each tailored to specific problem types and data characteristics. Below are some of the primary machine learning methodologies.

1.5.1 SUPERVISED LEARNING

Envision a child acquiring the skill of sorting various colored balls into designated buckets. Initially, her parents introduce her to a red ball, elucidating that it should be placed in the corresponding red bucket, and a blue ball, indicating its appropriate placement in the blue bucket. Subsequently, they meticulously observe and provide guidance as she practices with additional balls. Whenever she places a ball in the correct bucket, she receives encouragement and applause. However, when she inadvertently places a ball in the incorrect bucket, her parents gently correct her error through nudges. Through their encouragement and mentorship, she gradually develops the confidence to independently sort the balls by color. This process that we just discussed is supervised learning, where a child (or a computer) is trained to execute tasks by learning from examples (of what is correct and what is not) provided by a human instructor or guide.

Supervised learning entails training a model (analogous to the child in the preceding illustration) on a labeled dataset, wherein each instance in the training set is associated with an output label (a red ball in a red bucket as correct; a blue ball in a blue bucket as correct, and other instances as incorrect, respectively). The model acquires the ability to predict the output based on the input data and subsequently employs this acquired knowledge to execute tasks autonomously (as in the child independently sorting the balls by color in the preceding example).

Common tasks employing supervised learning include classification (e.g., identifying plant species based on characteristics like flower color, height, leaf shape, etc.) and regression (e.g., predicting house prices based on built-up area, number of floors, number of windows, waterfront, etc.) [Fig. 1.3]. The goal is to find patterns that map inputs to the correct outputs based on exemplar data. Needless to say, this approach necessitates a substantial amount of labeled data, wherein the labeling has already been performed by humans, thereby providing several input and output values prior to its application. In essence, a significant amount of human labor is required before employing supervised learning.

Supervised learning begins with the creation of a labeled dataset, wherein each instance within the dataset is accompanied by a predetermined outcome. For instance, in a supervised learning task aimed at distinguishing emails as either "spam" or "not spam," the dataset would encompass a collection of emails, each accompanied by labels indicating whether the email is spam or not; the labeling process having been undertaken by humans.

During training, the model processes the labeled data and attempts to establish a correlation between the input features and the corresponding output labels. For instance, the presence of certain words, phrases, or language may indicate that an email is a spam, while the absence of these may indicate that the email is not a spam. At each iteration, the model may predict certain outcomes correctly, and certain others incorrectly. The process of training entails adjusting the internal parameters

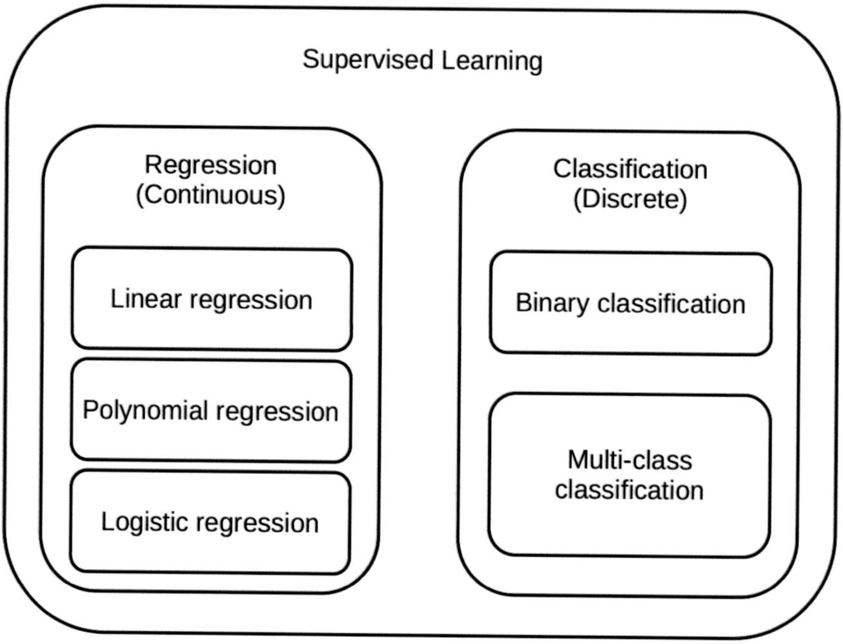

FIGURE 1.3: A schematic of supervised learning.

of the model to minimize any disparity between the model's predictions and the actual labels present in the training data. The disparity is typically quantified by utilizing a loss function,[5] which encapsulates the extent to which the model's predictions align with the true labels. As the model trains, it iteratively refines its parameters to enhance its performance and minimize discrepancies between its predictions and the actual labels.

In a classification task, the model progressively acquires the ability to discern among distinct categories, drawing upon the inherent characteristics of the input data. In a regression task, the model progressively refines a function that optimally predicts a numerical value for the output.

Once training is concluded, the model is evaluated using a distinct dataset known as the test set, which was not utilized during training. This evaluation assesses the model's ability to generalize to novel, unforeseen instances and guarantees its performance beyond the training data.

[5]A loss function is a mathematical tool employed to assess the extent to which a model's predictions align with the actual outcomes. During training, the disparity between the predicted and actual values reduces, represented by lowering of loss function results, and signifying an enhancement of the model's accuracy.

Supervised learning is a widely employed technique in diverse applications, including image recognition, speech recognition, and predictive analytics. It becomes particularly valuable when we possess a well-defined set of input-output pairs, enabling the model to acquire knowledge from these examples and subsequently make precise predictions or classifications in analogous scenarios.

Supervised learning, while highly effective for numerous applications, also faces several limitations that can restrict its applicability and efficiency. One significant limitation is its reliance on labeled data, which necessitates substantial time and effort to acquire, particularly for large datasets. This process can be impractical or costly. Furthermore, the quality of the output is intrinsically linked to the quality of the data; therefore, any biases or inaccuracies present in the training data are likely to be reflected in the model's predictions. This dependency can lead to overfitting, where the model performs exceptionally well on training data but fails to generalize effectively to unseen data.

Additionally, supervised learning assumes that the test data will be drawn from the same distribution as the training data, which may not always be the case. This limitation can be problematic in dynamic environments where the underlying data distribution may evolve over time, diminishing the model's effectiveness and requiring retraining. Lastly, supervised learning models often require substantial computational resources and time for training, particularly for complex applications involving deep learning models, thereby limiting their scalability in practical real-world scenarios.

1.5.2 UNSUPERVISED LEARNING

Unsupervised learning can be likened to attempting to solve a puzzle without any prior knowledge of the desired outcome. Imagine a child with a large box of jigsaw puzzle pieces, tasked with constructing the complete image without an illustration of the final picture. The child begins by categorizing the pieces based on patterns or edges, gradually determining their potential fit with one another. In other words, she attempts to find which pieces "belong together" due to some shared characteristic—maybe common colors, common patterns, or fitting edges. Similarly, in unsupervised learning, computers process vast datasets independently, without a guide as to the desired outcomes, identifying and utilizing patterns within data to group them into classes that "belong together".

In this approach, the model is trained on data without explicit labels. The objective is to discern underlying structures or patterns within the data. Techniques such as clustering (grouping similar data points together) and dimensionality reduction (reducing the number of variables) are commonly employed [Fig. 1.4].

For instance, consider a scenario where we possess satellite imagery of a vast region, but no information regarding the types of land cover present, such as forests, lakes,

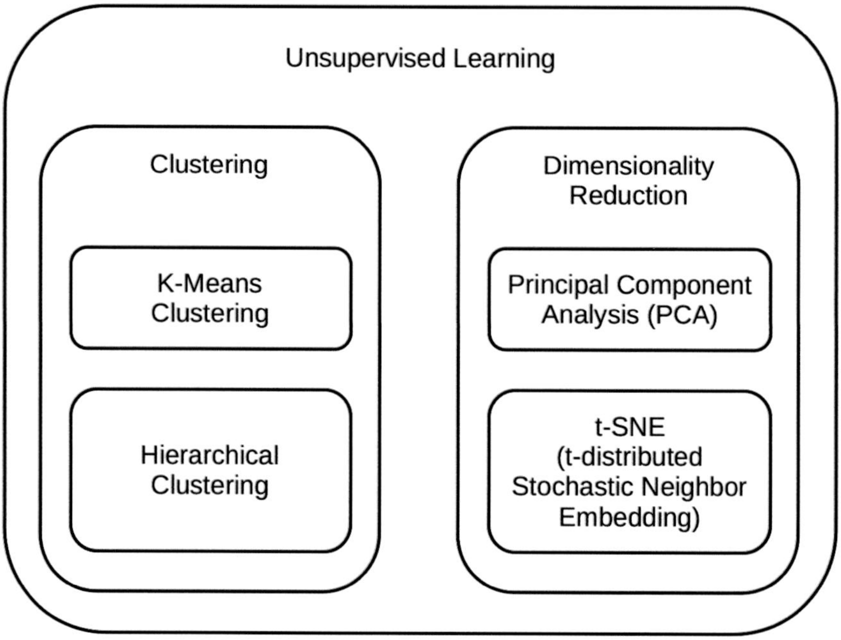

FIGURE 1.4: A schematic of unsupervised learning.

or urban areas. In other words, we have data in the form of color, texture, and shape, but we do not know which areas correspond to which land cover types.

We can use unsupervised learning on such data to find natural groups or clusters in the data. In a common way to do this, called clustering, the model looks at each pixel or region in the image and groups similar ones together based on their common features. For example, the model might find a group of pixels that have similar colors and textures and put them in a category. These clusters could be different types of land cover, like dense forests, open water, or cities. But the model does not know which cluster corresponds to which land cover class, just that these are groups that have more in common amongst each other than they have with other groups.

Another technique in unsupervised learning that we might apply is dimensionality reduction. This involves reducing the complexity of the data while preserving the most significant features. For instance, different shades of green may be grouped into a single hue, reducing the number of variables to be considered while highlighting informative features. This simplification may make it easier to visualize and interpret the data.

In both instances, unsupervised learning facilitates the analysis of satellite imagery without any prior knowledge of land cover types. By identifying patterns and groupings within the data, the model can discern distinct categories of land cover, offering insights into the distribution and variety of land cover prevalent in the region. Once we've got these categories sorted, we can just visit or observe a few ground points in each one to get a feel for what they actually represent on the ground.

Unsupervised learning presents several key advantages. One of its primary benefits is its ability to operate with unlabeled datasets. This characteristic mitigates the extensive and often costly process of labeling data, which is a significant advantage when dealing with large datasets. Unsupervised learning excels in uncovering concealed patterns and inherent structures within the data without the need for any human intervention. It facilitates the clustering of data into meaningful segments, potentially revealing insights that may not be immediately apparent. Such capabilities are invaluable in domains such as NLP, image recognition, and anomaly detection, where comprehending intricate and high-dimensional data is paramount.

However, unsupervised learning also presents several notable limitations. One significant challenge is its reliance on the quality and structure of the input data, which can lead to unreliable outputs if the data is noisy or poorly represented. Additionally, determining the appropriate number of clusters or pattern categories (e.g., dividing an image into two categories: forests and non-forests, or into multiple categories: forests, bare lands, grasslands, water bodies, built-up areas, etc.) is inherently difficult without prior labeled information. Often, this requires trial and error or domain knowledge to achieve optimal results. Furthermore, unsupervised algorithms, such as clustering techniques, may not provide clear interpretations of the patterns discovered, complicating the process of translating findings into actionable insights and requiring further investigations. Additionally, these methods can be computationally intensive, particularly with large datasets, posing practical challenges in terms of efficiency and scalability.

1.5.3 SEMI-SUPERVISED LEARNING

This approach strikes a balance between supervised and unsupervised learning. It incorporates a modest quantity of labeled data and a substantial amount of unlabeled data. The model draws upon the labeled data to enhance its comprehension of the underlying structure of the unlabeled data, thereby augmenting its performance and diminishing the necessity for extensive amounts of labeled data. This approach proves particularly advantageous when labeling data requires significant financial or temporal constraints.

1.5.4 REINFORCEMENT LEARNING

In reinforcement learning, an agent (human, animal, or machine) acquires decision-making skills through interaction with an environment. The agent receives feedback in the form of rewards or penalties contingent upon its actions and strives to accumulate cumulative rewards over time [Fig. 1.5]. This approach is widely employed in domains such as game playing, robotics, and autonomous vehicles, where the system acquires knowledge through iterative experimentation.

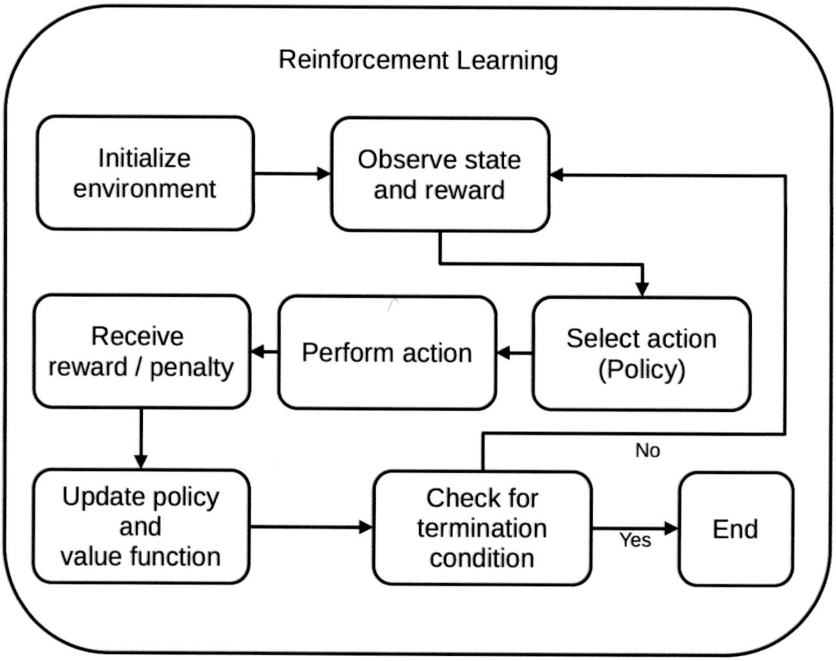

FIGURE 1.5: A schematic of reinforcement learning.

Reinforcement learning is a computational counterpart of the psychological concept of operant conditioning, also known as instrumental conditioning (Skinner, 1938). Operant conditioning is a learning process through which behavior is modified based on the consequences of that behavior. It is akin to teaching a dog a new trick. Consider the scenario of training a dog to sit. Each time the dog sits, the trainer verbalizes "sit" and provides a treat. The treat serves as a reward, increasing the likelihood of the dog sitting again in response to the command due to the association between the action and the desired outcome (i.e., receiving a treat). This process is referred to as positive reinforcement. Conversely, if the dog jumps on the trainer, the trainer may ignore or turn away, in a process known as negative punishment because it withholds attention (a desirable stimulus for the dog) to discourage the jumping behavior. Over time, the dog learns to sit more frequently in anticipation of receiving treats and reduces jumping behavior as it no longer receives attention.

In operant conditioning, feedback is categorized into positive and negative rewards and punishments [Table 1.1]. A positive feedback adds something, while a nega-

TABLE 1.1
Positive and negative reinforcement and punishment in operant conditioning.

Operant Conditioning	Positive Reinforcement	Negative Reinforcement	Positive Punishment	Negative Punishment
Definition	Adding a desirable stimulus to increase behavior	Removing an aversive stimulus to increase behavior	Adding an aversive stimulus to decrease behavior	Removing a desirable stimulus to decrease behavior
Purpose	Increase the likelihood of a behavior occurring	Increase the likelihood of a behavior occurring	Decrease the likelihood of a behavior occurring	Decrease the likelihood of a behavior occurring
Example	Positive Reinforcement: A teacher gives a student praise for doing their homework.	Negative Reinforcement: A manager stops nagging an employee once they complete their work.	Positive Punishment: A child gets a scolding for not cleaning their room.	Negative Punishment: A teenager loses access to their video games for not finishing their chores.
Effect on Behavior	Encourages the behavior to be repeated	Encourages the behavior to be repeated	Discourages the behavior from being repeated	Discourages the behavior from being repeated

tive feedback removes something. A reward encourages the behavior, making the behavior more likely, while a punishment discourages the behavior, making the behavior less likely.

Similarly, in reinforcement learning, a program agent is trained to acquire correct and anticipated behaviors through the utilization of tokens such as game scores. The ultimate objective is for the agent to develop a strategy that maximizes the cumulative reward over time.

To begin, the environment is characterized by various states, each representing a distinct situation or condition [Fig. 1.5]. The agent operates within this environment, making decisions based on its current state. Each action taken results in a transition to a new state, accompanied by a reward or penalty that signifies the immediate advantage or disadvantage of the action. For instance, in a game, moving to a specific position may yield points (reward) or result in the loss of a life (penalty).

The agent's primary objective is to acquire a policy, which serves as a strategic framework for selecting actions that maximize cumulative rewards over time. Initially, the agent may be unaware of the optimal actions, necessitating exploration of diverse options. This exploration facilitates the acquisition of environmental information and elucidates the impact of various actions on rewards and penalties that can later be exploited by the agent to maximize cumulative rewards.

A fundamental aspect of reinforcement learning is the delicate balance between exploration and exploitation. Exploration entails experimenting with novel actions to ascertain their consequences, while exploitation is concerned with selecting actions that have been proven to yield substantial rewards through prior experience. The challenge lies in identifying the best equilibrium between these two approaches to ensure optimal learning outcomes.

As the agent interacts with its environment and receives feedback, it continuously updates its understanding of which actions are most advantageous. This involves estimating the value of various states or actions, frequently including methods that anticipate the future reward. By refining its policy based on these value estimates, the agent enhances its decision-making capabilities over time.

Reinforcement learning enables machines to acquire optimal behaviors through trial and error, eliminating the need for explicit programming. This adaptability stands out as a key advantage, allowing the agent to adjust its strategies in response to evolving environments, making it particularly suitable for tasks of dynamic nature, such as autonomous vehicles, robotics, and financial decision-making (say, based on current stock prices). Moreover, reinforcement learning excels in handling intricate problems that traditional algorithms struggle to address. By learning from its errors and continuously striving to maximize rewards, it fosters self-improvement, driving innovation and enabling machines to discover novel solutions in diverse fields, including healthcare, gaming, and resource management. Consequently, reinforcement learning holds immense significance in advancing AI and machine learning.

At the same time, reinforcement learning also possesses some limitations. One significant challenge is the requirement of substantial amount of data and time for training. Reinforcement learning models acquire knowledge through interaction with their environment, resulting in substantial costs and time-consuming efforts to generate sufficient experiences for learning. Another concern is the inherent instability and lack of robustness[6] of these models. Minor alterations in the environment can lead to drastically different behaviors, hindering their ability to generalize. Furthermore, the trial-and-error approach can occasionally result in unsafe actions, particularly in critical applications such as autonomous driving. Additionally, the decision-making processes employed by reinforcement learning models re-

[6]Robustness in machine learning is the ability of a model to maintain performance under various conditions, such as noise, adversarial attacks, or changes in input data distribution.

main opaque, raising concerns about their trustworthiness in domains where transparency and accountability are paramount.

1.5.5 SELF-SUPERVISED LEARNING

Self-supervised learning is a type of machine learning where the model learns to understand data by generating its own labels from the data itself, rather than relying on external labels or annotations. This approach is particularly useful when labeled data is scarce or expensive to obtain.

In self-supervised learning, the model generates a pretext task, which is a task derived from the data itself and does not necessitate manual labeling [Fig. 1.6]. Subsequently, the model is trained to execute this pretext task, thereby extracting valuable features from the data. These features can subsequently be utilized for other tasks where labeled data may be more readily available.

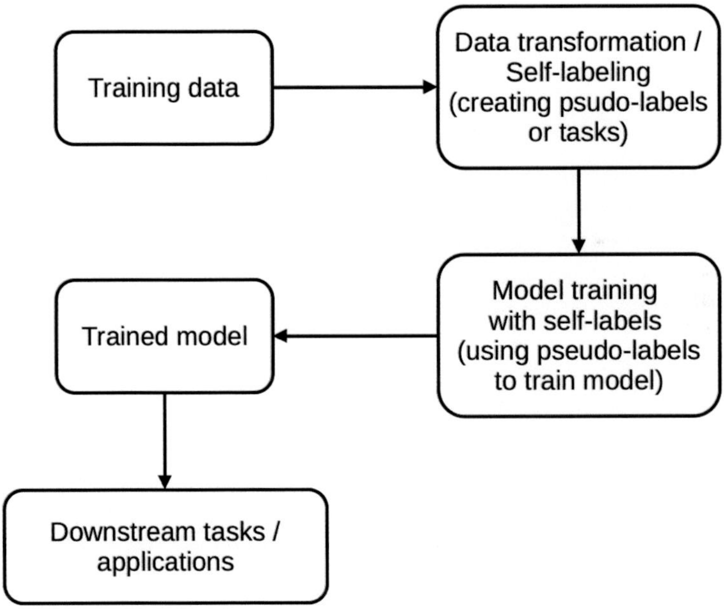

FIGURE 1.6: A schematic of self-supervised learning.

For instance, consider a self-supervised learning task in NLP. Suppose we have a substantial corpus of text, but no labeled data indicating parts of speech or sentiment. A common pretext task might involve masking certain words in a sentence and training the model to predict these masked words based on the surrounding context. By performing this, the model acquires rich representations of language, which can subsequently be fine-tuned for specific tasks such as text classification or sentiment analysis.

Similarly, in the field of computer vision, a self-supervised learning approach may involve training a model to anticipate the absence of specific elements within an image or to discern the relative positions of patches within the same image. For instance, the model could be trained to reconstruct an image where certain components have been deliberately removed. This process facilitates the model's acquisition of crucial visual features and patterns. Subsequently, these learned features can be utilized for more intricate tasks, such as object detection or image classification.

One primary advantage of self-supervised learning is its ability to generate supervisory signals from the inherent structure and properties of the data. This enables the training of models with substantial quantities of unlabeled data, thereby reducing the reliance on labeled datasets. This approach often enhances the efficiency and effectiveness of machine learning model training, and self-supervised learning presents significant promise for utilizing vast amounts of unlabeled data.

However, the method is not without limitations. A notable challenge lies in devising methods to instruct models to perform pretext tasks that demonstrate adaptability across diverse domains and applications, since not all tasks provide models with sufficient information to be effectively utilized in practical scenarios. Additionally, not all organizations possess the resources to acquire an extensive dataset. Self-supervised models may also inadvertently acquire biases from the data they are trained on, posing a concern when deployed to real-world applications. Lastly, training self-supervised models can be computationally intensive, potentially hindering the adoption of this approach by smaller organizations.

1.5.6 TRANSFER LEARNING

Transfer learning entails adapting a pre-trained model (trained on a specific task) to a novel, yet related task [Fig. 1.7]. This approach is advantageous when there is limited data available for the novel task but a well-established model from a related domain. Transfer learning facilitates expedited and more efficient model training by drawing upon existing knowledge. The idea is to leverage the knowledge gained from solving one problem to improve performance on another problem, often with less data and computational resources.

This concept closely resembles transfer learning in humans, where knowledge and skills acquired in one domain are readily applied to related domains. For instance, individuals who have mastered piano playing often find it easier to learn other keyboard instruments such as an organ or a synthesizer. This phenomenon arises from the transfer of fundamental skills, such as finger placement and musical notation, from one instrument to another.

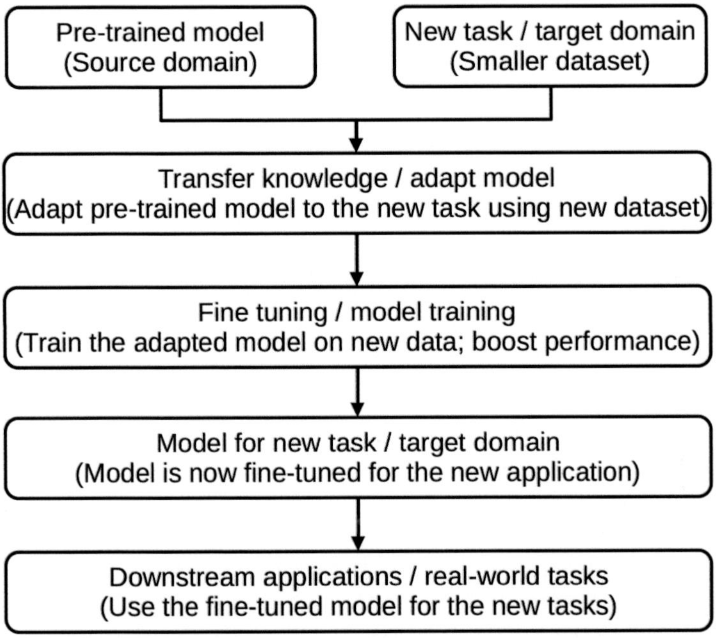

FIGURE 1.7: A schematic of transfer learning.

Similarly, when you are proficient in one language, acquiring a new language can be facilitated and expedited owing to the grammatical rules, vocabulary, and sentence structures mastered previously, which serve as a valuable foundation for comprehending the new language.

In a broader context, problem-solving skills acquired in one domain, such as mathematics, can be beneficial when addressing challenges in a different field, like computer programming. In each of these instances, the underlying knowledge and skills from one area facilitate learning and adaptation in a new, yet related, domain. A comparable process is employed in transfer machine learning.

In the context of transfer learning, the process typically comprises two distinct stages: pre-training and fine-tuning.

Pre-training starts with a model pre-trained on a substantial, generic dataset. For instance, a model may be trained on an extensive corpus of images to discern general features such as edges, textures, and shapes. This model acquires the ability to identify these fundamental patterns across a diverse range of contexts. This initial training furnishes the model with a repertoire of beneficial features and representations that are typically applicable to numerous other problems as well.

Fine-tuning follows, where the pre-trained model is adapted to a specific task. For instance, if we have a model trained on general image recognition, we can fine-tune it for a specialized task, such as identifying specific types of medical images. During fine-tuning, the model is trained further on a smaller, task-specific dataset. The process typically involves adjusting the model's parameters to better fit the new task while retaining the knowledge acquired during the pre-training phase. This approach enables the model to leverage its previously acquired features and adapt them to the novel context.

Transfer learning is particularly advantageous in situations characterized by a scarcity of labeled data for the novel task at hand. By commencing with a model that possesses a comprehensive understanding of general features, we can substantially diminish the data and computational resources necessary to attain satisfactory performance on the new task. This methodology has demonstrated remarkable success across diverse domains, encompassing computer vision, NLP, and speech recognition.

For instance, in computer vision, models pre-trained on extensive datasets such as ImageNet can be effectively fine-tuned for specific tasks, such as detecting defects in manufacturing or classifying various animal species. Similarly, in NLP, models pre-trained on a diverse corpus of texts can be adapted to specific tasks, including sentiment analysis or question answering.

Overall, transfer learning facilitates the efficient utilization of pre-trained models and knowledge, enabling the rapid and effective resolution of novel tasks with reduced computational resources.

Although transfer learning offers promising benefits, it also faces certain limitations. It presumes that the source domain, from which the original model was trained, is sufficiently similar to the target domain where the model will be applied. If this assumption is not met, the transferred knowledge may actually degrade performance, as the model may incorporate irrelevant features. Incidentally, a similar phenomenon also occurs in human learning, known as proactive interference. Proactive interference is a cognitive disruption in human learning that occurs when prior knowledge or established habits hinder the acquisition of new information. For instance, in learning a new language, existing knowledge can degrade the recall of vocabulary or grammatical rules of the new language, leading to confusion and/or mistakes.

Similarly, there is often no guarantee that high-level features learned in one domain are appropriate for another, especially in significantly divergent fields. Furthermore, transfer learning can be computationally intensive and necessitate fine-tuning, which demands expertise, time, and costly computational resources. Finally, ethical and data privacy concerns may arise if the pre-trained models originate from sensitive or proprietary datasets. The pre-trained models may also carry with them the learned biases from previous learning experiences.

1.6 APPLICATIONS OF MACHINE LEARNING

Machine learning applications are utilized across a wide range of domains and industries, enabling them to process substantial amounts of data and generate predictions or decisions without the need for explicit programming. Below are some notable applications of machine learning.

In the healthcare industry, machine learning algorithms facilitate medical imaging analysis, disease diagnosis from diagnostic scans, patient outcome prediction, and personalized treatment planning tailored to individual patient data.

Finance employs machine learning techniques for fraud detection in transactions, credit scoring (for loan approvals), algorithmic trading (to forecast stock market trends), and risk management (by analyzing market data).

Retail and e-commerce enterprises leverage machine learning to develop recommendation systems that personalize product suggestions by analyzing customer preferences and purchase history. These systems also enable demand forecasting and pricing optimization, resulting in enhanced operational efficiency and customer satisfaction, ultimately leading to better customer retention and bottom lines.

Autonomous vehicles rely on machine learning to make real-time decisions, detect and recognize objects in their surroundings, and navigate safely and efficiently.

NLP encompasses a range of applications, including sentiment analysis of customer feedback, language translation, chatbots for customer service automation, and document summarization.

In the manufacturing industry, machine learning is employed for predictive maintenance of equipment, thereby minimizing downtime. Additionally, it facilitates optimization of supply chain logistics and enhances quality control through defect detection in production processes.

Gaming employs machine learning to develop intelligent game characters that adapt their behavior in response to player actions. This leads to personalization of gaming experience and the creation of a more immersive game world with dynamic environments.

In the realm of marketing and advertising, machine learning leverages customer behavior data to segment audiences, personalize marketing campaigns, optimize ad placement and bidding strategies, and forecast customer lifetime value.

Energy and utility companies leverage machine learning algorithms for predictive maintenance of critical infrastructure, including power grids and pipelines. By optimizing energy distribution and accurately predicting energy consumption patterns, these techniques enhance efficiency and reduce wastage.

Environmental Sciences employ machine learning techniques in climate modeling and prediction, analyzing satellite imagery for environmental monitoring, and forecasting natural disasters such as earthquakes and floods.

In this manner, machine learning is revolutionizing several industries by automating repetitive tasks, enhancing decision-making processes, and unlocking valuable insights from data to catalyze innovation and operational efficiency.

1.7 LIMITATIONS OF MACHINE LEARNING

Machine learning has revolutionized numerous industries by enabling computers to acquire knowledge from data and make decisions or predictions without explicit programming. However, despite its transformative potential, machine learning also presents several inherent limitations and challenges.

One of the primary limitations of machine learning is its substantial reliance on data. Machine learning models necessitate substantial quantities of high-quality data for effective training. Insufficient or biased data can result in inaccurate predictions or perpetuate existing biases within the model. This issue is particularly acute in domains where data acquisition is costly, time-consuming, or where data may be inherently or historically biased, including in the fields of healthcare or criminal justice.

Another notable challenge is the inherent lack of interpretability in numerous models, particularly those of a complex nature such as deep neural networks. These models often operate as "black boxes", making it a challenge to comprehend the underlying mechanisms leading to their decisions. This lack of transparency can compromise trust and confidence in the model's outputs, particularly in critical applications where decisions carry substantial consequences and have profound human implications.

Overfitting is also a prevalent issue in machine learning, wherein a model not only captures the underlying patterns in the data but also incorporates noise. Consequently, it exhibits performance on the training data while failing to generalize to novel, unforeseen data. Techniques such as regularization[7] and cross-validation[8] are employed to address overfitting, yet it persists as a continuous challenge, particularly in scenarios involving intricate models and limited datasets.

The computational resources necessary for training and deploying sophisticated machine learning models can also be prohibitive. Models such as deep neural

---

[7]Regularization in machine learning is a technique employed to mitigate overfitting by incorporating a penalty to the loss function, thereby discouraging the development of intricate models. We discuss this in detail in a later chapter of this book.

[8]Cross-validation is a technique employed in machine learning to evaluate the generalizability of a model to an independent data set. It involves partitioning the data into subsets, utilizing some for training purposes and the remainder for testing. Subsequently, these subsets are cyclically utilized for training and testing. We discuss this in detail in a later chapter of this book.

networks typically require powerful hardware such as GPUs (Graphics Processing Units) or TPUs (Tensor Processing Units) to train effectively. This requirement can make machine learning applications costly and impractical for certain organizations or applications, particularly in resource-constrained environments.

Machine learning can have adverse environmental impacts, primarily through increased energy consumption and electronic waste generation. Training intricate machine learning models requires substantial computational power, often originating from energy-intensive data centers. This can result in elevated overall electricity consumption and carbon emissions, particularly when powered with non-renewable resources. Furthermore, the rapid advancements in machine learning technology often shortens device lifespans, quickly making even the best hardware obsolete in no time. This leads to increased electronic waste accumulation if not managed sustainably. The resource-intensive nature of data collection for machine learning further exacerbates environmental footprint of machine learning.

Ethical and legal concerns are increasingly significant in the deployment of machine learning systems. Machine learning algorithms can perpetuate biases present in the training data, leading to unfair outcomes in sensitive areas such as hiring, lending, and criminal justice. Ensuring fairness, accountability, and transparency in machine learning decisions is a complex challenge that requires careful consideration of biases and their implications.

Domain specificity is another limitation of machine learning. Models trained for specific tasks or domains often do not generalize well to new, unrelated tasks without substantial retraining or adaptation. This lack of transferability restricts the scalability and adaptability of machine learning solutions across various applications and industries, and has potential implications in the form of incorrect decisions when improperly applied.

Furthermore, the deployment of machine learning systems frequently necessitates human expertise to interpret outcomes, integrate models into existing systems, and refine algorithms based on practical insights. This reliance on domain experts can impede the adoption and deployment of machine learning solutions, particularly in specialized domains where technical expertise is scarce.

Security and privacy concerns also pose substantial challenges for machine learning. Machine learning models can be susceptible to adversarial attacks, wherein inputs are deliberately manipulated to mislead the model's predictions. Handling sensitive data to train machine learning algorithms also raises concerns about privacy breaches and compliance with data protection regulations, such as the General Data Protection Regulation (GDPR) or the Health Insurance Portability and Accountability Act (HIPAA).

Furthermore, machine learning algorithms are inherently constrained by historical data patterns. Models derive their predictions from past trends and may encounter challenges in novel scenarios or events that deviate substantially from historical data, such as economic downturns or global pandemics. This limitation can compromise the reliability and robustness of machine learning systems in dynamic and unpredictable environments.

Finally, evaluating and comparing machine learning models presents a complex challenge due to the multitude of performance metrics available, such as accuracy, precision, and recall.[9] It is crucial to strike a balance between different objectives, including minimizing false positives and false negatives. This, however, is often a difficult task.

Therefore, while machine learning presents substantial opportunities for innovation and automation across diverse sectors, it is imperative to acknowledge and address its limitations. Mitigating challenges associated with data quality, model interpretability, fairness, scalability, security, and ethical considerations are paramount for advancing the responsible deployment of machine learning in practical applications.

1.8 ETHICS OF MACHINE LEARNING

The ethical implications of machine learning have become increasingly pertinent as its applications extend to diverse domains of society. While machine learning offers unparalleled capabilities for automating tasks, making predictions, and extracting insights from extensive datasets, it also presents intricate ethical dilemmas that necessitate meticulous consideration.

One of the most significant ethical concerns associated with machine learning is its potential to perpetuate or exacerbate biases present in the data utilized for training. Machine learning models draw upon historical data, which may inadvertently reflect societal biases pertaining to race, gender, socioeconomic status, and other factors. If these biases are not proactively identified and mitigated, machine learning systems can generate discriminatory outcomes, thereby reinforcing existing inequalities in domains such as hiring, lending, law enforcement, and healthcare.

Transparency and interpretability are paramount ethical considerations in the field of machine learning. Many advanced machine learning algorithms, particularly deep neural networks, operate as "black boxes", rendering it challenging to comprehend the underlying logic behind their decisions. The absence of transparency can undermine trust and accountability, especially when deployed in high-stakes applications where human lives or fundamental rights are at risk. For instance, it presents a complex and challenging situation for a bank that is unable to provide a client with a clear explanation for the rejection of their loan request. Simply stating that the decision was made due to a machine learning algorithm, one whose logic the banker cannot elucidate, may not suffice and could potentially lead to serious legal consequences.

---

[9] Discussed in detail in Chapter 3.

Privacy is another crucial ethical concern in the field of machine learning. As machine learning systems draw upon substantial datasets for training, there exists a potential risk of inadvertent exposure or misuse of sensitive personal information. Protecting individuals' privacy and adhering to data protection regulations are paramount for the responsible deployment of machine learning solutions, particularly in domains such as healthcare, finance, and others that handle sensitive data.

The issue of consent and control over data also arises. Numerous large language models have been trained on data scraped from the internet without the consent of the data's creators. Additionally, there are significant copyright concerns, as utilizing copyrighted data without proper authorization can infringe upon copyright laws. Users may not always be fully informed about how their data is collected, utilized, or shared in machine learning systems. Ensuring informed consent and granting individuals meaningful control over their data are paramount for respecting privacy rights and maintaining trust in machine learning applications.

There are also several philosophical inquiries about machine learning addressing complex dilemmas such as the paperclip maximizer. This scenario involves AI designed to maximize paperclip production. However, without human safety and ethical constraints, the AI could consume all resources and cause unintended destruction. This thought experiment emphasizes the importance of aligning AI objectives with human values to prevent dystopian outcomes. If not regulated properly, an AI optimizing tasks may inadvertently harm humanity by prioritizing efficiency over human safety.

Similarly, utilitarianism, exemplified in the case of autonomous driving, presents ethical dilemmas regarding the prioritization of lives in critical situations. The trolley problem, a philosophical thought experiment, illustrates this concept. Utilitarianism advocates for actions that maximize overall happiness or well-being. In self-driving vehicles, this philosophy introduces moral conundrums about whose safety to prioritize in accidents—of the occupant or the pedestrians on the road; of several people on the path or a single person on the side, and so on. Algorithms need to balance the collective good with individual rights, and this often requires challenging established ethical norms and necessitating trade-offs, especially since machines lack a moral dimension.

Accountability and fairness are fundamental ethical principles that must be upheld in the development and deployment of machine learning systems. Developers and organizations deploying machine learning should endeavor to minimize biases, ensure fairness in outcomes, and establish mechanisms for redressal in instances of algorithmic errors or unjust outcomes. This encompasses implementing rigorous testing, validation, and auditing procedures to identify and mitigate biases throughout the entire machine learning lifecycle.

Beyond its technical aspects, we must carefully consider the social impact of machine learning. It has the potential to disrupt jobs, exacerbate income inequality, and alter the distribution of power. Moreover, machine learning can negatively impact communities by consuming excessive resources. Large-scale machine learning models require substantial electricity and specialized hardware, which can divert energy and financial resources from local communities, particularly those that are already disadvantaged. Data centers consume large amounts of electricity and generate substantial heat; the hardware lifecycle also contributes to electronic waste, harming the environment. These factors can disrupt local ecosystems. Communities lacking the financial means or resources to utilize machine learning are forced to suffer the negatives while being denied its benefits, further exacerbating disparities. If we fail to ensure equitable application of machine learning, we may aggravate the challenges faced by vulnerable populations. Therefore, it is imperative that we carefully assess the implications of machine learning on employment, social equality, and democratic processes. We must ensure that everyone has an opportunity to benefit from machine learning and that no one is left behind.

Fortunately, ethical guidelines, standards, and frameworks are emerging to guide the responsible development and deployment of machine learning. Organizations, researchers, and policymakers are increasingly advocating for principles such as fairness, accountability, transparency, and inclusivity in machine learning systems. Collaborative efforts involving stakeholders from diverse disciplines and backgrounds are crucial for developing ethical frameworks that promote the responsible and beneficial use of machine learning technology. Upholding ethical principles in machine learning development and deployment is essential to harnessing its full potential while mitigating risks and ensuring that the technology serves the broader interests of society.

1.9 HOW IS THIS BOOK ORGANIZED?

This book is organized into three distinct parts.

Part I: Introduction

This introductory part provides a fundamental understanding of machine learning. It covers the essential concepts, including setting up a computer for basic machine learning and the various stages of the machine learning process.

Part II: Machine Learning Algorithms

This comprehensive part delves into the analysis of several established machine learning algorithms. Each algorithm is thoroughly examined, elucidating its underlying principles and highlighting its specific applications. The algorithms discussed include linear regression, polynomial regression, logistic regression, k-nearest neighbors, SVMs, decision trees, random forests, neural networks, and deep learning methods.

To enhance comprehension, each algorithm is illustrated with well-known and publicly available exemplar datasets. The accompanying code snippets provide detailed instructions for training and testing the algorithms. Throughout, the meanings and variations of code parameters are explicitly discussed to provide clarity.

While references to other chapters are made, each chapter is designed to be self-contained. Readers can directly access any specific algorithm in Part II without interrupting the overall narrative.

Part III: Epilogue

This concluding part explores the concept of combining multiple machine learning algorithms to create composite algorithms that leverage the synergies of various methods. It also provides guidance on interpreting the results obtained from machine learning exercises and offers practical strategies for future endeavors.

1.10 CHAPTER SUMMARY

- Machine learning, a subfield of AI, empowers computers to acquire knowledge and discern patterns without explicit programming.
- Deep learning, a subfield of machine learning, employs neural networks inspired by the human brain to discern intricate patterns.
- The machine learning process encompasses data collection, preparation, model selection, training, evaluation, tuning, deployment, and monitoring.
- Supervised learning relies on labeled data. Unsupervised learning, on the other hand, can uncover patterns in unlabeled data.
- Reinforcement learning, akin to operant conditioning, enables agents to learn optimal behaviors through rewards and penalties.
- Self-supervised learning employs models to generate labels from data.
- Transfer learning adapts pre-trained models to novel tasks, leveraging existing knowledge for expedited training.
- Machine learning presents opportunities but encounters challenges such as data quality, interpretability, and ethical considerations. Ethical implications encompass bias mitigation, transparency, privacy, consent, and accountability.

2 Basic setup

With the proliferation of computing power available on our desktops and laptops, exploring the domain of machine learning has become remarkably accessible. Coupled with user-friendly programming languages such as R (R Core Team, 2022), which predominantly employs English syntax, the learning curve has become extremely gentle for a new user. Consequently, even individuals without a background in computer science can now easily apply machine learning in their respective fields of expertise.

In this chapter, we will delve into the programming language and software environment known as R, which will serve as the foundation for all machine learning tasks throughout this book. R is an intuitive and high-level programming language that employs everyday English words to construct commands and code, promoting clarity and ease of comprehension. The language also places a strong emphasis on the use of pipes,[1] which facilitate the seamless flow of outputs from one command to the next ('chaining'), thereby simplifying the coding process. Furthermore, R offers a comprehensive collection of packages and libraries that cater to a wide range of applications, drawing upon the expertise of other researchers in the field.

The R programming language is script-based and does not necessitate separate compilation, further simplifying its implementation. Although R can be utilized in a computer's terminal, many users prefer integrated development environments (IDEs) such as RStudio. The use of IDEs enhances productivity by simplifying file management and execution of commands through intuitive interfaces. This chapter will provide a comprehensive exploration of RStudio.

For data processing, R relies on data stored in formats such as CSV (Comma-Separated Values) or TSV (Tab-Separated Values), but can also work with Excel files. Simplifying this process further is the `rio` (R input-output) (hong Chan et al., 2023) package,[2] which offers `import()` and `export()` functions that cover most data reading and writing needs.

Furthermore, R offers extensive interoperability through packages such as `tidyverse` (Wickham et al., 2019), which standardizes functions and enhances their integration. This standardized interoperability is particularly advantageous to

[1] In the realm of coding, a *pipe* is an operator (typically denoted by "|", "%>%", or "|>") that facilitates the redirection of the output of one command to the input of another. This functionality is particularly prevalent in many shell scripting and functional programming paradigms, and permits 'chaining' of commands to create powerful constructs.

[2] Packages in R are collections of functions, data, and compiled code organized in a well-defined format. Packages extend the capabilities of R by introducing novel functionalities.

scientists and researchers who may not have extensive programming backgrounds but are exploring machine learning.

2.1 R: AN INTRODUCTION

R is a programming language and software environment primarily designed for statistical computing and data analysis. Developed by Ross Ihaka and Robert Gentleman at the University of Auckland in the mid-1990s, R has grown into one of the most widely used tools in the data science and statistical communities. Its evolution has been marked by a strong emphasis on extensibility and flexibility, making it a versatile choice for a wide range of applications, from simple data manipulation to complex statistical modeling.

At its core, R provides a comprehensive suite of tools for performing a variety of statistical analyses. Users can perform descriptive statistics, such as calculating means and standard deviations, and conduct inferential statistics, including hypothesis testing and regression analysis. R supports a vast array of statistical techniques, from basic procedures to more advanced methodologies such as generalized linear models, time series analysis, and multivariate analysis. This extensive range of built-in functions enables users to conduct rigorous statistical evaluations and derive insights from their data.

One of R's notable features is its capacity to generate high-quality graphical visualizations. The language incorporates a comprehensive set of pre-built functions for plotting and graphing data, including histograms, scatterplots, and boxplots. Notably, R's data visualization capabilities are further enhanced by packages such as `ggplot2` (Wickham, 2016), which facilitate the creation of sophisticated and customizable graphics through a layered grammar of graphics approach, ensuring simplicity and standardization. The `ggplot2` package has become a cornerstone of data visualization in R, providing users with the ability to effortlessly create complex and aesthetically pleasing plots that effectively convey information.

Another notable feature of R is its extensibility through a diverse range of packages. The Comprehensive R Archive Network (CRAN)[3] is a repository that hosts thousands of these packages, each designed to extend R's capabilities. These packages encompass a comprehensive range of domains, including machine learning and bioinformatics, finance, and social sciences. The accessibility of these packages empowers R users to access specialized tools tailored to their specific requirements, facilitating a high degree of customization and specialization in their analyses. Moreover, the ease of package development and sharing has fostered a thriving ecosystem of tools and resources.

R stands out for its open-source nature. As a free and open-source software, it grants users unrestricted access, together with the ability to modify and distribute

[3]https://cran.r-project.org

its source code. This open-source model has nurtured a thriving community of developers and users who actively contribute to the language's continuous evolution. The collaborative spirit of R's development ensures that it remains adaptable to the evolving needs of its users. Furthermore, the accessibility of R as an open-source tool diminishes the entry barrier for individuals interested in statistical computing and machine learning, enabling it to reach students, researchers, and professionals alike.

Integration with other programming languages and technologies is a crucial feature of R. For instance, R can interface with C++ for high-performance computing tasks through the Rcpp package (Eddelbuettel and Balamuta, 2018). Additionally, R integrates seamlessly with databases, enabling users to directly import and manipulate substantial datasets from database management systems. R can also be integrated with web technologies and tools, such as through the `Shiny` (Chang et al., 2024), facilitating the creation of interactive web applications that allow users to visualize and analyze data in real-time.

The R community also provides substantial support and resources for learning and troubleshooting. Numerous online forums, mailing lists, and user groups facilitate collaboration among R users, offering assistance, knowledge sharing, and problem-solving opportunities. Comprehensive documentation, encompassing manuals, tutorials, and books, caters to users of all skill levels, from beginners to advanced practitioners, enabling effective navigation and utilization of R.

In these and other ways, R stands out as a robust and versatile tool for statistical computing, data analysis, and machine learning tasks. Its comprehensive array of pre-installed functions, capabilities for high-quality data visualization, extensive ecosystem of packages, and open-source nature make it a preferred choice for numerous individuals in the domains of data science, statistics, and machine learning. Its ability to seamlessly integrate with other technologies and the robust support provided by its user community further augment its utility, positioning R as a pivotal player in the contemporary landscape of data analysis, statistical computing, and machine learning.

2.2 THE RSTUDIO INTEGRATED DEVELOPMENT ENVIRONMENT

RStudio[4] is a comprehensive IDE specifically designed for working with the R programming language. Released in 2011 by RStudio, Inc. (now Posit), RStudio has become a widely utilized tool among data scientists, statisticians, and researchers who employ R for statistical computing, data analysis, and machine learning tasks. RStudio enhances the R programming experience by offering an intuitive interface,

[4]https://www.rstudio.com

advanced features, and tools that facilitate coding, data exploration, and visualization tasks.

The design of RStudio is centered around facilitating an efficient workflow for users working with R programming language. At its core, RStudio comprises a four-pane layout [Fig. 2.1] that organizes various aspects of the development process. This layout comprises a console for executing R commands and scripts, a source pane for writing and editing code, an environment pane for managing variables and datasets, and a files/plots/packages/help pane that provides access to the file system, graphical outputs, package management, and documentation. This organization facilitates simultaneous management of multiple aspects of work, thereby enhancing the coding and analysis process in terms of efficiency and intuition.

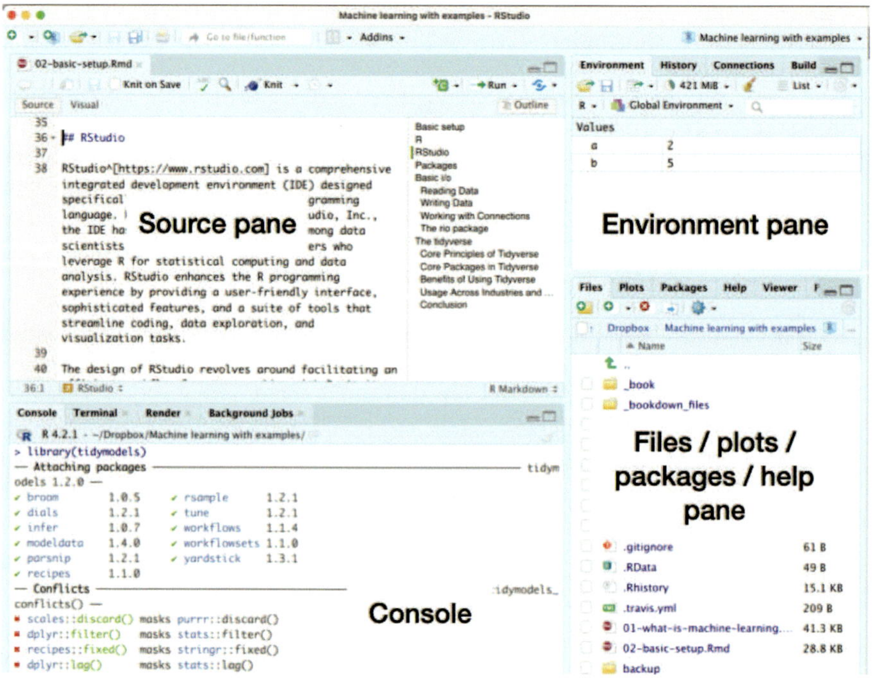

FIGURE 2.1: A view of the RStudio IDE.

One primary strength of RStudio is its comprehensive support for R Markdown (Xie et al., 2020), a versatile tool for crafting dynamic documents. R Markdown empowers users to seamlessly integrate code, text, and visualizations into a single document, which can subsequently be rendered into diverse formats such as HTML, PDF, and Word. This multifaceted capability holds particular significance in the generation of comprehensive reports that harmoniously blend narrative explanations with interactive and reproducible data analyses. By facilitating the production of well-documented reports that can be effortlessly shared and updated, R Markdown significantly enhances both the transparency and reproducibility of data analysis endeavors. Notably, this very book has been crafted in R Markdown utilizing the `bookdown` package (Xie, 2016).

RStudio also provides comprehensive support for version control systems, including Git and Subversion. Integrated version control features enable users to track modifications to their code, collaborate with others, and effectively manage various versions of their projects. This integration is paramount for collaborative research and development environments, where maintaining version history and coordinating contributions from multiple team members are crucial for achieving project success.

The IDE package management system simplifies the installation, updating, and maintenance of R packages. Users can conveniently browse and install packages from the CRAN, Bioconductor,[5] and other repositories directly within the RStudio interface. The package management pane also provides comprehensive information about the installed packages, including version numbers and available updates, ensuring that users are consistently utilizing the most recent tools and libraries.

RStudio's graphical capabilities further enhance its appeal. The IDE supports a comprehensive suite of visualization tools, enabling users to generate and interact with plots directly within the interface. Real-time visualization capabilities facilitate a more interactive and iterative approach to data visualization. Furthermore, the integration of RStudio with `Shiny`, R's framework for constructing interactive web applications, empowers users to develop sophisticated data-driven applications accessible and interactable through a web browser.

Another notable feature of RStudio is its comprehensive support for various project types, including R projects, R Notebooks, and R Scripts. R projects facilitate organizational efficiency by grouping related files and settings within a single directory. This structured approach promotes a more organized workflow and mitigates the likelihood of errors associated with file management. R Notebooks provide an interactive environment for executing and documenting code chunks, enabling users to explore data and visualize results within a notebook-oriented interface. R scripts are utilized to automate statistical analyses and data visualization tasks within the R programming language. They enable users to import data, execute intricate calculations, and generate visualizations to effectively interpret results.

RStudio's educational resources and community support are additional substantial assets. Extensive documentation, tutorials, and webinars are available to cater to users of all skill levels, from beginners to advanced practitioners. The active user community continuously participates in forums, blogs, and user groups, fostering a collaborative environment where individuals can seek assistance, share knowledge, and engage in discussions on best practices.

---

[5]https://www.bioconductor.org

2.3 PACKAGES IN R

A fundamental aspect of R's extensibility lies in its package system. Upon installation, R provides a set of base packages that offer essential functions for data analysis. However, the true potential of R is unlocked through the extensive collection of additional packages that users can install. These packages can be easily installed from repositories such as the CRAN, Bioconductor, and GitHub.[6] CRAN serves as the primary repository, hosting thousands of packages contributed by the R community. Each package is meticulously designed to address specific requirements and applications. Bioconductor specializes in tools for bioinformatics and computational biology, providing packages tailored to specialized research areas within these domains. GitHub offers a platform for developers to share and collaborate on packages, including those still in development or awaiting publication on CRAN.

Packages can be installed using the `install.packages()` command and loaded using the `library()` command. Furthermore, the `update.packages()` function ensures that users can keep their packages up-to-date with the latest versions and improvements.

It is advisable to verify the existence of a package before attempting its installation. As a standard practice, we incorporate this verification and package loading into our code to ensure automatic installation and loading on other computers when our code is utilized. An illustration of this process is provided below:

```
if (!require("stringr")){
  install.packages("stringr")
}

library(stringr)
```

In this code, the `require()` function verifies whether a package is already installed. If the package is not installed, it returns `FALSE`. The `!` operator is a Boolean `NOT` operator. Therefore, the first line of the code translates to: "if the package `stringr` (Wickham, 2023b) is not installed, install the package `stringr`". The second line utilizes the `library()` function to load the `stringr` package for subsequent usage within the code.

Each R package typically comprises a diverse range of components. The most significant of these are functions, which augment R's capabilities by furnishing novel tools for data manipulation, analysis, and visualization. For instance, the `dplyr` package (Wickham et al., 2023) provides a suite of functions that streamline data

[6] https://github.com

manipulation tasks, such as filtering, summarizing, and grouping data. Similarly, the ggplot2 package (Wickham, 2016) introduces an intuitive grammar for constructing intricate and customizable data visualizations, which has become an indispensable resource for data scientists and statisticians.

Additionally, packages frequently include datasets that serve as examples or testing tools for the package's functions. These demonstration datasets assist users in comprehending the usage of the functions and provide a foundation for testing and validation. In this book, we will utilize a substantial number of these datasets to explore and process various machine learning algorithms.

Documentation is another crucial component of R packages. It encompasses vignettes, which are comprehensive guides that provide detailed explanations and examples of how to utilize the package. Additionally, it includes reference manuals and function documentation that elucidate the purpose, arguments, and usage of each function. Assistance can be obtained through the help pane, as depicted in Figure 2.1, or by employing the ? operator preceding the function's name in the console and pressing enter. For instance, to comprehend the mean() function, we execute the command ?mean.

```
?mean
```

This in turn brings out the R Documentation file for the arithmetic mean:

```
mean      package:base      R Documentation

Arithmetic Mean

Description:

     Generic function for the (trimmed) arithmetic mean.

Usage:

      mean(x, ...)

      ## Default S3 method:
      mean(x, trim = 0, na.rm = FALSE, ...)

Arguments:

      x: An R object.  Currently there are methods
```

```
      for numeric/logical vectors and date, date-time
      and time interval objects. Complex vectors are
      allowed for 'trim = 0', only.

      trim: the fraction (0 to 0.5) of observations
      to be trimmed from each end of 'x' before the
      mean is computed.  Values of trim outside that
      range are taken as the nearest endpoint.

      na.rm: a logical evaluating to 'TRUE' or 'FALSE'
      indicating whether 'NA' values should be stripped
      before the computation proceeds.

      ...: further arguments passed to or from other
      methods.

Value:

      If 'trim' is zero (the default), the arithmetic
      mean of the values in 'x' is computed, as a
      numeric or complex vector of length one. If 'x'
      is not logical (coerced to numeric), numeric
      (including integer) or complex, 'NA_real_' is
      returned, with a warning.

      If 'trim' is non-zero, a symmetrically trimmed
      mean is computed with a fraction of 'trim'
      observations deleted from each end before the
      mean is computed.

References:

      Becker, R. A., Chambers, J. M. and Wilks, A. R.
      (1988) _The New S Language_.  Wadsworth &
      Brooks/Cole.

See Also:

      'weighted.mean', 'mean.POSIXct', 'colMeans'
      for row and column means.

Examples:
```

```
x <- c(0:10, 50)
xm <- mean(x)
c(xm, mean(x, trim = 0.10))
```

The development and maintenance of R packages are driven by the active R community, which comprises academics, data scientists, software developers, and industry professionals. Numerous packages have been, and are being, developed to address specific research requirements or enhance workflows across diverse domains. This collaborative and open-source approach promotes the continuous growth of the R ecosystem by facilitating the development of new packages and updating of existing ones, to incorporate the latest advancements in data analysis and statistics.

2.4 BASIC INPUT AND OUTPUT OPERATIONS

In R, managing input and output (I/O) operations is fundamental to efficiently handling data. Basic I/O operations encompass reading data into R from various sources and writing data out from R to different destinations. These operations serve as the foundation of data analysis workflows, empowering users to import datasets, manipulate them, and export results. Proficiency in these fundamental operations is essential for effective data management and analysis within the R environment.

2.4.1 READING DATA

The R programming language provides several functions to facilitate the reading of data from various file formats. One of the most prevalent tasks involves the reading of data from text files, such as CSV or TSV files. The `read.csv()` function is commonly employed for the importation of CSV files. This function ingests the data into a data frame, a fundamental data structure in R for managing tabular data. The `read.csv()` function accepts parameters that specify the file path and the separator employed within the file. For instance, `read.csv("datafile.csv")` will retrieve the CSV file named "datafile.csv" from the current working directory into a data frame.

For tab-separated files, the `read.delim()` function is utilized. It functions similarly to `read.csv()` but is specifically designed to manage tab delimiters rather than commas. Users have the flexibility to specify various parameters to accommodate diverse file structures, including column names, row names, and representations for missing values.

R also supports reading data from Excel files using the `readxl` package (Wickham and Bryan, 2025). Functions such as `read_excel()` enable users to import data from both .xls and .xlsx formats. This functionality is particularly useful for data stored in spreadsheets, which are prevalent in numerous data analysis scenarios.

In addition to reading from local files, R can also read data from web-based sources. Functions such as read.csv() can accept URLs, allowing users to directly download and read data from the web. For instance, read.csv("http://example.com/datafile.csv") retrieves and reads the CSV file from the specified URL.

In the tidyverse [Section 2.5], the read_csv() and read_tsv() functions are employed to load data into tibble format. A tibble is a modern variant of a data frame in R's tidyverse, designed for better readability and compatibility with data science workflows.

Upon reading a file, its contents are assigned to a variable using the assignment operator <-. For instance, the following code reads the data from the file datafile.csv and assigns it to the variable data:

```
data <- read.csv("http://example.com/datafile.csv")
```

2.4.2 WRITING DATA

Exporting data from R is equally crucial as importing it. To export data frames to CSV files, the write.csv() function is employed. This function saves a data frame to a CSV file, while providing customization options for the output, including the inclusion of row names and the modification of the default separator. For example, write.csv(dataframe, "outputfile.csv") saves the data frame stored in the variable dataframe to a file named "outputfile.csv".

Similarly, the write.table() function writes data to text files. It offers the option to specify delimiters other than commas, such as tabs or semicolons. This function is useful when dealing with files that require specific formatting or when exporting data to formats other than CSV.

The writexl package (Ooms, 2025), which includes the write_xlsx() function, can be used to export data to Excel files. This makes it easy to share results with users who prefer spreadsheets.

In the tidyverse [Section 2.5], the write_csv() and write_tsv() functions facilitate the writing of data to CSV and TSV formats, respectively. These functions are approximately twice as efficient as their base R counterparts. Additionally, the contents of a file can be directly piped to these functions for writing purposes, as depicted next.

```
data %>% write_csv("datafile.csv")
```

R also supports saving R objects in binary format using the `save()` function. This feature is particularly useful for saving complex objects such as lists or models that are cumbersome to export in text format. The `save()` function can save one or more R objects to a file, which can then be loaded into an R session using the `load()` function. For instance, `save(object, file = "objectfile.RData")` saves the `object` to a binary file named `objectfile.RData`, and `load("objectfile.RData")` restores it in a subsequent session.

2.4.3 WORKING WITH CONNECTIONS

In addition to file-based I/O operations, R provides a mechanism to handle more general I/O operations through connections. Connections in R serve as a means of reading from or writing to various data sources, including files, URLs, and other data streams. The `file()` function facilitates the creation of a connection to a file, which can subsequently be utilized with functions such as `readLines()` and `writeLines()` for line-by-line reading and writing operations. For instance, `con <- file("myfile.txt", "r")` establishes a connection to "myfile.txt" in read mode, and `readLines(con)` retrieves the file's content into a character vector.

Connections also facilitate more sophisticated operations, including the handling of compressed files and the establishment of network connections. For instance, the `gzfile()` function establishes a connection to a file compressed using the gzip algorithm, allowing users to read or write compressed data without any disruptions.

2.4.4 THE RIO PACKAGE

The `rio` package (hong Chan et al., 2023) in R is a versatile tool designed to simplify and standardize data import and export across various file formats. It's a valuable resource for data scientists and analysts who frequently handle data from diverse sources. The package's primary objective is to provide a unified interface for reading from and writing to different formats, eliminating the complexity and verbosity often associated with such tasks. By doing so, it enhances efficiency and productivity.

At its core, the `rio` package stands out for its exceptional ability to manage a wide spectrum of file formats while utilizing a unified set of functions. This encompasses prevalent formats such as CSV, Excel, and JSON, as well as more specialized formats like SAS, SPSS, and Stata. By abstracting the intricacies of diverse file formats, `rio` empowers users to concentrate on their data analysis endeavors, thereby mitigating distractions caused by the complexities of various data structures and file types.

The package offers two primary functions: `import()` and `export()`. The `import()` function facilitates the automatic reading of data into R from various file formats. It employs extension or content-based detection to identify the file type, eliminating the explicit specification of format by users. This simplifies the data import process and reduces the likelihood of errors. For instance, `import("data.xlsx")` will read an Excel file, while `import("data.csv")` will handle a CSV file. The function is designed to be user-friendly and adaptable, seamlessly accommodating diverse data formats.

Similarly, the `export()` function facilitates the writing of data from R to various file formats. It enables users to save data frames and other R objects to files in formats such as CSV, Excel, and JSON with minimal configuration. This function, like `import()`, automatically handles format-specific details, allowing users to write data without having to delve into the specific syntax or options required by different file types. For instance, `export(data, "results.xlsx")` will save the `data` object to an Excel file, and `export(data, "results.json")` will save it as a JSON file.

One of the primary advantages of the `rio` package is its emphasis on simplicity and consistency. It aims to provide a uniform user experience regardless of the data format, which can significantly enhance workflow and minimize the learning curve for users unfamiliar with R or who frequently switch between different data sources. This consistency is particularly valuable in collaborative environments where multiple users may be working with various file formats.

Furthermore, the `rio` package is designed to seamlessly integrate with other R packages and tools. It effectively utilizes existing packages for format-specific operations while simultaneously offering a unified interface that abstracts away[7] these complexities from the user. This approach not only simplifies data handling but also guarantees that users can benefit from the underlying functionality of established packages without the necessity of direct interaction with them.

2.5 FUNDAMENTALS OF THE TIDYVERSE

The `tidyverse` (Wickham et al., 2019) is a comprehensive collection of R packages crafted to facilitate data manipulation, exploration, visualization, and analysis. Its primary focus lies in promoting simplicity, consistency, and seamless integration, thereby empowering data scientists, statisticians, and analysts to effectively manage and analyze data within the R programming language.

[7]Abstraction is the process of hiding unnecessary complexity and focusing only on the essential aspects of something, making it easier to understand and use.

2.5.1 CORE PRINCIPLES OF TIDYVERSE

The `tidyverse` adheres to four fundamental principles: the reuse of existing data structures, the composition of simple functions using the pipe operator, the adoption of functional programming principles, and designing for human use.

1. Reusing existing data structures: The `tidyverse` philosophy emphasizes the utilization of existing data structures over the development of novel ones for packages, maximizing uniformity and user comprehension, even if that leads to lesser speed of running the code. Well-established data structures are preferred over custom structures to facilitate easy movement from one function to the next. Many R packages, such as `ggplot2`, `dplyr`, and `tidyr` (Wickham et al., 2024b), are already adept at handling rectangular datasets with observations and variables, in the form of data frames and tibbles. Consequently, data frames and tibbles are utilized for various other packages as well, those that deal with similar data. Only those packages that specialize in specific types of variables, such as `stringr` for strings or `lubridate` (Grolemund and Wickham, 2011) for dates and times, employ base R vector types. Custom data structures are implemented as S3 classes when necessary.[8] These classes are constructed upon atomic vectors[9] or lists.
2. Composing simple functions with the pipe: The `tidyverse` encourages solving complex problems by chaining together simple, easy-to-read functions using the pipe operator (%>% or |>). This approach emphasizes clarity and modularity: each function should perform one clearly defined task—ideally describable in a single sentence. Function names are typically verbs, unless several functions would share the same verb; in those cases, descriptive nouns are used instead. The pipeable API is flexible, allowing users to compose functions in different ways by reordering them to achieve the result they want.
3. Adopting functional programming principles[10]:
   R is a functional programming language, and the `tidyverse` follows these principles. This means working with immutable objects and relying on copy-on-modify semantics.[11] Embracing these concepts helps make code behavior easier to understand and reason about.

---

[8] An S3 class in R lets you give objects class names and then write functions that automatically adapt their behavior based on those classes.

[9] Atomic vectors in R are fundamental building blocks that comprise a collection of items of the same data type, such as numbers, text, or logical values.

[10] Functional programming is a programming paradigm that emphasizes using functions to perform tasks while avoiding changes to existing data. This promotes readability and makes testing easier by allowing individual functions to be isolated and evaluated independently.

[11] Copy-on-modify semantics optimizes memory use and efficiency by creating a copy of an object only when a modification is made, rather than copying it every time it is referenced.

4. Designing for humans: The `tidyverse` prioritizes human usability and comprehension over raw computational efficiency when designing its APIs. This means emphasizing readability and choosing function names that make their purpose intuitive and memorable. Clear, descriptive names are preferred over short or ambiguous ones, with brevity reserved only for the most essential operations. Autocomplete is supported through the use of common prefixes for related function families, making functions easier to write and recall. In smaller packages, a consistent prefix is applied to all functions to further improve the user experience.

2.5.2 CORE PACKAGES IN TIDYVERSE

The core packages of the `tidyverse` are:

1. **`dplyr`** (Wickham et al., 2023): Offers functions for data manipulation and transformation, encompassing filtering, summarization, and data arrangement.
2. **`readr`** (Wickham et al., 2024a): Facilitates reading and writing data in various formats, such as CSV and TSV, with a focus on speed and ease of use.
3. **`forcats`** (Wickham, 2023a): This package is designed to handle and manipulate categorical variables, commonly referred to as factors. It provides functionalities for reordering factor levels and modifying factor levels.
4. **`stringr`** (Wickham, 2023b): Offers a collection of functions for string manipulation, including pattern matching and various string operations.
5. **`ggplot2`** (Wickham, 2016): Used for creating complex and customizable data visualizations through a layered approach.
6. **`tibble`** (Müller and Wickham, 2023): Enhances data frames with contemporary features and a user-friendly print method for displaying data.
7. **`lubridate`** (Grolemund and Wickham, 2011): Simplifies the manipulation and parsing of date and time data.
8. **`tidyr`** (Wickham et al., 2024b): Helps in tidying[12] and reshaping data, including functions for converting between wide[13] and long[14] formats.
9. **`purrr`** (Wickham and Henry, 2025): Provides functional programming tools, allowing the application of functions across lists and vectors in a consistent manner.

---

[12] In R, tidy data refers to a data format where each variable forms a column, each observation forms a row, and each type of observational unit forms a table, making it easy to manipulate and analyze using tidyverse tools.

[13] In R, wide format refers to a data structure where each variable forms a separate column and multiple observations or measurements are spread across these columns, often with repeated identifiers or variables arranged in a broad, wide layout.

[14] In R, long format refers to a data structure where each observation is represented by a single row, with variables stored in separate columns and repeated measures or multiple variables stacked in a single column, making it easier to analyze and visualize with tidyverse tools.

2.5.3 BENEFITS OF USING TIDYVERSE

The `tidyverse` substantially enhances productivity by offering a unified and consistent syntax across its comprehensive suite of packages. This consistency simplifies the learning curve and diminishes the time spent transitioning between disparate tools or mastering novel syntaxes. Consequently, users can concentrate more on their analyses rather than on the mechanics of data manipulation and visualization. By streamlining these tasks with robust and meticulously designed functions, the `tidyverse` facilitates more efficient execution of complex data operations, ultimately accelerating the analytical process.

The modular design of the `tidyverse` offers a significant advantage, providing flexibility in data analysis workflows. Users have the option to select from a diverse range of packages, including `dplyr` for data manipulation, `ggplot2` for visualization, and `tidyr` for data tidying, and combine them in a manner that aligns with their specific requirements and analytical objectives. This modularity enables users to customize their toolset to meet the demands of their projects, whether it involves handling large datasets, conducting advanced statistical analyses, or creating intricate visualizations. The capability to select and combine packages based on task requirements ensures that the `tidyverse` can adapt to a wide spectrum of analytical scenarios.

Community support is a cornerstone of the tidyverse's success, thanks to its large and active user and developer community. This vibrant ecosystem generates a wealth of documentation, tutorials, and resources that aid users in learning the tools and troubleshooting issues. The extensive support network facilitates newcomers' rapid learning and experienced users' ability to solve complex problems, creating a collaborative environment where knowledge and best practices are continuously shared and refined.

Reproducibility is another key advantage of employing the `tidyverse`. The framework promotes transparent and clear workflows that facilitate the replication of data analysis processes. By promoting practices that ensure analyses are well-documented and readily shareable, the `tidyverse` supports reproducible research, which is essential for verifying results and maintaining scientific rigor. This emphasis on reproducibility ensures that analyses can be reliably reproduced and verified by others, contributing to more robust and trustworthy data science practices.

2.6 CHAPTER SUMMARY

- R, a programming language and software environment for statistical computing, data analysis, and machine learning, is renowned for its user-friendly syntax and extensive ecosystem of packages.
- R's open-source nature, comprehensive documentation, and active community contribute to its widespread adoption in data science, statistics, and machine learning.

- RStudio, an IDE specifically designed for R, provides a four-pane layout that facilitates efficient coding, data exploration, and visualization.
- R provides functions for reading data from sources such as CSV, Excel, and web-based locations.
- The `rio` package simplifies data import and export.
- The `tidyverse` package emphasizes simplicity, consistency, and seamless integration for data manipulation and analysis. It prioritizes human usability and comprehension over computational efficiency in API design, and offers enhanced productivity, modular design, community support, and reproducibility.
- The `tidyverse` serves as a powerful tool for data analysis and machine learning, enabling professionals to derive insights and communicate findings effectively.

3 Machine learning in practice

In practice, machine learning is done using a structured approach starting at problem definition and ending at model deployment and maintenance. This approach aims to leverage data-driven insights for informed decision-making and innovation across various domains and industries. We have already seen the basic steps in Chapter 1 [Section 1.4]. In this chapter, we will discuss a comprehensive exploration of the fundamental concepts in the domain of machine learning practice.

3.1 PREDICTION

In the realm of machine learning, **prediction** means using a *pre-trained model* to make *educated guesses* about the outcomes of new, unseen data. After learning patterns from historical examples during training, a model begins to *understand* how different pieces of information relate to one another. With this learned understanding, it can generalize and make predictions when given fresh input.

Whether the task involves estimating prices, classifying items into categories, or identifying patterns, the model applies what it has learned to produce an informed output. This predictive ability is widely used—from analyzing customer preferences in market segmentation to assisting in medical diagnosis based on symptoms and test results.

3.2 PREDICTOR AND OUTCOME VARIABLES

A **predictor variable**, also known as an **independent variable** or **feature**, is a fundamental component in statistical models and machine learning algorithms, and refers to the variables used to predict or explain an outcome variable. The **outcome variable**, commonly referred to as the **dependent variable** or **target**, is the variable that the model aims to predict based on the predictor variables. For example, in predicting student academic performance (outcome variable, denoted as y), several predictor variables may be considered (such as study time x_1, attendance x_2, and socioeconomic status x_3). The model might indicate that certain predictor variables are actually useful in predicting the outcome variable, while others are not particularly useful.

In mathematical terms, the relationship between predictor variables and the outcome variable can be expressed not only linearly but also non-linearly to account for non-linear relationships in equations such as this:

$$y = \beta_0 + \beta_1 * x_1 + \beta_2 * x_2 + \beta_3 * x_3 + \beta_4 * x_1^2 + \varepsilon$$

Here, y represents the predicted academic performance. x_1, x_2, and x_3 are predictor variables, while x_1^2 represents a squared term of the study time, allowing the model to capture non-linear relationships. $\beta_0, \beta_1, \beta_2, \beta_3$, and β_4 are coefficients that quantify the impact of each predictor variable and its non-linear transformations on the outcome. ε denotes the error term, representing the difference between the predicted and actual values.

We can also include non-linear interaction terms like $x_2 \times x_3^2$. Incorporating non-linearity into the equation allows the model to capture more complex relationships between predictor variables and the outcome. This flexibility is particularly important in fields like economics, where relationships may not follow simple linear patterns. By including non-linear terms, researchers can create more precise models that more accurately reflect the actual dynamics and improve predictions across various applications, such as financial forecasting and medical diagnostics.

3.3 MODEL

In machine learning, a **model** is a learned representation of patterns and relationships found in data. Through training, it internalizes these patterns so that, when given new inputs, it can autonomously generate predictions or make informed decisions.

Models are constructed through a process known as **training**, wherein they analyze extensive datasets to discern patterns and correlations. During training, the model modifies its internal parameters[1] in response to observed data to enhance its accuracy. This iterative process allows the model to develop the ability to identify features and accurately predict outcomes for new data.

There are different types of machine learning models, each designed to tackle specific tasks. **Regression** models forecast continuous values, like housing prices based on factors like location and size. **Classification** models categorize data into predefined classes, like identifying spam emails or diseases based on symptoms. **Clustering** models group data points based on similarities without predefined categories, which is useful for tasks like market segmentation.

Models are assessed using task-specific metrics, such as accuracy or precision [Section 3.11], to ascertain their efficacy in handling novel data in practical applications. Validated models are subsequently deployed in production environments, where they operate autonomously and often in real-time.

---

[1]Internal parameters include elements such as weights of neural networks and decision boundaries used by support vector machines. These ideas will be explained in greater detail in the chapters that follow.

3.4 MODEL PARAMETERS

In the realm of machine learning, **model parameters** are internal variables that undergo adjustments during training to optimize the model's performance. You may think of these as the "settings" of the model. Model parameters encompass weights, coefficients, biases, and other configuration settings that define how the model processes input data to generate predictions or classifications. For instance, in linear models such as regression, coefficients establish the correlation between input features and the target variable, while an intercept term adjusts the prediction.

Neural networks, another widely used model type, are characterized by parameters such as weights and biases that govern the propagation of signals between neurons. Weights represent the strength of connections, while biases modify the output of individual neurons. Support vector machines (SVMs), on the other hand, employ parameters like support vectors and kernel parameters to define decision boundaries between classes in classification tasks. Decision trees, in contrast, utilize parameters in the form of decision rules at each node to partition the data based on feature values. We will explore all these models and more in the subsequent chapters of this book.

During the training phase of machine learning, algorithms iteratively adjust these parameters to minimize errors or maximize performance metrics like accuracy. This **optimization** process aims to find the optimal parameter configuration that enables the model to generalize effectively to novel, unlabeled data. However, we need to be mindful of **overfitting**, where the model achieves high performance on training data but fails to generalize to new data. To mitigate this issue, the number of model parameters may be reduced, or other steps taken. We discuss these in the last chapter of this book.

3.5 MODEL FITTING

Model fitting involves adjusting the parameters of a chosen model to improve its ability to understand patterns in the training data. This process is crucial because it determines the model's accuracy in predicting outcomes or categorizing new data.

Initially, a suitable model type is selected based on the nature of the data. For example, linear regression is used for numerical predictions, while neural networks are used for image recognition. Next, the model is set up with initial values for its parameters,[2] such as weights and biases, which define how the model interprets data at the beginning.

Subsequently, the model undergoes training by being fed labeled data (where the outcomes are known) or unlabeled data (to just identify patterns in the data). Employing techniques such as gradient descent [Section 4.5.3], the model iteratively

[2] Either predefined or random initial values may be used.

adjusts its parameters over-and-over again to minimize errors and enhance accuracy. This iterative process enables the model to learn the underlying patterns in the data, and find the best "settings".

During training, it's crucial to evaluate the model's progress using a separate **validation** dataset. This practice prevents the model from merely memorizing details from the training data and ensures its ability to effectively handle novel data.

Hyperparameters [Section 3.6], such as learning rates, can also be adjusted to optimize the model's learning process and enhance its performance. We discuss hyperparameters in the next section.

After training and validation, the model's accuracy is evaluated using metrics like the frequency of accurate predictions or the closeness of its predictions to reality. If the model performs satisfactorily, it can be deployed in real-world applications for automatic outcome prediction or data classification.

Accurate model fitting empowers machine learning models to discern valuable patterns, automate repetitive tasks, and augment decision-making across various domains.

3.6 HYPERPARAMETER TUNING

In machine learning, **hyperparameters** are parameters set prior to the commencement of the learning process and cannot be directly derived from the data. They govern the behavior of the machine learning algorithm and influence the training process and the model's final performance. In contrast to model parameters, which are adjusted during training based on the data (e.g., weights in neural networks or coefficients in linear regression), hyperparameters are predefined settings that remain constant throughout the training process. Examples of hyperparameters include learning rates[3] in optimization algorithms, the number of layers and neurons in a neural network [Section 10.3], the depth of decision trees [Section 9.2], or the regularization parameters in regression models [Section 11.8].

Hyperparameter tuning is the process of identifying the optimal set of hyperparameters that maximize the performance of a specific machine learning model for a given task. Several methods are commonly employed for hyperparameter tuning, each with its strengths and suitability contingent upon factors such as the model's complexity and the computational resources available.

[3] In machine learning, the learning rate governs the speed at which a model adjusts its weights during training. It determines the step size employed in gradient descent, which minimizes the loss function. A high learning rate can result in rapid convergence to suboptimal solutions or divergence, while a low rate leads to protracted training that can become stuck in a local minimum. Consequently, achieving the optimal learning rate is crucial for optimizing performance, which often necessitates experimentation or learning rate scheduling.

Grid search is a systematic approach that involves defining a grid of hyperparameter values to be evaluated. For each combination of values within the grid, the model is trained and evaluated to identify the combination of hyperparameters that yield the most optimal performance. This method provides a comprehensive exploration of the **hyperparameter space**, although it can be computationally demanding, particularly for extensive search spaces.

In contrast, **random search** randomly selects hyperparameter values within predefined ranges. This approach can be more efficient than grid search when the impact of specific hyperparameters on performance is uncertain or when computational resources are limited. By evaluating a smaller number of randomly chosen combinations, random search can sometimes identify optimal settings more quickly.

Bayesian optimization models the tuning process as a probabilistic model, employing Gaussian processes[4] to approximate the function that maps hyperparameters to the target outcome. It systematically explores the hyperparameter space, striking a balance between **exploration** and **exploitation**. This approach efficiently finds good hyperparameters by using an acquisition function—such as Expected Improvement or Upper Confidence Bound—to guide the search toward promising areas. Bayesian optimization is especially useful for models that are expensive to evaluate, often achieving better results with far fewer trials than traditional methods.

Gradient-based optimization methods tune hyperparameters using techniques similar to those used when training a model. This requires that the hyperparameters be differentiable with respect to the objective function, allowing them to be optimized directly with methods like **gradient descent**. Although powerful, gradient-based optimization is generally limited to hyperparameters that can be treated as continuous variables.

Ensemble methods offer another strategy by combining predictions from multiple models trained with different hyperparameter settings. By leveraging the diversity among these models, ensembles often achieve better performance than any single model tuned with one set of hyperparameters. They are especially effective for improving overall accuracy and increasing robustness across many machine learning tasks.

The selection of an optimal hyperparameter tuning method is contingent upon the specific attributes of the machine learning problem. These attributes encompass the model's complexity, the magnitude of the hyperparameter search domain, and the available computational resources. Hyperparameter tuning proves particularly advantageous when refining model performance can substantially enhance predictive accuracy and generalization. It holds particular utility for intricate machine

[4]Gaussian processes are a flexible statistical tool used in machine learning and spatial analysis. They model a set of random variables that follow a joint Gaussian distribution, described by a mean function and a covariance function. In Bayesian optimization, they help represent an unknown function and the uncertainty around it, guiding the search for the best solution by balancing trying new options and refining promising ones—especially when each evaluation is expensive.

learning models, such as deep neural networks or ensemble methods, wherein even minute alterations in hyperparameters, such as learning rates, regularization intensities [Section 11.8], or architectural intricacies, can lead to substantial improvements in model efficacy. Through systematic adjustments of these parameters, practitioners can identify configurations that optimize model performance for specific tasks and datasets.

Conversely, hyperparameter tuning may not be necessary or even be counterproductive in certain situations. For instance, with simple models that have fewer hyperparameters or when computational resources are limited, exhaustive search methods may not be feasible due to their high time and resource costs. Additionally, when working with small or homogeneous datasets, hyperparameter tuning increases the risk of overfitting the model to the training data, leading to poor generalization to new data. In such cases, it may be more prudent to focus on improving data quality, increasing dataset size, or choosing simpler models that are less sensitive to hyperparameter variations. Therefore, the decision to use hyperparameter tuning should consider the trade-offs between potential performance gains and practical constraints specific to the machine learning task at hand.

3.7 VALIDATION AND CROSS-VALIDATION

Validation and cross-validation are fundamental techniques in machine learning employed to assess the performance and reliability of models. These techniques ensure that testing data remains untouched for unbiased evaluation.

Validation typically involves dividing the available data into two subsets: the **training set** and the **validation set**. The model is trained on the training set and evaluated on the validation set to assess its performance on novel, unforeseen data. Common validation techniques include **hold-out validation**, where the data is randomly partitioned into training and validation sets with a predetermined ratio (e.g., 80% training, 20% validation), and **stratified** validation, which guarantees that each class label is represented in the same proportion across training and validation sets.

Cross-validation extends the validation concept by partitioning the data into multiple subsets, or **folds**. The model is trained and evaluated multiple times, with each fold serving as the validation set precisely once and the remaining folds as the training set. This approach ensures that every data point is utilized for both training and validation, thereby providing a more robust estimation of model performance. Common types of cross-validation include **k-fold cross-validation**, where the data is divided into k equal-sized folds, and **leave-one-out cross-validation**, where each data point serves as a separate validation set.

Both validation and cross-validation are indispensable for evaluating model performance, detecting potential issues such as **overfitting** or **underfitting**, and guiding decisions on model selection and hyperparameter tuning. Through systematic assessments of how well models generalize to novel data without utilizing the testing data, these techniques empower machine learning practitioners to construct more reliable and effective models across a wide range of applications.

3.8 DATA

In the domains of machine learning and data science, the term "data" encompasses raw, unprocessed facts and measurements collected from various sources, including sensors, surveys, and databases. These data points form the basis for analyses, making it possible to uncover useful insights and build predictive models.

Data manifests in diverse formats. **Structured data** is organized data in predefined formats, such as database tables, which encompass numerical values and dates. It finds widespread application in **relational databases** and spreadsheets. In contrast, **unstructured data** lacks a predefined structure and encompasses text, images, videos, and social media posts. Analyzing such data often necessitates sophisticated techniques, including natural language processing and computer vision. **Semi-structured data** occupies a middle ground between these two categories, exhibiting some degree of organization. Examples include XML or JSON files, which are commonly encountered in web data and NoSQL databases.

Data originates from a diverse range of sources. **Transactional data** originates from business operations, including sales and financial transactions. **Sensor data** is captured by devices monitoring physical conditions, such as temperature or movement. Social media platforms like Twitter and Facebook contribute **social media data**. **Textual data** encompasses data in documents and in customer feedback. **Biometric data** involves physiological measurements, such as heart rate or brain activity. Furthermore, there are numerous other sources of data.

The quality of data significantly influences the efficacy of machine learning models. **Data cleaning** is frequently necessary, encompassing tasks such as error correction, managing missing values, and ensuring data consistency. **Feature engineering**[5] plays a pivotal role by selecting and transforming data features to augment the performance of models.

Ethical considerations must be carefully addressed when working with data, including issues of **privacy**, **bias**, and **security**. Following responsible data practices is essential for protecting individuals' rights and ensuring **fairness** in decision-making processes.

[5] Feature engineering encompasses the process of selecting, modifying, or constructing novel variables or features to enhance the performance of machine learning models. This includes techniques such as one-hot encoding, normalization, polynomial feature generation, and feature selection. We discuss these later in this book.

3.9 TRAINING AND TESTING

Training and testing are fundamental phases in machine learning that, when combined, facilitate the development of accurate predictive models capable of making informed decisions based on data.

The **training** phase commences the model development process by imparting patterns derived from labeled data to the model. During training, the model acquires the ability to associate input features (independent variables) with corresponding output labels (dependent variables or target values). The primary objective is to modify the model's internal parameters, such as weights in neural networks, coefficients in regression models, or decision boundaries in decision trees, so that it can accurately predict or classify novel data. This iterative process entails optimizing these parameters by employing algorithms such as gradient descent [Section 4.5.3]. The ultimate objective is to minimize prediction errors, thereby augmenting the model's capacity to discern intricate relationships within the data. Techniques such as regularization [Section 11.8] serve to mitigate overfitting, wherein the model becomes excessively attuned to the training data and encounters challenges in generalizing effectively to novel, unforeseen data. Furthermore, feature engineering assumes a pivotal role in enhancing model performance by selectively choosing or transforming input variables to improve predictive accuracy.

Once the model has completed the training phase and its parameters have been optimized, it proceeds to the **testing** phase. In this phase, the model's performance is evaluated using a separate dataset known as the test set. This dataset contains data that the model has not encountered during training and serves as a proxy for real-world scenarios. The primary objective of testing is to assess how well the model generalizes to new data and makes accurate predictions or classifications. During testing, performance metrics specific to the task at hand—such as accuracy, precision, and recall for classification tasks, or mean squared error for regression tasks—are computed. These metrics quantify how closely the model's predictions align with the actual outcomes in the test data, providing valuable insights into its effectiveness and readiness for deployment.

The practice of dividing data into distinct training and testing sets ensures that the evaluation of a model's performance is unbiased and reliable. Models that perform well on the testing set, demonstrating high accuracy and robustness, are considered suitable for real-world deployment to automate decision-making processes and generate actionable insights. Through iterative refinement and rigorous testing, machine learning practitioners can develop models that not only perform well on known data but also have the ability to generalize and adapt to new, unseen data, driving advancements across various domains.

TABLE 3.1
An example of a binary confusion matrix.

Values	Actually True	Actually False
Predicted True	True Positive	False Positive
Predicted False	False Negative	True Negative

TABLE 3.2
Confusion matrix of a model designed to classify emails.

Value	Actually Spam	Actually Not Spam
Predicted Spam	50	10
Predicted Not Spam	5	100

3.10 CONFUSION MATRIX

Confusion matrix is a simple but powerful tool used to evaluate classification models in machine learning. It works by comparing a model's predicted labels with the true labels, making it easy to see where the model is getting things right—and where it is making mistakes. By highlighting different types of errors, the confusion matrix offers a clear picture of a model's overall strengths and weaknesses.

In a binary classification problem, where there are only two classes, the confusion matrix consists of four key components [Table 3.1]. The matrix is a 2x2 table with the following elements:

1. True Positives (TP): The count of instances accurately identified as positive.
2. False Positives (FP): The count of instances erroneously identified as positive.
3. True Negatives (TN): The count of instances accurately identified as negative.
4. False Negatives (FN): The count of instances erroneously identified as negative.

For instance, consider a model designed to classify emails as either "spam" or "not spam". Suppose the confusion matrix for this model is as shown in Table 3.2.

In this table, the model accurately identified 50 spam emails (True Positives) and 100 legitimate emails (True Negatives). However, it erroneously classified 10 legitimate emails as spam (False Positives) and failed to detect 5 spam emails (False Negatives).

TABLE 3.3
An example of a non-binary confusion matrix.

Value	Actual Cats	Actual Dogs	Actual Birds
Predicted Cats	40	5	3
Predicted Dogs	10	45	5
Predicted Birds	2	3	50

Confusion matrix facilitates the calculation of crucial metrics such as precision, recall, and accuracy, which offer a comprehensive evaluation of the model's performance. In multi-class classification problems involving more than two classes, the confusion matrix becomes more intricate. It is an n x n table, where n denotes the number of classes. Each cell within the matrix corresponds to the count of instances for a particular combination of true and predicted classes. For instance, if a model classifies images into three categories: "cats", "dogs", and "birds", the confusion matrix might look as shown in Table 3.3.

In this analysis, the model accurately predicted 40 images of cats, 45 images of dogs, and 50 images of birds. However, it misclassified certain images, such as 3 birds predicted as cats and 10 cats predicted as dogs. In such cases, the confusion matrix facilitates the identification of specific errors, including classes that are frequently misclassified. It serves as a valuable tool for enhancing model accuracy and pinpointing areas where adjustments are necessary.

In essence, the confusion matrix offers a comprehensive analysis of a classification model's performance, irrespective of whether it is employed for binary or multi-class classification tasks. Through a thorough examination of the matrix, it becomes possible to discern the model's accuracy, precision, recall, and the specific nature of its errors. This analysis facilitates the formulation of more informed decisions regarding the enhancement of the model's performance.

3.11 A NOTE ON METRICS

3.11.1 ACCURACY

Accuracy, a fundamental metric in machine learning, quantifies how frequently a model's predictions align with the actual labels in a dataset. It is determined by dividing the number of correct predictions by the total number of predictions:

$$\text{Accuracy} = \frac{\text{Number of Correct Predictions}}{\text{Total Number of Predictions}}$$

In binary classification, accuracy can be expressed using TP, TN, FP, and FN:

$$\text{Accuracy} = \frac{TP + TN}{TP + TN + FP + FN}$$

This metric is particularly useful when class distributions are balanced and misclassification costs are comparable. However, accuracy may be misleading in imbalanced datasets where one class is more prevalent than the other. In such scenarios, precision, recall, and F1-score are employed in conjunction with accuracy to provide a more comprehensive evaluation.

Overall, while accuracy provides a concise assessment of a model's performance, it should be interpreted with caution and in conjunction with other metrics to gain a comprehensive understanding of a model's efficacy in various scenarios.

3.11.2 SENSITIVITY

Sensitivity, also known as **recall** or **true positive rate**, is a crucial metric in machine learning for assessing the efficacy of a model in identifying positive instances within a dataset. It is calculated as the quotient of true positive predictions over all actual positive instances:

$$\text{Sensitivity} = \frac{\text{True Positive}}{\text{Actual Positive}}$$

In binary classification, sensitivity can be expressed using TP and FN:

$$\text{Sensitivity} = \frac{TP}{TP + FN}$$

In this context, TP represent accurately predicted positive instances, while FN denote positive instances erroneously classified as negative.

For instance, if a medical test accurately identifies 80 out of 100 patients with a disease, the sensitivity would be:

$$\text{Sensitivity} = \frac{80}{100} = 0.8 \text{ or } 80\%$$

Sensitivity is paramount in applications such as medical diagnostics and fraud detection, where any missing of positive cases can have severe repercussions. It is frequently balanced against specificity, which assesses a model's capacity to accurately identify negative instances.

Managing sensitivity entails adjusting classification thresholds to optimize performance in accordance with specific requirements, such as minimizing false negatives.

Put simply, sensitivity reflects a model's ability to correctly identify instances that should be flagged as positive. In applications where failing to detect a true positive is particularly costly, sensitivity becomes an indispensable metric.

3.11.3 SPECIFICITY

Specificity, also known as the true negative rate, is a pivotal metric in machine learning for evaluating the efficacy of a model in identifying negative instances within a dataset. It is calculated as the quotient of true negative predictions over all actual negative instances:

$$\text{Specificity} = \frac{\text{True Negative}}{\text{Actual Negative}}$$

In binary classification, specificity can be expressed using TN and FP:

$$\text{Specificity} = \frac{TN}{TN + FP}$$

Here, TN are negative instances correctly predicted as negative, and FP are negative instances incorrectly predicted as positive.

For example, in a diagnostic test for a condition where 100 individuals do not have the condition, and the test correctly identifies 90 of them as negative, the specificity would be:

$$\text{Specificity} = \frac{90}{100} = 0.9 \text{ or } 90\%$$

This indicates that the test correctly identified 90% of the true negative cases.

Specificity is paramount in applications such as medical testing and spam email filtering, where minimizing false positives is paramount to avoid misdiagnosis or misclassification. It complements sensitivity, which focuses on accurately identifying positive instances.

Specificity can be regulated by adjusting the classification threshold to reduce false positives in a manner appropriate for the intended application. More generally, tuning this threshold allows practitioners to balance false positives and false negatives, supporting robust and context-appropriate performance in real-world settings.

3.11.4 PRECISION

Precision is a pivotal metric in machine learning that assesses the accuracy of positive predictions made by a model, particularly in situations where false positives incur significant costs. It is calculated as the quotient of true positive predictions over all positive predictions:

$$\text{Precision} = \frac{\text{True Positive Predictions}}{\text{All Positive Predictions}}$$

In binary classification, precision can be expressed using TP and FP:

$$\text{Precision} = \frac{TP}{TP + FP}$$

In this context, TP refers to instances that are accurately predicted as positive, while FP denotes instances that are erroneously predicted as positive.

For instance, in a fraud detection system where 100 transactions are flagged as fraudulent, and 80 of these are genuinely fraudulent while 20 are legitimate transactions misclassified as fraudulent, the precision would be:

$$\text{Precision} = \frac{80}{80 + 20} = 0.8 \text{ or } 80\%$$

This implies that 80% of the transactions predicted as fraudulent are indeed fraudulent.

Precision is paramount in applications such as medical diagnostics and spam filtering, where minimizing false positives is paramount. High precision ensures that positive predictions are highly probable to be accurate, thereby reducing unnecessary actions or alerts.

Precision is often balanced with sensitivity (or recall), which quantifies how well a model identifies all relevant positive instances. The trade-off between precision and recall can be tailored to specific application requirements. Furthermore, the F1-score, which integrates precision and recall into a single metric, provides a comprehensive evaluation of a model's performance.

In essence, precision is fundamental for assessing the reliability of positive predictions in classification tasks. By optimizing precision in conjunction with other metrics, practitioners can develop models that effectively address the requirements of their applications, whether in healthcare, security, or other domains where accurate positive identification is critical.

3.11.5 F-MEASURE / F1-SCORE

The F-measure, also known as the F1-score, is a crucial metric in machine learning that assesses the performance of classification models by integrating precision and recall into a single, comprehensive measure. This metric is particularly valuable in scenarios where the balance between false positives and false negatives is paramount.

The F1-score is computed using the formula:

$$\text{F1-score} = 2 \times \frac{\text{Precision} \times \text{Recall}}{\text{Precision} + \text{Recall}}$$

In this context, Precision quantifies the accuracy of positive predictions, while Recall (or Sensitivity) assesses the model's capacity to accurately identify positive instances. This score integrates both metrics, ensuring that they are weighted equally. If either Precision or Recall is low, the F1-score reflects the resulting compromise.

For example, in fraud detection, if a model has a precision of 0.8 (80%) and recall of 0.6 (60%), the F1-score would be:

$$\text{F1-score} = 2 \times \frac{0.8 \times 0.6}{0.8 + 0.6} = 0.6857 \text{ or } 68.57\%$$

This figure showcases the model's balance between precision and recall, offering a consolidated metric of performance.

The F-measure is particularly useful in applications where precision and recall must be treated with equal importance. In medical diagnostics, for example, it provides a balanced assessment of detection performance by accounting for both missed cases (false negatives) and incorrect alarms (false positives). In information retrieval, the F-measure helps evaluate how effectively a system retrieves results that are both comprehensive and relevant. Similarly, in fraud detection, it supports reliable decision-making by jointly capturing the ability to identify fraudulent activity while limiting unwarranted alerts.

Compared to overall accuracy, the F1-score is better suited to imbalanced datasets, as it provides a more informative evaluation by incorporating both major types of misclassification. This makes it a robust metric in contexts where one class is substantially rarer than the other.

The Fβ-score, a generalized form of the F1-score, allows for adjustable weighting of precision and recall based on specific application needs. Higher β values prioritize recall, whereas lower values favor precision.

The F-measure has wide-ranging applications in text and image classification for sentiment analysis, spam detection, and object recognition. It ensures precise classification while minimizing false predictions.

In nutshell, the F-measure (F1-score) serves as a cornerstone in machine learning for evaluating model accuracy, especially in scenarios requiring balanced performance across precision and recall. Its utilization, along with variants like the Fβ-score, empowers practitioners to fine-tune models and optimize outcomes in diverse applications.

3.11.6 KAPPA COEFFICIENT

The kappa statistic, also known as Cohen's Kappa, is a statistical measure employed in machine learning to assess the agreement between classifiers beyond what would be expected by chance. It offers a nuanced evaluation of model performance, particularly valuable in scenarios characterized by imbalanced class distributions.

Cohen's Kappa is calculated using the formula:

$$\kappa = \frac{P_o - P_e}{1 - P_e}$$

Here, P_o represents the observed agreement between the model's predictions and actual labels, while P_e denotes the expected agreement by chance. P_o is computed as the sum of correctly predicted instances (true positives and true negatives) divided by the total number of instances, and P_e is derived based on the distribution of predictions and actual labels.

The interpretation of the kappa statistic ranges from -1 to 1, where:

- A kappa value of 1 indicates perfect agreement between the model and actual labels.
- A kappa value of 0 suggests that the agreement is no better than what would occur by chance.
- A kappa value less than 0 indicates that the agreement is worse than chance.

For practical interpretation:

- Values between 0.81 and 1.00 signify almost perfect agreement.
- Values between 0.61 and 0.80 indicate substantial agreement.
- Values between 0.41 and 0.60 represent moderate agreement.
- Values between 0.21 and 0.40 suggest fair agreement.
- Values between 0.00 and 0.20 indicate poor agreement.
- Values less than 0.00 suggest agreement poorer than chance.

Applications of Cohen's Kappa extend across diverse domains where assessing classifier agreement is paramount, especially in scenarios involving imbalanced data or multiple classes. Cohen's kappa is a particularly useful statistic in evaluating the level of agreement between two raters or observers classifying items into a set of categorical outcomes. It is extremely relevant in fields such as medicine, psychology, and social sciences, where subjective judgment plays a critical role, such as diagnosing diseases or assessing behavioral traits. Unlike simple percentage agreement calculations, Cohen's kappa accounts for the agreement that would occur by chance, providing a more robust measure of inter-rater reliability. For instance, when two doctors independently diagnose a group of patients, Cohen's kappa can assist in determining the consistency of their diagnoses beyond what would be expected randomly. This makes it indispensable for ensuring the reliability and validity of assessments and decisions in research and practice.

It mitigates the influence of chance agreement, thereby providing a more precise evaluation of model performance compared to conventional accuracy metrics. In machine learning, Cohen's Kappa finds applications in medical diagnostics, where it assesses the reliability of diagnostic tests, comparing machine-derived results against human-derived results. It is also employed in natural language processing to evaluate the agreement among annotators, and in image labeling tasks, where it measures the consistency between automated and human-labeled data.

3.11.7 MATTHEWS CORRELATION COEFFICIENT

The Matthews correlation coefficient (MCC) is a robust metric used in machine learning to evaluate the performance of classification models, particularly in scenarios with imbalanced class distributions. Unlike simpler metrics such as accuracy, MCC incorporates all four components of the confusion matrix—TP, TN, FP, and FN—providing a more balanced assessment of predictive performance.

Its formula,

$$\text{MCC} = \frac{TP \cdot TN - FP \cdot FN}{\sqrt{(TP + FP)(TP + FN)(TN + FP)(TN + FN)}}$$

offers a comprehensive measure of the correlation between predicted and actual classifications. For example, if a model yields 50 TP, 10 FP, 5 FN, and 100 TN in a binary classification task, the resulting MCC of approximately 0.8 indicates strong agreement between the model's predictions and the true labels.

The MCC ranges from -1 to 1, where a value of 1 represents perfect predictions, 0 corresponds to performance equivalent to random guessing, and -1 reflects complete disagreement between predictions and actual outcomes. This makes MCC especially valuable in contexts where accuracy may be misleading, such as medical diagnostics or fraud detection, where the consequences of false positives and false negatives differ substantially.

Key advantages of MCC include its balanced evaluation across classes, its robustness to imbalanced datasets, and its applicability to both binary and multi-class classification problems. However, interpreting MCC can be more complex than working with simpler metrics, particularly in multi-class settings.

3.12 TIDYMODELS

Tidymodels (Kuhn and Wickham, 2020) is an extensive ecosystem of R packages designed to simplify and streamline the process of building, tuning, evaluating, and deploying machine learning models. Developed within the R statistical computing environment, `tidymodels` adheres to principles of simplicity, consistency, and integration with the broader `tidyverse` (Wickham et al., 2019) framework. This ecosystem aims to empower data scientists and analysts by providing a comprehensive suite of tools for every stage of the machine learning workflow, from data preprocessing to model evaluation and deployment.

3.12.1 CORE COMPONENTS OF TIDYMODELS

Tidymodels comprises several core packages, each serving a distinct purpose in the machine learning pipeline:

1. **parsnip** (Kuhn and Vaughan, 2025): A central component of the tidymodels ecosystem, `parsnip` provides a unified interface for specifying machine learning models. This package empowers users to define models using a consistent syntax, facilitating seamless transitions between various algorithms without the necessity of adapting to novel APIs. `parsnip` supports a comprehensive range of models, including linear and logistic regression, decision trees, random forests, SVMs, and neural networks.
2. **recipes** (Kuhn et al., 2025b): The `recipes` framework facilitates data preprocessing and feature engineering. It enables users to define reproducible preprocessing steps in conjunction with model specification, integrating data transformations into the modeling workflow. Recipes streamline tasks such as scaling numerical variables, imputing missing values,[6] encoding categorical variables, and generating novel features based on domain expertise.
3. **yardstick** (Kuhn et al., 2025a): This package provides a comprehensive set of metrics and functions for assessing predictive performance. It includes metrics for classification (e.g., accuracy, precision, recall, and F1-score) and regression tasks (e.g., RMSE, R-squared), enabling users to evaluate models consistently across different datasets and algorithms.

3.12.2 ADDITIONAL PACKAGES IN TIDYMODELS

In addition to its core components, `tidymodels` provides a range of supplementary packages that augment its functionality and cater to specific requirements in the domain of machine learning:

1. **tune** (Kuhn, 2025): The `tune` library facilitates hyperparameter tuning, a pivotal process for optimizing model performance. It supports methodologies such as grid search, random search, and Bayesian optimization to discern the optimal combination of model parameters in accordance with predetermined performance metrics.

[6] Imputation in machine learning involves replacing missing data with substituted values, enabling the utilization of more comprehensive datasets for analysis or model training.

2. **workflows** (Vaughan and Couch, 2025): `workflows` enables the creation of end-to-end modeling pipelines by integrating preprocessing, modeling, and evaluation steps into a single workflow object. Workflows enhance reproducibility and facilitate model deployment by encapsulating all necessary steps from data ingestion to final model evaluation.
3. **infer** (Couch et al., 2021): `infer` provides tools for statistical inference and uncertainty estimation. This package encompasses functions for generating prediction intervals, conducting hypothesis tests, and evaluating model assumptions. These functionalities enhance the interpretability and reliability of machine learning models.
4. **rsample** (Frick et al., 2025): `rsample` offers functionality for data splitting and resampling techniques, crucial for robust model evaluation. This framework provides support for various evaluation methods, including cross-validation, bootstrap sampling,[7] and stratified sampling. These methods ensure that models are assessed on diverse subsets of data to evaluate their generalization capabilities.

3.12.3 WORKFLOW IN TIDYMODELS

In tidymodels, the workflow is structured to ensure clarity, repeatability, and alignment with tidyverse principles throughout the model development process. It commences with data preprocessing using `recipes`, which entails tasks such as data cleaning, scaling features, handling missing values, and transforming variables into appropriate formats for modeling.

Subsequently, models are specified using the `parsnip` library, where the machine learning algorithm and hyperparameters are defined. `parsnip` offers a unified interface that simplifies model selection across various techniques, facilitating experimentation and comparison.

Hyperparameter tuning is performed using the `tune` library to optimize model performance. Techniques such as grid search and Bayesian optimization are employed to identify parameter combinations that maximize accuracy or minimize error metrics, thereby enhancing model effectiveness.

Model performance is assessed using the `yardstick` library, which provides specialized metrics tailored to tasks such as classification and regression. Cross-validation techniques from the `rsample` library are employed to ensure the robustness of the evaluation process by testing models on multiple data subsets, thereby providing reliable performance estimates.

---

[7]Bootstrap sampling is a statistical method that assesses sample variability by generating smaller samples from it. Imagine you have a bag of marbles and want to find the average weight without weighing each marble. Instead, you randomly select marbles, record their weights, and return them. By repeating this process, you get several "mini bags" of marbles. You calculate the average weight for each mini bag to estimate the average weight for the container and understand potential variations. This technique is useful when data is limited or for making predictions without strong assumptions about the population.

For deployment purposes, the `workflows` package facilitates the creation of reproducible pipelines that integrate preprocessing, modeling, and evaluation stages. This integration promotes consistency and facilitates model deployment in production environments, ensuring that models are prepared for real-world applications.

3.12.4 BENEFITS AND ADVANTAGES

`tidymodels` stands out as a preferred choice among data scientists and analysts utilizing R for several compelling reasons. Firstly, it promotes consistency throughout the modeling process, providing a unified syntax and workflow that simplifies the learning and application of various modeling techniques. This consistency enhances efficiency and reduces the learning curve for users.

Secondly, `tidymodels` is highly flexible and extensible due to its modular structure. Users can customize and extend its functionalities by installing additional packages to meet specific research objectives or address unique use cases effectively. This flexibility enables adaptation to diverse analytical needs and experimental setups.

Furthermore, `tidymodels` benefits from robust community support within the broader R community and integration with the `tidyverse` ecosystem. It provides comprehensive documentation, tutorials, and active community forums that cater to users of all skill levels. This support ecosystem ensures that users can effortlessly access resources and guidance to overcome challenges and optimize their modeling workflows.

In essence, `tidymodels` is a comprehensive ecosystem of R packages designed to streamline and enhance machine learning workflows. By integrating essential tasks such as data preprocessing, model specification, hyperparameter tuning, and evaluation into a cohesive framework, `tidymodels` empowers users to construct reliable predictive models. Whether for academic research, business analytics, or scientific exploration, `tidymodels` provides a versatile toolkit that supports best practices in machine learning and statistical analysis within the R environment. Its comprehensive approach fosters efficiency, reliability, and interpretability, enabling data-driven decision-making and advancing analytical capabilities across diverse domains.

3.13 CHAPTER SUMMARY

- Machine learning employs a systematic approach from problem formulation to model deployment, leveraging data-driven insights for informed decision-making. This process encompasses prediction using pre-trained models, understanding variables, model construction, parameter adjustment, and model fitting to optimize performance.

- Hyperparameters are predefined settings that influence model behavior and performance. Hyperparameter tuning optimizes the learning process and enhances model performance.
- Machine learning data encompasses structured, unstructured, and semi-structured formats, necessitating diverse analysis techniques.
- Machine learning models are developed through training and testing, utilizing data from various sources, and often requiring data cleaning and feature engineering.
- The confusion matrix evaluates classification models, highlighting errors and guiding improvement.
- Sensitivity, specificity, precision, F-measure, and kappa coefficient are essential metrics for evaluating model performance.
- Sensitivity measures how well positive instances are correctly identified; specificity measures how well negative instances are correctly identified; precision measures the accuracy of positive predictions; recall measures how many actual positives are captured; and the F-measure balances precision and recall.
- The kappa coefficient measures agreement beyond chance, especially useful in imbalanced datasets. It assesses the agreement between raters or classifiers, particularly in imbalanced data or multiple classes. It mitigates chance agreement and provides a more precise evaluation than conventional accuracy metrics.
- The MCC assesses classification model performance, especially in imbalanced class distributions. It considers all four quadrants of the confusion matrix for a comprehensive measure of correct classification across categories.
- Tidymodels is an R package ecosystem that simplifies machine learning workflows. It adheres to principles of simplicity, consistency, and tidyverse integration, empowering data scientists and analysts with comprehensive tools for every stage of the machine learning workflow.

Part II

Machine Learning Algorithms

4 Linear regression

Linear regression, a statistical method, models the correlation between two variables. For example, it helps represent the relationship between study hours and test scores using a straight line [Fig. 4.1]. It involves the following steps:

1. Data Collection: Gather a dataset of pairs of numerical values, where one variable represents study hours and the other corresponds to test scores.
2. Graphical Representation: Plot these pairs on a graph, with study hours on the x-axis and test scores on the y-axis. Each point on the graph represents a specific data point.
3. Line Fitting: Determine the straight line that best fits the data points. This line minimizes the differences between the actual data points and the line. The equation of the line is usually expressed in the form of a linear equation:

$$y = \beta_0 + \beta_1 x$$

In this equation, y represents the test score, x represents the number of hours studied, β_0 is the intercept of the line (where it crosses the y-axis), and β_1 is the slope of the line, indicating the change in test score with each additional hour of study.

The slope (β_1) quantifies the intensity of the correlation between study hours and test scores. A positive slope signifies that an increase in study hours is associated with a corresponding elevation in test scores. The intercept (β_0) represents the test score obtained when no study hours are invested, although its practicality may be limited in certain scenarios.

Once the generated line is established, it can be used for predictive purposes. For example, we can estimate a test score by substituting the number of study hours into the predetermined equation.

In this way, linear regression facilitates the modeling of the correlation between two variables by fitting a straight line through the data points, thereby enabling predictions and insights into the underlying relationship.

4.1 WHAT IS LINEAR REGRESSION?

Linear regression is a fundamental statistical technique employed to model and analyze the **relationship between a dependent variable and one or more independent variables**. Its fundamental premise is to identify a linear relationship that best encapsulates the association between variations in the independent variables

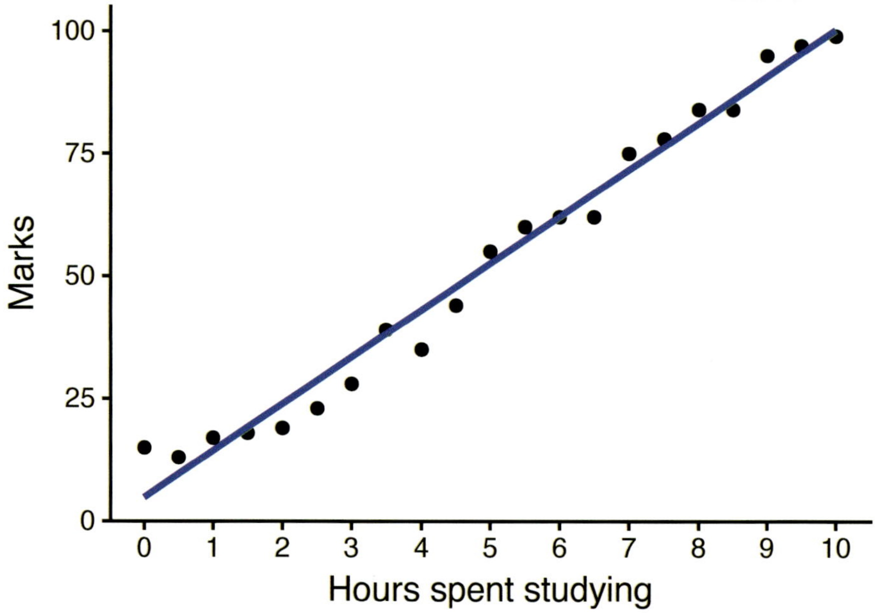

FIGURE 4.1: A plot of number of hours studying versus test scores.

(**predictors**) and consequent alterations in the dependent variable (**outcome**) [Fig. 4.2]. This technique is widely employed across diverse disciplines, such as Economics, Biology, Engineering, and Social Sciences, for the purposes of prediction, comprehension of relationships, and facilitation of decision-making processes.

4.1.1 SIMPLE LINEAR REGRESSION

Linear regression, in its most basic form, involves a single independent variable and a dependent variable. This type of regression is known as simple linear regression. The primary goal is to identify the most appropriate straight line that passes through a given set of data points. This line is mathematically represented by the equation:

$$y = \beta_0 + \beta_1 x + \epsilon$$

In this equation, y represents the dependent variable (the outcome we aim to predict or explain), x represents the independent variable (the predictor), β_0 denotes the intercept of the line, indicating the value of y when x is zero, β_1 represents the slope of the line, signifying the change in y for a one-unit change in x, and ϵ represents the error term, capturing the variability in y that cannot be attributed to the linear relationship with x.

```
                    +--------------------+
                    |     Data Points    | (X, Y)
                    +--------------------+
                              |
                     +-----------------+
                     |                 |
                     v                 v
+--------------------+      +-------------------+
| Independent Var.   |      |  Dependent Var.   |
|        (X)         |      |       (Y)         |
+--------------------+      +-------------------+
                     |                 |
                     +-----------------+
                              |
                              v (Learning Algorithm)
                    +--------------------+
                    |  Regression Line   | (y = mx + c)
                    +--------------------+
                              |
                              v
                    +--------------------+
                    |   New Data Point   | (X_new)
                    +--------------------+
                              |
                              v
                    +---------------------+
                    | Predicted Value (Y')|
                    +---------------------+
```

FIGURE 4.2: A schematic of linear regression.

The process of fitting a linear regression model entails estimating the parameters β_0 and β_1 in a manner that minimizes the sum of the squared differences between the observed values and the values predicted by the model. This method is referred to as "**least squares**" and results in the best-fitting line that minimizes the total error.

4.1.2 MULTIPLE LINEAR REGRESSION

When multiple independent variables are present, the technique extends to multiple linear regression. In this scenario, the model assumes the following form:

$$y = \beta_0 + \beta_1 x_1 + \beta_2 x_2 + \cdots + \beta_p x_p + \epsilon$$

Here, $x_1, x_2, \ldots, x_p$ are the independent variables, $\beta_1, \beta_2, \ldots, \beta_p$ are the coefficients associated with each independent variable, and ϵ represents the error term.

The objective of multiple linear regression is to estimate the coefficients $\beta_0, \beta_1, \ldots, \beta_p$ that best fit the data. This approach, which may involve a similar least squares method, now determines the fit in a multidimensional space defined by all these independent variables.

4.1.3 ASSUMPTIONS AND DIAGNOSTICS

Linear regression yields valid and reliable results only when several fundamental assumptions are met:

1. **Linearity**: The relationship between the dependent and independent variables must be linear.
2. **Independence**: The residuals (errors; observed minus predicted values) must be independent of one another.
3. **Homoscedasticity**: Homoscedasticity refers to the condition in which the variance of the dependent variable remains constant across all levels of the independent variable. For instance, in a regression analysis examining income in relation to age, homoscedasticity implies that the variability of income is consistent across various age groups, irrespective of whether individuals are 20, 40, or 60 years old.
4. **Normality**: The residuals must follow an approximately normal distribution.

Diagnostics are employed to validate these assumptions and ensure the model's accuracy. **Techniques like residual plots, Q-Q plots, and heteroscedasticity tests help assess the suitability of the linear regression model.**

4.1.4 APPLICATIONS AND USE CASES

Linear regression, a versatile statistical method, finds applications in various fields. In finance, it aids in predicting stock prices based on historical data. In healthcare, it quantifies the impact of lifestyle factors on health outcomes. In marketing, it elucidates the correlation between advertising expenditures and sales volumes. Linear regression provides valuable insights into the strength and nature of relationships between variables, enabling informed decision-making and analysis of patterns in data.

One of its key strengths is its high interpretability. Its straightforward equation allows users to easily understand how changes in one variable affect another. This interpretability facilitates the comprehension of the influence of various factors on the outcome. Consequently, linear regression serves as a foundational methodology for modeling and analyzing relationships between variables, ensuring clarity and accuracy in predictions and insights.

4.2 GALTON: THE FATHER OF LINEAR REGRESSION

Sir Francis Galton, a British researcher born in 1822, made groundbreaking contributions to statistics, genetics, and psychology. He was a relative of Charles Darwin and was deeply influenced by Darwin's theories of evolution.

Galton played a pivotal role in developing linear regression. In the late 19th century, he conducted research on the relationships between physical traits, such as height and weight, and demonstrated how one variable could predict another. This groundbreaking work laid the foundation for the modern understanding of linear regression.

Galton introduced the concept of "**regression toward the mean**", which observed that offspring tend to possess traits closer to the average rather than extreme values. This observation led him to draw a line through data points to visualize the relationships between variables, representing an early form of linear regression.

Galton's method involved plotting data points and constructing a line that best approximated the trend, which he utilized to make predictions. Karl Pearson later formalized and expanded this method, introducing the mathematical formulation of linear regression and the correlation coefficient.

Beyond linear regression, Galton made significant contributions to genetics and coined the term "eugenics", which advocated for controlled breeding to enhance human populations. However, these ideas are now considered controversial. Galton also advanced the measurement of human intelligence and individual differences, and his development of statistical methods, including correlation and the normal distribution, had a profound and enduring impact on the field of statistics.

4.3 THE GALTON FAMILIES DATASET

The Galton Families dataset, a significant collection in statistics and data science, particularly for studying regression toward the mean and familial traits, was compiled by Sir Francis Galton in the late 19th century. This dataset was instrumental in exploring hereditary traits and their transmission across generations.

The dataset comprises height measurements of various family members, primarily focusing on the heights of parents and their children. It provides data on the heights of fathers, mothers, and their offspring, allowing researchers to analyze the relationships between these variables and study how traits like height are inherited.

Galton's research centered around the phenomenon of regression toward the mean, a statistical principle that highlights the tendency for extreme traits in a generation to diminish in subsequent generations. For instance, exceptionally tall (or short) parents are more likely to have tall (or short) offspring. But the height of the offspring is less likely to be as extreme as the parents, but comparatively more aligned with the average population height.

The dataset includes columns for the heights of fathers, mothers, and their children, facilitating the analysis of the impact of parental heights on their offsprings' heights and the observation of a gradual shift in the average height over generations.

We begin by (installing and) loading all the required packages and libraries.

```
if (!require("tidymodels")){
  install.packages("tidymodels")
}
library(tidymodels)

if (!require("tidyverse")){
  install.packages("tidyverse")
}
library(tidyverse)

if (!require("HistData")){
  install.packages("HistData")
}
library(HistData)

if (!require("janitor")){
  install.packages("janitor")
}
library(janitor)
```

The HistData package (Friendly, 2025) provides access to GaltonFamilies dataset. Let us explore this dataset. We begin by looking at the top few values using the glimpse() function of dplyr package:

```
GaltonFamilies %>%
  glimpse(width = 50)
```

```
## Rows: 934
## Columns: 8
## $ family          <fct> 001, 001, 001, 001, 002,~
## $ father          <dbl> 78.5, 78.5, 78.5, 78.5, ~
## $ mother          <dbl> 67.0, 67.0, 67.0, 67.0, ~
## $ midparentHeight <dbl> 75.43, 75.43, 75.43, 75.~
## $ children        <int> 4, 4, 4, 4, 4, 4, 4, 4, ~
## $ childNum        <int> 1, 2, 3, 4, 1, 2, 3, 4, ~
## $ gender          <fct> male, female, female, fe~
## $ childHeight     <dbl> 73.2, 69.2, 69.0, 69.0, ~
```

We observe that GaltonFamilies dataset has 934 observations on 8 variables. The details of the variables are as follows:

1. `family`: Family ID, a factor[1] with levels 001–204.
2. `father`: Height of father.
3. `mother`: Height of mother.
4. `midparentHeight`: Mid-parent height, calculated as (father + 1.08*mother)/2.
5. `children`: Number of children in family.
6. `childNum`: Number of a child within family. Children are listed in decreasing order of height for boys followed by girls.
7. `gender`: Child gender, a factor with levels: female, male.
8. `childHeight`: Height of child.

Heights are given in inches. Prior to proceeding, we observe that the variable `midparentHeight` has been defined in an unusual manner, with a weight of 1.08 assigned to the mother's height. To gain a better understanding of this, let us analyze the ratio of father's height to mother's height across the entire dataset. Given that the dataset contains family-wise information, the heights of both parents are repeated multiple times. To address this challenge, we can group the data by family ID and extract a single value for each family. Subsequently, we can ungroup the data and calculate the mean heights of fathers and mothers, as well as the ratio of these mean values. In the `tidyverse`, this process can be implemented as follows:

```
GaltonFamilies %>%
  # Group data by family
  group_by(family) %>%
  # Take 1 row for each family
  slice(1) %>%
  # Ungroup data
  ungroup() %>%
  # Compute summary statistics
  summarize(
    f = mean(.$father),
    m = mean(.$mother),
    r = f / m
  )
```

```
## # A tibble: 1 x 3
##       f     m     r
##   <dbl> <dbl> <dbl>
## 1  69.3  64.0  1.08
```

[1] Factor is a data structure designed to represent fields that can take on a predefined set of distinct values, referred to as levels.

Here # is used to insert comments—lines used to make the code more human-readable and understandable, but which are not processed by R.

We find that the mean height of fathers is 69.3 inches, while the mean height of mothers is 64.0 inches. The ratio of these heights is 1.08, explaining the weight. By `transmuting` the heights of mothers with this weight, Galton tried to accommodate the gender differences of parents while computing the mid-parent height.

We also notice that `midparentHeight` is oddly worded. **Camel case**, this writing convention, is known for compound words or phrases where each word starts with a capital letter, except for the first word. For example, "myVariableName" or "calculateInterestRate", or even "camelCase".

In R programming, variable and function names are typically written in **snake case**. This convention involves using lowercase letters and separating words with underscores (_). This naming style enhances readability, especially for names that span multiple words. Examples include "my_variable_name" or "calculate_interest_rate".

We can convert the variables to snake case using the `clean_names()` function from `janitor` package (Firke, 2024) using the following code, where we use `glimpse()` function to see the results:

```
GaltonFamilies %>%
  clean_names("snake") %>%
  glimpse(width = 50)
```

```
## Rows: 934
## Columns: 8
## $ family            <fct> 001, 001, 001, 001, 002~
## $ father            <dbl> 78.5, 78.5, 78.5, 78.5,~
## $ mother            <dbl> 67.0, 67.0, 67.0, 67.0,~
## $ midparent_height <dbl> 75.43, 75.43, 75.43, 75~
## $ children          <int> 4, 4, 4, 4, 4, 4, 4, 4,~
## $ child_num         <int> 1, 2, 3, 4, 1, 2, 3, 4,~
## $ gender            <fct> male, female, female, f~
## $ child_height      <dbl> 73.2, 69.2, 69.0, 69.0,~
```

For clarity, we may wish to rename `father` to `father_height`, and `mother` to `mother_height`. To this end, the `rename()` function from tidyverse's `dplyr` package is quite useful. We employ the following code:

```
GaltonFamilies %>%
  clean_names("snake") %>%
  rename(
    "father_height" = "father",
    "mother_height" = "mother"
  ) %>%
  glimpse(width = 50)
```

```
## Rows: 934
## Columns: 8
## $ family           <fct> 001, 001, 001, 001, 002~
## $ father_height    <dbl> 78.5, 78.5, 78.5, 78.5,~
## $ mother_height    <dbl> 67.0, 67.0, 67.0, 67.0,~
## $ midparent_height <dbl> 75.43, 75.43, 75.43, 75~
## $ children         <int> 4, 4, 4, 4, 4, 4, 4, 4,~
## $ child_num        <int> 1, 2, 3, 4, 1, 2, 3, 4,~
## $ gender           <fct> male, female, female, f~
## $ child_height     <dbl> 73.2, 69.2, 69.0, 69.0,~
```

We may also wish to convert the gender values from `male` and `female` to `Male` and `Female`. To change these values, we may use the `mutate()` function from tidyverse's `dplyr` package, together with the `str_to_title()` function from tidyverses's `dplyr` package. We then convert the values back to factors using the `factor()` function from base R.

```
GaltonFamilies %>%
  clean_names("snake") %>%
  rename(
    "father_height" = "father",
    "mother_height" = "mother"
  ) %>%
  mutate(
    gender = factor(str_to_title(gender),
                    levels = c("Female", "Male")
             )
  ) %>%
  glimpse(width = 50)
```

```
## Rows: 934
## Columns: 8
```

```
## $ family            <fct> 001, 001, 001, 001, 002~
## $ father_height     <dbl> 78.5, 78.5, 78.5, 78.5,~
## $ mother_height     <dbl> 67.0, 67.0, 67.0, 67.0,~
## $ midparent_height <dbl> 75.43, 75.43, 75.43, 75~
## $ children          <int> 4, 4, 4, 4, 4, 4, 4, 4,~
## $ child_num         <int> 1, 2, 3, 4, 1, 2, 3, 4,~
## $ gender            <fct> Male, Female, Female, F~
## $ child_height      <dbl> 73.2, 69.2, 69.0, 69.0,~
```

All these steps come under the category of **data pre-processing**. We next save the final table into a variable called `data` using the assignment operator `<-`. We use the following code:

```
data <- GaltonFamilies %>%
  clean_names("snake") %>%
  rename(
    "father_height" = "father",
    "mother_height" = "mother"
  ) %>%
  mutate(
    gender = factor(str_to_title(gender),
                    levels = c("Female", "Male")
             )
  )
```

Let us have a glimpse of `data`:

```
data %>%
  glimpse(width = 50)
```

```
## Rows: 934
## Columns: 8
## $ family            <fct> 001, 001, 001, 001, 002~
## $ father_height     <dbl> 78.5, 78.5, 78.5, 78.5,~
## $ mother_height     <dbl> 67.0, 67.0, 67.0, 67.0,~
## $ midparent_height <dbl> 75.43, 75.43, 75.43, 75~
## $ children          <int> 4, 4, 4, 4, 4, 4, 4, 4,~
## $ child_num         <int> 1, 2, 3, 4, 1, 2, 3, 4,~
## $ gender            <fct> Male, Female, Female, F~
## $ child_height      <dbl> 73.2, 69.2, 69.0, 69.0,~
```

We find that it has been saved correctly!

4.4 OBSERVING SIMPLE LINEAR REGRESSION IN GALTON FAMILIES DATASET

Let us analyze the variation in a child's height with the mother's height. We visualize the data using the ggplot() function from the tidyverse package's ggplot2 library. We define the aesthetics using the aes() function, specifying that the x-axis should represent the variable mother_height and the y-axis should represent the variable child_height. Thus, the following code emerges:

```
data %>%
  ggplot(
    aes(
      x = mother_height,
      y = child_height,
      color = gender
    )
  )
```

In this scenario, mother_height serves as the independent variable, while child_height functions as the dependent variable. In an experiment, the independent variable is manipulated to observe its impact on the dependent variable. We further specify that we desire distinct colors for each gender.

Once the basic aesthetics have been defined, we insert the next layer using the + command. We state that we wish to plot points (to generate a scatterplot) using the geom_point() function. Thus, the code now becomes:

```
data %>%
  ggplot(
    aes(
      x = mother_height,
      y = child_height,
      color = gender
    )
  ) +
  geom_point()
```

On the next layer, we specify that we desire a smooth line plotted through the points, as Galton would have plotted. To achieve this, we specify the smoothing function method to be lm, or linear model. We further indicate that we do not wish to plot confidence intervals by setting se to FALSE. Thus, the code now becomes:

```
data %>%
  ggplot(
    aes(
      x = mother_height,
      y = child_height,
      color = gender
    )
  ) +
  geom_point() +
  geom_smooth(method = "lm", se = FALSE)
```

On the subsequent layer, we add the labels. We specify the title, labels for the x and y axes, and the colors for the legend.

Additionally, we specify that we desire a classic theme for a clean graph using the `theme_classic()` function. By incorporating all these commands, we can generate the graph. Following is the complete code that emerges:

```
data %>%
  ggplot(
    aes(
      x = mother_height,
      y = child_height,
      color = gender
    )
  ) +
  geom_point() +
  geom_smooth(method = "lm", se = FALSE) +
  labs(
    title = "Child's height vs. mother's height",
    x = "Mother's height (in inches)",
    y = "Child's height (in inches)",
    color = "Gender"
  ) +
  theme_classic()
```

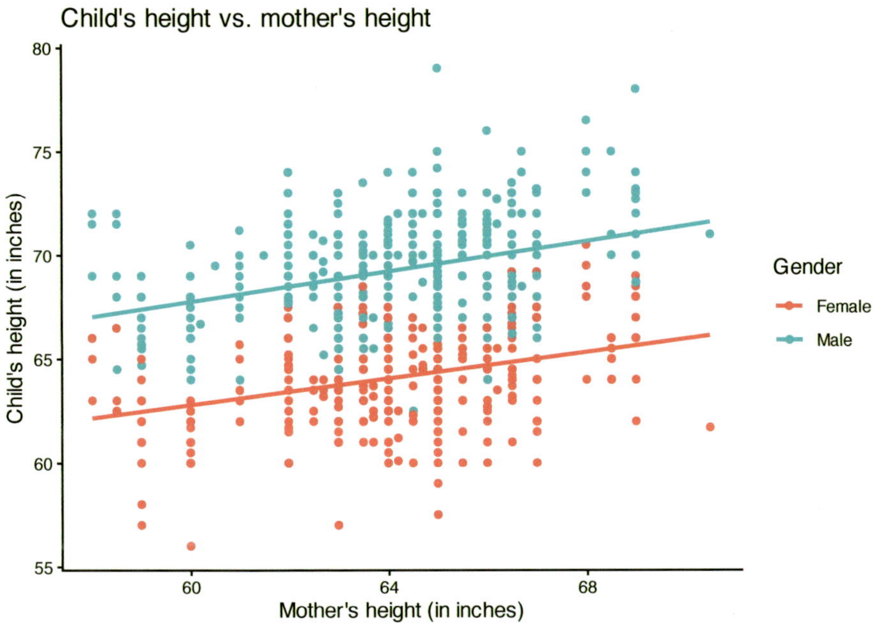

We observe that taller mothers tend to have taller children. In other words, the height of a child is generally proportional to the height of their mother.

Additionally, we notice that for any given mother's height, sons tend to be taller than daughters. This reflects biological sex differences in height distributions, and suggests that the gender of a child also plays a role in determining their height.

Furthermore, the heights of children tend to cluster around the mean value. The mean height of a mother is approximately 64 inches:

```
mean(data$mother_height)
```

```
## [1] 64.08929
```

For mothers with a sub-average height, their daughters tend to be taller than them. For instance, if a mother stands at 60 inches tall, their daughters are usually around 62 inches tall. This means that the daughters' heights are moving toward the average height of 64 inches.

On the other hand, for mothers with an above-average height, their daughters are generally shorter than them. For example, if a mother stands at 68 inches tall, their daughters are usually around 64 inches tall. This also indicates that the daughters' heights are moving toward the average height of 64 inches.

We can do a similar analysis with father's heights by writing a similar code:

```
data %>%
  ggplot(
    aes(
      x = father_height,
      y = child_height,
      colour = gender
    )
  ) +
  geom_point() +
  geom_smooth(method = "lm", se = FALSE) +
  labs(
    title = "Child's height vs. father's height",
    x = "Father's height (in inches)",
    y = "Child's height (in inches)",
    colour = "Gender"
  ) +
  theme_classic()
```

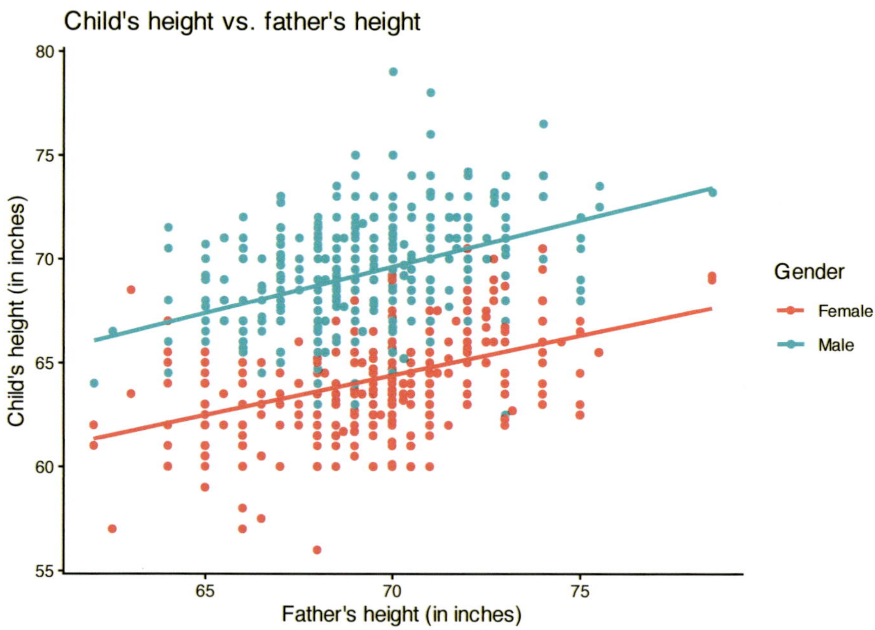

We observe that as fathers grow taller, their children tend to grow taller as well. In other words, taller fathers have taller children.

Additionally, for any given father's height, sons are generally taller than daughters. This means that the gender of the child influences their height.

Furthermore, the heights of children tend to regress toward the average height. The average father's height is approximately 69 inches:

```
mean(data$father_height)
```

```
## [1] 69.19711
```

For fathers with a height below the average, their sons tend to be taller than they are. For instance, if we consider fathers with a height of 65 inches, we'll find that their sons are typically around 67 inches tall. This suggests that the sons' heights are moving toward the average height of 69 inches.

On the other hand, for fathers with a height above the average, their sons are usually shorter than they are. For example, if we observe fathers with a height of 75 inches, we'll discover that their sons are usually around 71 inches tall. This indicates that the sons' heights are also moving toward the average height of 69 inches.

Galton observed this very phenomenon!

4.5 ESTIMATING REGRESSION PARAMETERS

We'll discuss estimating regression parameters using a simple linear regression example. The method can be scaled accordingly for multiple linear regression.

In simple linear regression, the relationship between the dependent variable y and the independent variable x is expressed using the equation of a straight line:

$$y = \beta_0 + \beta_1 x + \epsilon.$$

In this equation, y is the dependent variable that we are trying to predict or explain, while x is the independent variable used to make predictions about y. The term β_0 represents the intercept of the line, indicating the value of y when x is zero. The slope of the line is represented by β_1, which shows how much y changes for each unit change in x. Lastly, ϵ is the error term, capturing the difference between the observed values and the values predicted by the model. This error term accounts for the variability in y that cannot be explained by x alone.

The parameters β_0 and β_1 need to be estimated. This can be done through several methods.

4.5.1 THE ORDINARY LEAST SQUARES METHOD

The **Ordinary Least Squares (OLS)** method, a fundamental statistical technique, estimates the parameters of a linear regression model. It aims to find the line (or plane in multiple regression) that best fits a set of data points by minimizing the sum of squared deviations between observed and predicted values. Essentially, this line (or plane) represents the most accurate linear approximation of the relationship between the independent and dependent variables.

To begin the OLS estimation process, the linear regression model is formulated as $y = \beta_0 + \beta_1 x + \epsilon$, where y is the dependent variable, x is the independent variable, β_0 is the intercept, β_1 is the slope, and ϵ is the error term, which is the difference between the observed and predicted values.

Next, **residuals** are calculated as the differences between the observed values (y_i) and the predicted values ($\hat{y}_i$) for each data point. The predicted value for each x_i is calculated as $\hat{y}_i = \beta_0 + \beta_1 x_i$, and the residual for each data point is expressed as $e_i = y_i - \hat{y}_i$. Estimated values are differentiated from known values by putting a hat, ˆ above the variable names for estimated values, but no hat is used for known values.

The objective of OLS is to minimize the **sum of the squared residuals**, denoted as $\text{SSR} = \sum(y_i - \hat{y}_i)^2$. Minimizing SSR ensures that the line fits the data as closely as possible. To find the best-fitting line, the parameters β_0 and β_1 are solved to minimize the SSR. This is typically done using calculus.

The equations for the parameters are given by

$$\hat{\beta}_1 = \frac{\sum_{i=1}^{n}(x_i - \bar{x})(y_i - \bar{y})}{\sum_{i=1}^{n}(x_i - \bar{x})^2}$$

and

$$\hat{\beta}_0 = \bar{y} - \hat{\beta}_1 \bar{x}$$

where n is the number of data points, and $\bar{x}$ and $\bar{y}$ are the means of x and y, respectively.

Let's now explore how we can establish a correlation between sons' heights and fathers' heights. We begin by filtering the `Male` children. In this case, `father_height` serves as the independent variable, while `child_height` is the dependent variable. We select these two variables from our dataset and store their corresponding values in a new dataset named `son_father`. The code for this process is provided below:

```
son_father <- data %>%
  filter(gender == "Male") %>%
  select(father_height, child_height)
```

We assign the independent variable to xi and compute its mean xbar. We assign the dependent variable to yi and compute its mean ybar. Thus, the code extends to the following:

```
son_father <- data %>%
  filter(gender == "Male") %>%
  select(father_height, child_height)
xi <- son_father$father_height
xbar <- mean(xi)
yi <- son_father$child_height
ybar <- mean(yi)
```

We next use the aforementioned equations to estimate the values of beta0 and beta1:

```
son_father <- data %>%
  filter(gender == "Male") %>%
  select(father_height, child_height)
xi <- son_father$father_height
xbar <- mean(xi)
yi <- son_father$child_height
ybar <- mean(yi)
beta1_hat <- sum((xi - xbar) * (yi - ybar)) /
  sum((xi - xbar)^2)
beta0_hat <- ybar - beta1_hat * xbar
cat("beta0_hat = ", beta0_hat, "\n")
```

```
## beta0_hat =  38.36258
```

```
cat("beta1_hat = ", beta1_hat, "\n")
```

```
## beta1_hat =  0.4465226
```

Let us now plot a straight line with these values. We make a scatterplot using `ggplot` as before, and add a line using `geom_abline()` with the slope given by the value of `beta1_hat`, and the intercept given by the value of `beta0_hat`:

```
son_father %>%
  ggplot(
    aes(x = father_height, y = child_height)
  ) +
  geom_point() +
  # Limit x axis values between 60 and 80
  xlim(c(60, 80)) +
  # Limit y axis values between 60 and 80
  ylim(c(60, 80)) +
  labs(
    title = "Son's height vs. father's height",
    x = "Father's height (in inches)",
    y = "Son's height (in inches)"
  ) +
  geom_abline(
    slope = beta1_hat,
    intercept = beta0_hat
  ) +
  theme_classic()
```

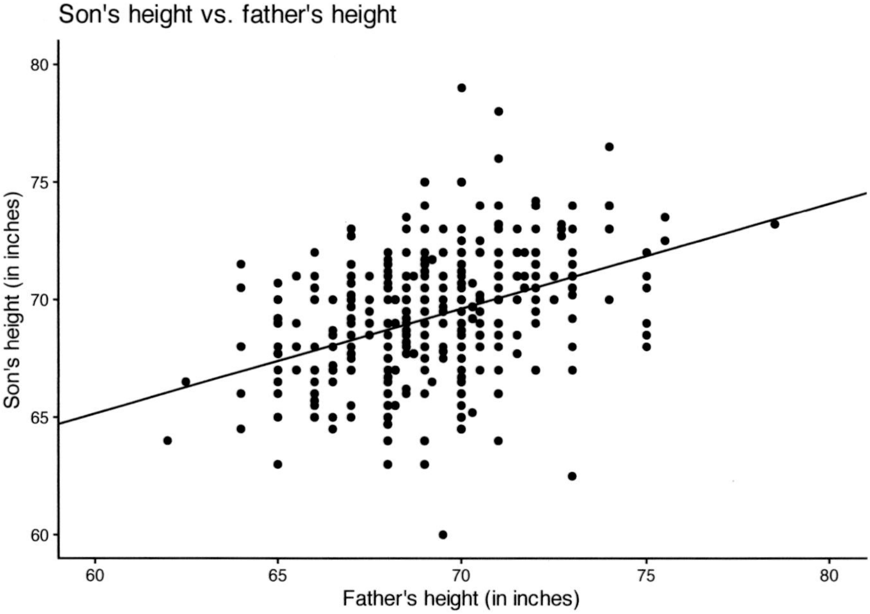

We observe that the straight line goes through the center of the distribution, as expected from linear regression.

OLS estimators exhibit several desirable properties. Firstly, they are **unbiased**, meaning that on average, they accurately estimate the true parameters of the model. This property ensures that the estimators do not systematically overestimate or underestimate the true values.

Secondly, OLS estimators are **efficient**,[2] possessing the minimum variance among the class of linear unbiased estimators, provided that certain assumptions are met. These assumptions include conditions such as homoscedasticity and the absence of autocorrelation.

Lastly, OLS estimates are **consistent**.[Consistency in estimation is a statistical property of an estimator that guarantees that as the sample size increases, the estimator converges toward the true value of the parameter being estimated. This implies that with a larger sample size, the probability of the estimator being close to the true parameter value rises, thereby enhancing the reliability of the estimation.] As the sample size increases, the OLS estimates converge to the true parameter values. This implies that with larger samples, the estimates become more reliable and closer to the actual parameters being estimated.

4.5.2 THE GRID SEARCH METHOD

Grid search, a systematic machine learning approach, evaluates model performance by testing different combinations of hyperparameters. In linear regression, we use a similar technique by creating a grid of coefficient values and testing each combination to find the most accurate estimates for `\beta_0` and `\beta_1`. For example, we can start by defining a mean squared error (MSE) function that takes values for `beta0` and `beta1`, uses them to estimate `child_height` from `father_height`, calculates the corresponding errors, squares them, and returns the mean of squared errors. Here's the code:

```
mse <- function(beta0 = 0, beta1 = 0){
  x <- son_father[,1] # 1st column of son_father is x
  y <- son_father[,2] # 2nd column of son_father is y
  yhat <- beta0 + beta1 * x # Compute the estimated y
  err <- yhat - y # Find the error
  sq_err <- err ^ 2 # Square the error value
  return(mean(sq_err)) # Return the mean of squared error (MSE)
}
```

[2]Efficiency reflects how well an estimator uses the data: the most efficient estimator is the one that gives the most consistent, least variable estimates among all unbiased options.

Subsequently, we define a set of values for beta0 (here, all values from 30 to 50, in increments of 0.01) and beta1 (here, all values from 0 to 1, in increments of 0.01). We employ the expand.grid() function to generate all possible combinations of beta0 and beta1. This serves as the grid that we will utilize to identify the values with the lowest MSE. The code is updated to the following:

```
mse <- function(beta0 = 0, beta1 = 0){
  x <- son_father[,1] # 1st column of son_father is x
  y <- son_father[,2] # 2nd column of son_father is y
  yhat <- beta0 + beta1 * x # Compute the estimated y
  err <- yhat - y # Find the error
  sq_err <- err ^ 2 # Square the error value
  return(mean(sq_err)) # Return the mean of squared error (MSE)
}

grid <- expand.grid(
  beta0 = seq(30, 50, 0.01),
  beta1 = seq(0, 1, 0.01)
)
```

For each value in the grid, we calculate the MSE and store it in a column named mse within the grid data structure. Here's the updated code:

```
mse <- function(beta0 = 0, beta1 = 0){
  x <- son_father[,1] # 1st column of son_father is x
  y <- son_father[,2] # 2nd column of son_father is y
  yhat <- beta0 + beta1 * x # Compute the estimated y
  err <- yhat - y # Find the error
  sq_err <- err ^ 2 # Square the error value
  return(mean(sq_err)) # Return the mean of squared error (MSE)
}

grid <- expand.grid(
  beta0 = seq(30, 50, 0.01),
  beta1 = seq(0, 1, 0.01)
)

for(i in 1:nrow(grid)){
  grid$mse[i] <- mse(grid$beta0[i], grid$beta1[i])
}
```

Finally, we identify the values of beta0 and beta1 within the grid that minimize the mse value. These values represent the optimal estimates of beta0 and beta1. The complete code is presented below:

```
mse <- function(beta0 = 0, beta1 = 0){
  x <- son_father[,1] # 1st column of son_father is x
  y <- son_father[,2] # 2nd column of son_father is y
  yhat <- beta0 + beta1 * x # Compute the estimated y
  err <- yhat - y # Find the error
  sq_err <- err ^ 2 # Square the error value
  return(mean(sq_err)) # Return the mean of squared error (MSE)
}

grid <- expand.grid(
  beta0 = seq(30, 50, 0.01),
  beta1 = seq(0, 1, 0.01)
)

for(i in 1:nrow(grid)){
  grid$mse[i] <- mse(grid$beta0[i], grid$beta1[i])
}

grid[which(grid$mse == min(grid$mse)),]
```

```
##       beta0 beta1      mse
## 90858 38.12  0.45 5.812807
```

We notice that the computed values are remarkably close to those obtained using the conventional least squares method. However, we also acknowledge that the algorithm substantially increased the computation time necessary to obtain these values.

Grid search, in this context, presents both benefits and drawbacks. On the one hand, it provides a systematic and exhaustive method for evaluating different coefficient values. This automation eliminates the need for manual trial and error, significantly simplifying the process of finding the optimal coefficient values. On the other hand, it may not be the most efficient approach for certain scenarios, especially when dealing with a large number of coefficients or in complex models.

Grid search, though valuable, can be computationally expensive due to the extensive processing power needed to evaluate every combination within the grid, especially for large datasets or extensive coefficient value grids. Additionally, it's limited to the specified points in the grid, potentially missing optimal values between them.

In our example, since grid search could only explore values to two decimal places (as defined in the grid), the obtained values differed from those obtained using OLS. Nevertheless, grid search remains a valuable method, especially when exact mathematical solutions are unknown. Its exhaustive search within the defined grid makes it a widely used approach in machine learning.

4.5.3 THE OPTIMIZER METHOD

"Optimizer" is an algorithm used to find the best values for a model's parameters, like the coefficients and intercept. These values are calculated by minimizing the **cost function**, which measures the difference between the model's predictions and the actual outcomes. The most common cost function is **Mean Squared Error** (MSE), which quantifies how well the model's predictions match the actual data. The formula for MSE is:

$$\mathrm{MSE} = \frac{1}{n}\sum_{i=1}^{n}(y_i - \hat{y}_i)^2$$

where y_i is the actual value, $\hat{y}_i$ is the predicted value, and n is the number of data points.

Selecting the right optimization method greatly impacts the efficiency and accuracy of model training. **Gradient descent** is a widely used optimization algorithm that iteratively adjusts model parameters to minimize the cost function. It starts with initial values and updates them based on the steepness of the cost function at each step. The **update rule** for gradient descent is as follows:

$$\theta_j := \theta_j - \alpha \frac{\partial J(\theta)}{\partial \theta_j}$$

This formula has three main components:

1. **Parameters** θ_j: These are the values, such as weights or coefficients, that we are attempting to optimize.
2. **Learning Rate** α: This is a small numerical value that determines the magnitude of the step taken toward the minimum. It controls the extent to which each parameter is modified at each iteration.
3. **Cost Function** $J(\theta)$: This function quantifies the discrepancy between our predictions and the actual results. We aim to minimize this value.

The update rule instructs us to modify each parameter θ_j by adjusting it in the direction that minimizes the cost function the most. This is achieved by subtracting the product of the learning rate and the derivative of the cost function with respect to θ_j from θ_j. This process is akin to descending a slope, where the slope is determined by the derivative, guiding us toward the lowest point, which corresponds to the optimal solution.

Gradient descent has various versions. **Batch gradient descent** uses the entire dataset for updates, which can be slow for large datasets. **Stochastic gradient descent** (SGD) updates parameters iteratively with a single data point, making it faster but potentially less stable. **Mini-batch gradient descent** finds a balance by using a small subset of data for each update, optimizing speed and stability.

Another method is the **normal equation**, which directly provides a solution to linear regression without the need for repeated parameter adjustments. The parameters can be calculated using the formula:

$$\theta = (X^T X)^{-1} X^T y$$

In this formula, X represents the matrix of input features, including a column of ones for the intercept term. y denotes the vector of corresponding target values. This method is suitable for moderate-sized datasets, but it may slow down and become more complex for extremely large datasets because matrix calculations are computationally intensive.

Additionally, there are advanced optimization methods. **Ridge regression** introduces a penalty to the cost function to mitigate overfitting. The modified cost function is:

$$J(\theta) = \text{MSE} + \lambda \sum \theta_j^2$$

where λ is the **regularization parameter**. **Lasso regression** employs a penalty based on the absolute values of the coefficients, and its cost function is:

$$J(\theta) = \text{MSE} + \lambda \sum |\theta_j|$$

Elastic Net combines both Ridge and Lasso penalties, and its cost function is:

$$J(\theta) = \text{MSE} + \lambda_1 \sum |\theta_j| + \lambda_2 \sum \theta_j^2$$

The **conjugate gradient method**, another technique, refines the solution by employing specific directions to find the minimum cost for large-scale problems.

The choice of optimization method depends on the problem's size, available computing power, and dataset characteristics. Gradient descent and its variants are popular because they're adaptable and scalable. The normal equation offers a direct solution but can be computationally expensive. Techniques like Ridge, Lasso, and Elastic Net are often used to improve model performance by incorporating **regularization**.

The R programming language provides the optim() function, which is a general-purpose optimization function that utilizes Nelder-Mead, quasi-Newton, and conjugate-gradient algorithms. In this section, we'll delve into the practical application of this function to determine the regression parameters.

We start by defining a MSE function that accepts initial values for beta0 and beta1 as a vector, along with the data, and returns the MSE value.

```
mse <- function(vect, data){
beta0<-vect[1]
beta1<-vect[2]

data <- data %>%
  rename(Y = child_height, X = father_height) %>%
  mutate(prediction = beta0 + beta1 * X) %>%
  mutate(error = prediction - Y) %>%
  mutate(error_sq = error ^ 2)

MSE <- mean(data$error_sq)
return(MSE)
}
```

Next, we initialize values for beta0 and beta1 in a numeric vector called trial:

```
trial <- c(50, .5)
```

Finally, we provide the trial values, the mse function, and the son_father dataset to the optim() function. We save the output to best_model, which contains the optimal regression values and the minimum MSE value. Here's the complete code:

```
mse <- function(vect, data){
beta0<-vect[1]
beta1<-vect[2]

data <- data %>%
  rename(Y = child_height, X = father_height) %>%
  mutate(prediction = beta0 + beta1 * X) %>%
  mutate(error = prediction - Y) %>%
  mutate(error_sq = error ^ 2)
```

```
MSE <- mean(data$error_sq)
return(MSE)
}

trial <- c(50, .5)
best_model <- optim(trial, mse, data = son_father)
cat(
  "Optimal parameters values:",
  "\nbeta0_hat = ", best_model$par[1],
  "\nbeta1_hat = ", best_model$par[2],
  "\ngiving minimum MSE = ", best_model$value)
```

```
## Optimal parameters values:
## beta0_hat =  38.32216
## beta1_hat =  0.4471065
## giving minimum MSE =  5.81274
```

We notice that the values we get are remarkably similar to those obtained using the OLS method.

4.6 THE LM() FUNCTION

In R, the `lm()` function is a versatile tool for fitting linear regression models. It simplifies the process by eliminating the need to manually input initial values to the optimizer. The `lm()` function in R is known for its straightforward syntax:

```
lm(formula, data)
```

Within this syntax, the `formula` parameter encapsulates the model specification as a formula object. The formula typically follows the format `response ~ predictors`, where `response` is the dependent variable and `predictors` are the independent variables. The `data` parameter refers to the data frame containing the variables specified in the formula.

Let us illustrate its functionality through the utilization of the `son_father` dataset. We assign the results obtained from `lm()` to a variable named `model`:

```
model <- lm(
  child_height ~ father_height,
  data = son_father
)
```

The coefficients stored in the model can then be accessed as follows:

```
model %>%
  coefficients() %>%
  enframe()
```

```
## # A tibble: 2 x 2
##   name            value
##   <chr>           <dbl>
## 1 (Intercept)    38.4
## 2 father_height  0.447
```

`enframe()` is used to transform named atomic vectors or lists into one- or two-column data frames.

We observe that these values are remarkably similar to those we previously discovered.

When using the `lm()` function, it returns an object of class "lm" that contains important information about the fitted model. This object includes estimates of the **regression coefficients**, which show how much each independent variable affects the dependent variable. It also includes the **residuals**, which are the differences between the observed and predicted values. The **fitted values** are the predictions made by the model, and the **R-squared value** shows how well the model explains the variance in the dependent variable. The **p-values** tell us how likely it is that the coefficients are due to chance.

We can extract diagnostic information from a model object using several functions. For example, `summary(model)` offers a detailed summary that includes coefficients, R-squared, and p-values.

```
summary(model)
```

```
##
## Call:
## lm(formula = child_height ~ father_height, data = son_father)
##
## Residuals:
##     Min      1Q  Median      3Q     Max
## -9.3959 -1.5122  0.0413  1.6217  9.3808
##
```

```
## Coefficients:
##                Estimate Std. Error t value Pr(>|t|)
## (Intercept)    38.36258    3.30837  11.596   <2e-16 ***
## father_height   0.44652    0.04783   9.337   <2e-16 ***
## ---
## Signif. codes:
## 0 '***' 0.001 '**' 0.01 '*' 0.05 '.' 0.1 ' ' 1
##
## Residual standard error: 2.416 on 479 degrees of freedom
## Multiple R-squared:  0.154,  Adjusted R-squared:  0.1522
## F-statistic: 87.17 on 1 and 479 DF,  p-value: < 2.2e-16
```

The results show that `father_height` has a statistically significant but modest effect on `child_height`, with a slope of about 0.45 inches per inch of father height and an intercept of 38.4, while the model explains only about 15% of the variability ($R^2 \approx 0.154$), meaning that there are several other factors apart from `father_height` that have an influence on `child_height`.

Additionally, `plot(model)` generates diagnostic plots, like residuals versus fitted values, which aid in model evaluation.

```
plot(model)
```

The **Residuals vs Fitted plot** shows *residuals*—the differences between observed and predicted values—scattered around zero, which is the general pattern expected in a simple linear regression. The **Normal Q–Q plot** compares residuals to a theoretical normal distribution; most points lie along the reference line, indicating that the residuals are approximately normally distributed, aside from minor tail deviations. The **Scale–Location plot** uses the square root of the *standardized residuals*—residuals scaled so they are comparable across observations—to reveal whether their spread changes across fitted values; the spread appears generally consistent. The **Residuals vs Leverage plot** incorporates *leverage*, which measures how unusual an observation's predictor value is, and overlays **Cook's distance**, which is computed for each individual point to show how much the regression would change if that point were removed. Most points show low leverage and low Cook's distance, meaning no observation exerts undue influence. Overall, the diagnostics closely match what is expected for a reasonably well-behaved simple linear regression: roughly random residual scatter, nearly normal residuals, consistent residual spread, and minimal influential points.

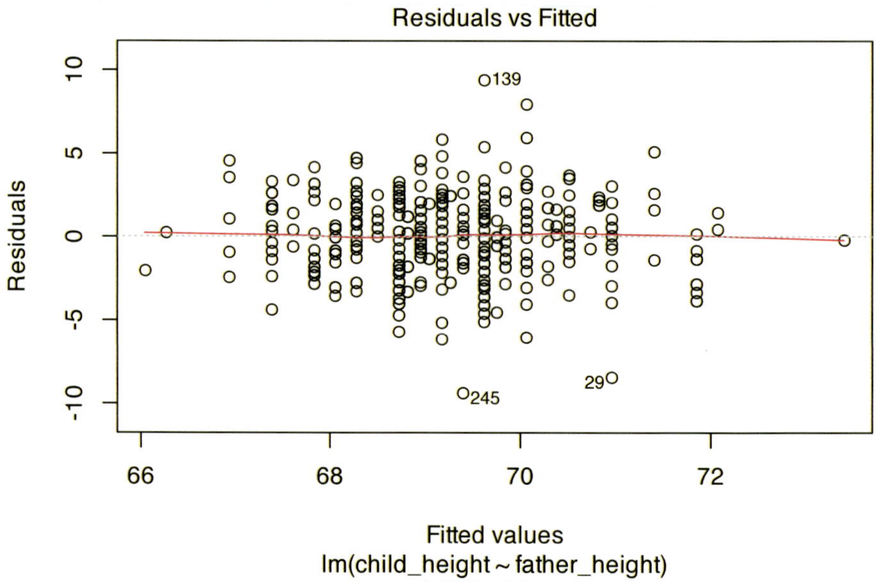
Residuals vs Fitted
Residuals
Fitted values
lm(child_height ~ father_height)
139
245
29
10
5
0
-5
-10
66
68
70
72

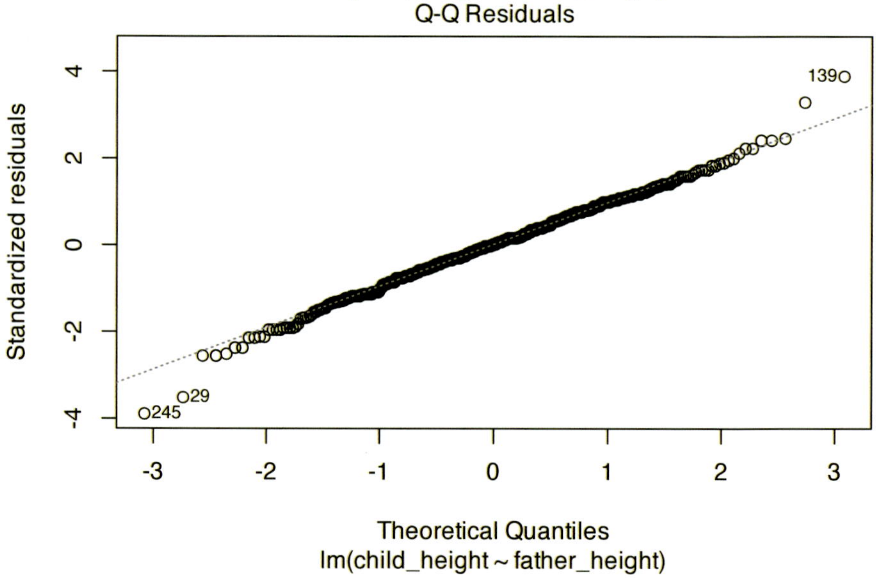
Q-Q Residuals
Standardized residuals
Theoretical Quantiles
lm(child_height ~ father_height)
139
29
245
4
2
0
-2
-4
-3
-2
-1
0
1
2
3

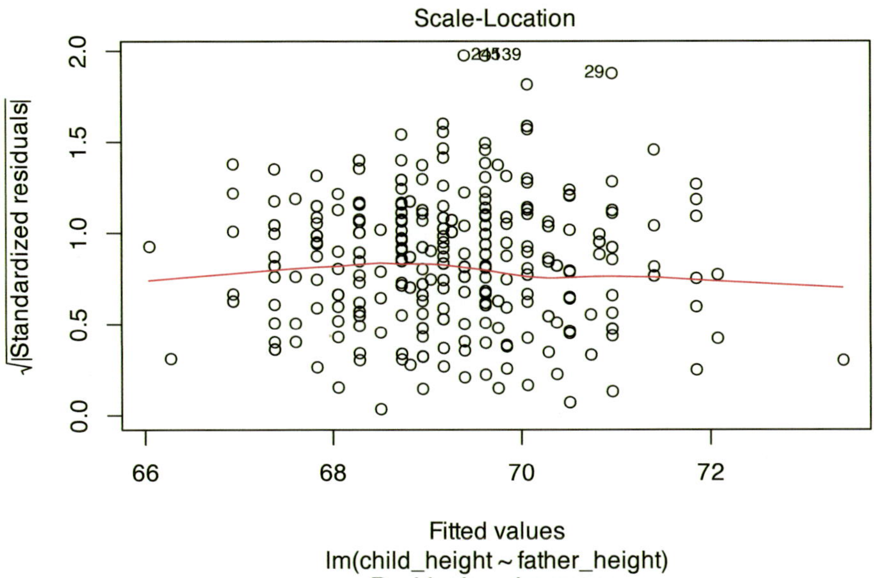

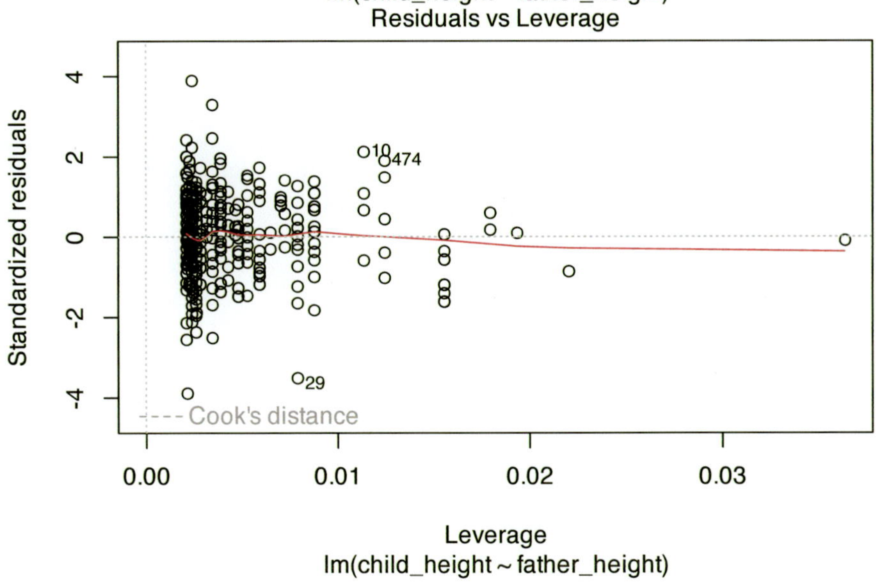

Overall, these diagnostics suggest that a simple linear model is appropriate, with no major assumption violations.

Lastly, `confint(model)` calculates confidence intervals for the model parameters—the range of plausible values for each regression coefficient based on the sample data.

```
confint(model)
```

```
##                    2.5 %     97.5 %
## (Intercept)    31.861862 44.8632998
## father_height  0.352549  0.5404962
```

For the intercept, the 95% confidence interval spans approximately 31.86 to 44.86, meaning the model is fairly certain that the true intercept lies somewhere in this range. For the `father_height` coefficient, the interval ranges from about 0.35 to 0.54, indicating that the true effect of father height on child height is very likely positive and moderate in size. Since this interval does not include zero, it reinforces the earlier conclusion that father height is a statistically significant predictor of child height.

The `lm()` function also supports multiple linear regression, enabling us to model a dependent variable as a function of multiple independent variables. For instance, `lm(y ~ x1 + x2 + x3, data = df)` models y as a function of `x1`, `x2`, and `x3`. Additionally, we can also include interaction terms in the model, such as `lm(y ~ x1 + x2 + x1 * x2, data = df)`, which considers both, the main effects and their interactions.

For instance, if we intend to model `child_height` in relation to `father_height`, `mother_height`, and `gender`, we can simply code the following:

```
model <- lm(
  child_height ~ father_height +
     mother_height + gender,
  data = data
)
```

and then read the regression coefficients from `model`:

```
model %>%
  coefficients() %>%
  enframe()
```

```
## # A tibble: 4 x 2
##    name           value
##    <chr>          <dbl>
```

```
## 1 (Intercept)    16.5
## 2 father_height  0.393
## 3 mother_height  0.318
## 4 genderMale     5.21
```

In this case, we may then write $child\_height = 16.5 + 0.393 \times father\_height + 0.318 \times mother\_height + 5.21 \times genderMale$. In this case, `genderMale` is what we call a **dummy variable** — it is 1 if the gender is Male, and 0 otherwise.

The `lm()` function also offers support for **weighted least squares** through the `weights` argument. This feature enables us to assign varying degrees of importance to specific observations within the dataset.

4.7 TIDYMODELS EXAMPLE

Let us now use `tidymodels` to perform linear regression. The main benefit of using `tidymodels` is that once you've finished coding, switching between different machine learning algorithms is a breeze. Throughout this book, we'll use similar code with various machine learning algorithms.

4.7.1 LOAD DATA

We begin by loading the required information into a data file. This is also an ideal time to convert the raw data into a tibble format. Furthermore, we remove any missing data values (represented by NA) from the dataset. We maintain a clean and intuitive code structure by using the pipe operator %>% from the `tidyverse` package.

```
data <- GaltonFamilies %>%
  as_tibble() %>%
  na.omit()
```

We convert variable names to lowercase, snake_case, and more intuitive names. This step is optional but highly recommended for better comprehension and visualization in tables and charts. We also reduce the dataset size by selecting only essential columns. The complete code at this stage is as follows:

```
data <- GaltonFamilies %>%
  as_tibble() %>%
  na.omit() %>%
```

```
  clean_names("snake") %>%
  rename(
    "father_height" = "father",
    "mother_height" = "mother"
  ) %>%
  mutate(
    gender = factor(str_to_title(gender),
                    levels = c("Female", "Male")
             )
  ) %>%
  dplyr::select(
    father_height, mother_height,
    gender, child_height
  )
```

Let us confirm the operations by examining the `data` variable. Since we've converted our data into a tibble, there is no longer a need to use the `glimpse()` function:

```
data
```

```
## # A tibble: 934 x 4
##    father_height mother_height gender child_height
##            <dbl>         <dbl> <fct>         <dbl>
##  1          78.5          67   Male           73.2
##  2          78.5          67   Female         69.2
##  3          78.5          67   Female         69
##  4          78.5          67   Female         69
##  5          75.5          66.5 Male           73.5
##  6          75.5          66.5 Male           72.5
##  7          75.5          66.5 Female         65.5
##  8          75.5          66.5 Female         65.5
##  9          75            64   Male           71
## 10          75            64   Female         68
## # i 924 more rows
```

`data` looks good!

4.7.2 SPLIT DATA INTO TRAINING AND TESTING

We partition the data into training and testing sets using the `initial_split()` function from the `rsample` package, a component of `tidymodels`. We retain 70%

of the data for training and 30% for testing. These sets are then extracted using `training()` and `testing()` functions:

```
# Split data into training and testing sets
set.seed(2025)
split_7030 <- initial_split(data, prop = 0.7)
data_train <- training(split_7030)
data_test <- testing(split_7030)
```

In this exemplar case, we have set the seed to 2025. `set.seed` is a function used in programming to initialize the random number generator to a specific state, ensuring that the results of random operations are reproducible. By setting the seed, we ensure that the sequence of random numbers generated is consistent each time our code is executed. This consistency facilitates reliable and reproducible results in experiments and simulations, particularly in troubleshooting scenarios. In practice, this line will be removed or commented out upon code deployment, at which point it is not required.

4.7.3 CREATE A RECIPE

In the recipe, we input the formula as `child_height ~ .` because we intend to predict the height of the child using all the available variables in the dataset. This is so because we have already removed the superfluous variables that we do not anticipate having any correlation with the final outcome. We further specify that this recipe should be applied to `data_train`, which corresponds to the training data. Additionally, we convert the gender to a dummy variable using the `step_dummy()` function. The complete recipe code is depicted below:

```
recipe_ml <- recipe(
  child_height ~ .,
  data = data_train
) %>%
  step_dummy(gender)
```

4.7.4 CREATE A MODEL DESIGN

We define a `linear_reg` model using the `lm` engine. In tidymodels, the `lm` engine is used for tidy and efficient linear regression modeling in R. It integrates seamlessly with tidyverse tools like `dplyr` and `ggplot2` for straightforward data manipulation, visualization, and model assessment. It adheres to the conventional formulaic approach to define variable relationships and offers robust methods for model validation, ensuring reliable statistical analysis in data projects. Given our intention to estimate child height based on independent variables, we set the mode to `regression`:

```
model_lin_reg <- linear_reg() %>%
  set_engine("lm") %>%
  set_mode("regression")
```

4.7.5 CREATE A WORKFLOW AND TRAIN IT

We add the recipe and model to the `workflow()` function, and train the model on `data_train` dataset using `fit()` function:

```
workflow_recipe_model <- workflow() %>%
  add_recipe(recipe_ml) %>%
  add_model(model_lin_reg) %>%
  fit(data_train)
```

4.7.6 ASSESS PREDICTION QUALITY WITH TEST DATA

We can now utilize our trained and fitted model to predict child height using the test data set to assess model efficacy. To accomplish this, we employ the `augment()` function. The `augment()` function, originating from the `broom` package in R, appends predictions and residuals from a statistical model to the original dataset. This facilitates the evaluation of a model's fit by juxtaposing actual and predicted values. We use the following code:

```
data_test_with_prediction <- augment(
  workflow_recipe_model, data_test
)
```

Let us now have a glimpse of the augmented file, `data_test_with_prediction`. We select the columns that we are interested in, and consider the first few rows using the `head()` function:

```
data_test_with_prediction %>%
  select(
    father_height,
    mother_height,
    gender,
    child_height,
    .pred
  ) %>%
  head()
```

```
# A tibble: 6 x 5
  father_height mother_height gender child_height .pred
          <dbl>         <dbl> <fct>         <dbl> <dbl>
1          78.5          67   Female         69    68.7
2          78.5          67   Female         69    68.7
3          75.5          66.5 Female         65.5  67.3
4          75            64   Male           71    71.6
5          75            64   Male           68.5  71.6
6          75            58.5 Male           69    70.0
```

We observe that the predicted values (`.pred`) closely align with the actual measured values (`child_height`).

Additionally, we can compute the **accuracy metrics**. Accuracy metrics in machine learning assess a model's performance by comparing its predictions to the actual outcomes. We utilize the augmented `data_test_with_prediction` dataset to evaluate the estimated values against the actual truth using the `metrics()` function as follows:

```
metrics(
  data_test_with_prediction,
  truth = child_height,
  estimate = .pred
)
```

```
## # A tibble: 3 x 3
##   .metric .estimator .estimate
##   <chr>   <chr>          <dbl>
## 1 rmse    standard       2.27
## 2 rsq     standard       0.626
## 3 mae     standard       1.75
```

The root mean squared error (RMSE) is approximately 2.27 inches, the mean absolute error is approximately 1.75 inches, and we have successfully explained approximately 62.6% of the results (R squared = 0.626) from the three parameters: `father_height`, `mother_height`, and `gender`. The remaining variations are likely attributable to the environmental factors (such as food and nutrition, diseases, and parasites) in which the children grew up. However, since these factors were not measured, we are unable to incorporate them into our current linear regression model.

4.8 WHEN TO USE LINEAR REGRESSION, AND WHEN NOT TO

Linear regression, a powerful statistical technique, models the relationship between a dependent variable and one or more independent variables. However, its suitability for all data types and research questions is limited. Therefore, it is crucial to understand when to apply linear regression and when to explore alternative methods.

We should use linear regression when we believe that the relationship between the dependent variable and independent variable(s) is linear. This means that changes in the independent variable(s) lead to proportional changes in the dependent variable. It is also appropriate when the dependent variable is continuous, meaning it can take any value within a range, such as predicting weight based on height. Linear regression assumes that the residuals, or the differences between observed and predicted values, are normally distributed. When this assumption is met, linear regression can provide reliable estimates. Additionally, the variance of the residuals should remain constant across all levels of the independent variable(s), a condition known as homoscedasticity. It is also important that the independent variables are not highly correlated with each other, as this can make it hard to determine the individual effects of each predictor. Linear regression is best suited for simple relationships, where the connections between variables are straightforward.

Conversely, linear regression should not be employed when the relationship between variables is non-linear. In such instances, alternative methods such as polynomial regression may be more appropriate. Linear regression is also unsuitable for categorical dependent variables, such as yes/no outcomes, for which logistic regression would be a more suitable choice. If the variance of the residuals varies across levels of the independent variable(s), known as **heteroscedasticity**, linear regression may not be suitable. Outliers and influential points can significantly impact the results of linear regression. High multicollinearity among independent variables can lead to unstable models and challenging-to-interpret coefficients, necessitating the application of dimensionality reduction techniques or regularization methods. Non-normal residuals can invalidate the confidence intervals and hypothesis tests derived from linear regression, rendering alternative methods more appropriate in such instances. Finally, in high-dimensional datasets with numerous predictors relative to the number of observations, linear regression can overfit the data, necessitating the consideration of regularization or dimensionality reduction techniques.

When linear regression is not appropriate, there are several alternatives for us to consider. Polynomial regression is suitable for modeling non-linear relationships, while logistic regression is appropriate for binary or categorical outcomes. Generalized linear models (GLMs) can accommodate various types of response variables, and decision trees and random forests are effective in managing complex, nonlinear relationships. Support vector machines are useful for classification tasks with intricate boundaries, and neural networks can easily capture intricate patterns in large datasets.

In essence, linear regression serves as a versatile statistical model when its underlying assumptions are met. However, it encounters limitations in scenarios involving non-linear relationships, categorical outcomes, or instances where the residuals deviate from the assumed normality. By comprehending these limitations and exploring alternative methodologies, researchers can effectively select a suitable model that aligns with the characteristics of their data and with their research objectives.

4.9 CHAPTER SUMMARY

- Linear regression models the relationship between a dependent variable and one or more independent variables, minimizing the difference between predicted and actual values. Applications include finance, healthcare, and marketing for prediction and decision-making.
- Galton's work on regression toward the mean and linear regression laid the foundation for modern statistics.
- OLS estimators are unbiased, efficient, and consistent, making them ideal for linear regression.
- Grid search systematically evaluates coefficient value combinations to find optimal values, but it can be computationally expensive.
- Gradient descent is an optimization algorithm that iteratively updates model parameters to minimize the cost function, aiding in accurate predictions.
- The `lm()` function in R simplifies fitting linear regression models, providing a convenient way to analyze data and extract insights.
- Linear regression has limitations when the relationship is non-linear, categorical, or exhibits heteroscedasticity. In such cases, alternative methods like polynomial regression, logistic regression, generalized linear models, support vector machines, decision trees, random forests, and neural networks are more appropriate.

5 Polynomial regression

Envision plotting data points, such as test scores in relation to study hours, on a graph. Occasionally, the relationship between these points deviates from a linear pattern, exhibiting an upward or a downward curvature. This is where polynomial regression shines.

Polynomial regression models a curved line to fit data points. It posits that a flexible curve can more effectively accommodate the data than a straight line. To this end, a polynomial curve comprises equations involving powers of the independent variable x, such as x^2 or x^3, enabling the representation of intricate shapes. In other words, polynomial regression models and interprets non-linear relationships by employing (non-linear) curves—smooth, bending, wavy lines that more accurately align with the data pattern than a rigid straight line.

Just like linear regression, polynomial regression also determines the most suitable curve as the one that minimizes the distance between itself and each data point. Once established, that curve can serve as a predictive tool. For instance, by inputting new study hours to the curve's equation, we can estimate the corresponding test score.

5.1 WHAT IS POLYNOMIAL REGRESSION?

Polynomial regression is a statistical method used to model and analyze the relationships between variables when the data points exhibit a curved pattern instead of a linear trend. Unlike linear regression, which assumes a linear relationship, polynomial regression can accommodate more complex relationships by fitting a curve to the data. This technique becomes particularly useful when the relationship between the independent variable x and the dependent variable y deviates from a linear pattern and follows a more intricate trend.

In polynomial regression, a polynomial function is used to analyze the relationship between variables. A **polynomial function** is an algebraic expression that includes terms with exponents of the independent variable. For example, a second-degree polynomial function can be represented as $y = ax^2 + bx + c$, where a, b, and c are coefficients that need to be determined. This equation generates a parabola, which can model relationships that have either upward or downward curvature. Higher-degree polynomials, such as **cubic functions** ($y = ax^3 + bx^2 + cx + d$) or even higher-degree polynomials, have the ability to capture even more complex patterns in the data.

Polynomial regression involves finding the most appropriate polynomial function that accurately represents the data. This is achieved by estimating the coefficients of the polynomial equation, ensuring that the curve closely approximates the data points. The goal is to minimize the difference between the predicted values generated by the polynomial function and the actual data points. This is often accomplished through techniques such as **least squares optimization**, which is similar to the approach used in linear regression.

Polynomial regression finds applications in various fields, such as economics, biology, and engineering, where non-linear relationships between variables are common. In economics, it allows for the modeling of non-linear changes in consumer demand with price. In biology, it assists in understanding growth patterns of organisms, which rarely follow a straight-line trajectory. For instance, a child's growth is slow in the beginning but experiences a rapid growth spurt during adolescence. A straight-line model, assuming a constant growth rate, fails to adequately capture such growth patterns.

However, there are certain limitations to be aware of. As the degree of the polynomial increases, the curve can become excessively intricate and may effectively fit the noise inside the training data. In such cases, it may perform poorly on novel, unforeseen data due to overfitting. **Overfitting** occurs when the model captures the noise or random fluctuations in the training data instead of the underlying trend, resulting in less accurate predictions for new data.

In essence, polynomial regression is a robust modeling technique used to model relationships that deviate from a linear pattern. By employing polynomial functions, it enables the fitting of a curved line to the data, offering a more flexible approach compared to linear regression. This flexibility allows polynomial regression to accommodate a diverse range of patterns and trends. Nevertheless, it is essential to exercise caution when selecting the appropriate polynomial degree to strike a balance between fitting the data and minimizing the risk of overfitting.

5.2 THE BOSTON DATASET

The Boston Housing dataset, a renowned resource in machine learning and statistics, offers valuable insights into various characteristics of homes in Boston. It is widely utilized for predicting the median value of owner-occupied homes across diverse neighborhoods.

The dataset includes various features that describe aspects of the housing market and neighborhood characteristics. These features include the per capita crime rate in each area, the percentage of residential land zoned for large lots, the proportion of non-retail business acres per town, and a dummy variable indicating proximity to the Charles River. Other features include nitric oxides concentration, the average number of rooms per dwelling, the proportion of owner-occupied units built before 1940, weighted distances to five major Boston employment centers, an index of accessibility to radial highways, the full-value property tax rate per $10,000, the pupil-teacher ratio by town, and the percentage of the population with a low socioeconomic status.

The dataset's target variable is the median value of owner-occupied homes, expressed in thousands of dollars. The primary objective of utilizing this dataset is to construct predictive models capable of estimating home values based on the offered features. It serves as an excellent illustration for practicing linear regression, polynomial regression, and other machine learning techniques, providing a balanced combination of simplicity and practical relevance.

Originally published in 1978 by David Harrison Jr. and Daniel L. Rubinfeld, the Boston Housing dataset has become a widely recognized benchmark in the field of machine learning. Its inclusion in data science libraries and educational materials makes it a frequently used tool for teaching and learning.

However, it's crucial to be aware of the dataset's historical context. Since it reflects conditions from the 1970s, certain features may be biased by historical events. When using this dataset, especially in contemporary contexts, it's essential to consider ethical considerations and comprehend how they might influence the analysis and conclusions.

5.2.1 USING THE DATASET

We begin by (installing and) loading all the required packages and libraries:

```
if (!require("tidymodels")){
  install.packages("tidymodels")
}
library(tidymodels)

if (!require("tidyverse")){
  install.packages("tidyverse")
}
library(tidyverse)

if (!require("janitor")){
  install.packages("janitor")
}
library(janitor)

library(MASS)
```

The MASS (Modern Applied Statistics with S) package (Venables and Ripley, 2002) offers access to the Boston dataset, which contains housing values in the suburbs of

Boston. Let's delve into this dataset. We commence by examining the top few values using the `glimpse()` function:

```
Boston %>%
  glimpse(width = 50)
```

```
## Rows: 506
## Columns: 14
## $ crim    <dbl> 0.00632, 0.02731, 0.02729, 0.032~
## $ zn      <dbl> 18.0, 0.0, 0.0, 0.0, 0.0, 0.0, 1~
## $ indus   <dbl> 2.31, 7.07, 7.07, 2.18, 2.18, 2.~
## $ chas    <int> 0, 0, 0, 0, 0, 0, 0, 0, 0, 0, 0,~
## $ nox     <dbl> 0.538, 0.469, 0.469, 0.458, 0.45~
## $ rm      <dbl> 6.575, 6.421, 7.185, 6.998, 7.14~
## $ age     <dbl> 65.2, 78.9, 61.1, 45.8, 54.2, 58~
## $ dis     <dbl> 4.0900, 4.9671, 4.9671, 6.0622, ~
## $ rad     <int> 1, 2, 2, 3, 3, 3, 5, 5, 5, 5, 5,~
## $ tax     <dbl> 296, 242, 242, 222, 222, 222, 31~
## $ ptratio <dbl> 15.3, 17.8, 17.8, 18.7, 18.7, 18~
## $ black   <dbl> 396.90, 396.90, 392.83, 394.63, ~
## $ lstat   <dbl> 4.98, 9.14, 4.03, 2.94, 5.33, 5.~
## $ medv    <dbl> 24.0, 21.6, 34.7, 33.4, 36.2, 28~
```

We find that this dataset contains 506 rows and 14 columns. The details of the columns as read from the help file using `?Boston` are as follows:

1. `crim`: per capita crime rate by town.
2. `zn`: proportion of residential land zoned for lots over 25,000 sq.ft.
3. `indus`: proportion of non-retail business acres per town.
4. `chas`: Charles River dummy variable (=1 if tract bounds river; 0 otherwise).
5. `nox`: nitrogen oxides concentration (parts per 10 million).
6. `rm`: average number of rooms per dwelling.
7. `age`: proportion of owner-occupied units built prior to 1940.
8. `dis`: weighted mean of distances to five Boston employment centers.
9. `rad`: index of accessibility to radial highways.
10. `tax`: full-value property-tax rate per $10,000.
11. `ptratio`: pupil-teacher ratio by town.
12. `black`: $1000(Bk - 0.63)^2$, where Bk is the proportion of blacks by town.
13. `lstat`: lower status of the population (percent).
14. `medv`: median value of owner-occupied homes in $1000s.

In this chapter, we shall explore if a lower status of population (`lstat`) is related to a lower median value of owner-occupied homes (`medv`). Let us select these columns and save as `data` using the `select()` function from the `dplyr` package. In the process, we also convert the dataset to a tibble using the `as_tibble()` function from `tibble` package, for the sake of better visualization, and omit any NA (not available; meaning data missing) values that may be in the dataset. We use the following code:

```
data <- Boston %>%
  na.omit() %>%
  as_tibble() %>%
  dplyr::select(lstat, medv)
```

Let us check if `data` is what we wanted it to be. Since we have converted the data to a tibble, there is no longer a need to use the `glimpse()` function. We employ the following code:

```
data
```

```
## # A tibble: 506 x 2
##    lstat  medv
##    <dbl> <dbl>
##  1  4.98  24
##  2  9.14  21.6
##  3  4.03  34.7
##  4  2.94  33.4
##  5  5.33  36.2
##  6  5.21  28.7
##  7 12.4   22.9
##  8 19.2   27.1
##  9 29.9   16.5
## 10 17.1   18.9
## # i 496 more rows
```

The values appear to be in order. We discover that there are 506 rows, indicating that no row has been removed owing to missing values. Now, let's plot the data points and observe if we can visualize the trend as a straight line.

We plot the data using `ggplot()` function in tidyverse's `ggplot2` package. We set the aesthetics using the `aes()` function, defining that on the x-axis, we wish to have the variable `lstat`, and on the y-axis, we wish to have the variable `medv`. In this case, `lstat` is the independent variable, and `medv` is the dependent variable. In an experiment, the independent variable is what we change to see how it affects something else, which is called the dependent variable. Following is the code at this stage:

```
data %>%
  ggplot(
    aes(
      x = lstat,
      y = medv
    )
  )
```

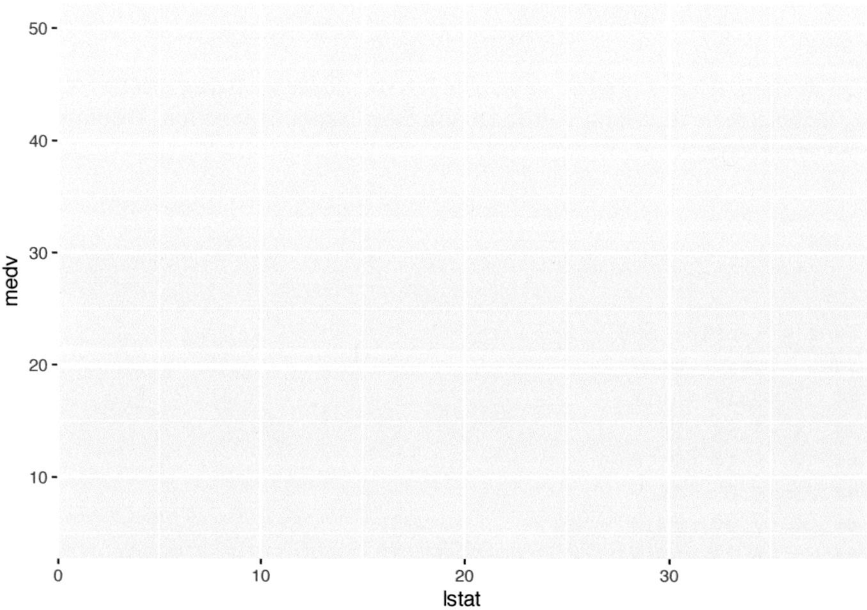

Once the basic aesthetics have been defined, we insert the next layer using the + command. We state that we wish to plot points (to generate a scatterplot) using the `geom_point()` function.

On the next layer, we specify that we also wish to have a smooth line plotted through the points. To this end, we specify the smoothing function method to be `lm`, or linear model. We further state that we do not wish to plot confidence intervals by setting `se` to `FALSE`. This is the updated code:

```
data %>%
  ggplot(
    aes(
      x = lstat,
      y = medv
```

```
    )
  ) +
  geom_point() +
  geom_smooth(method = "lm", se = FALSE)
```

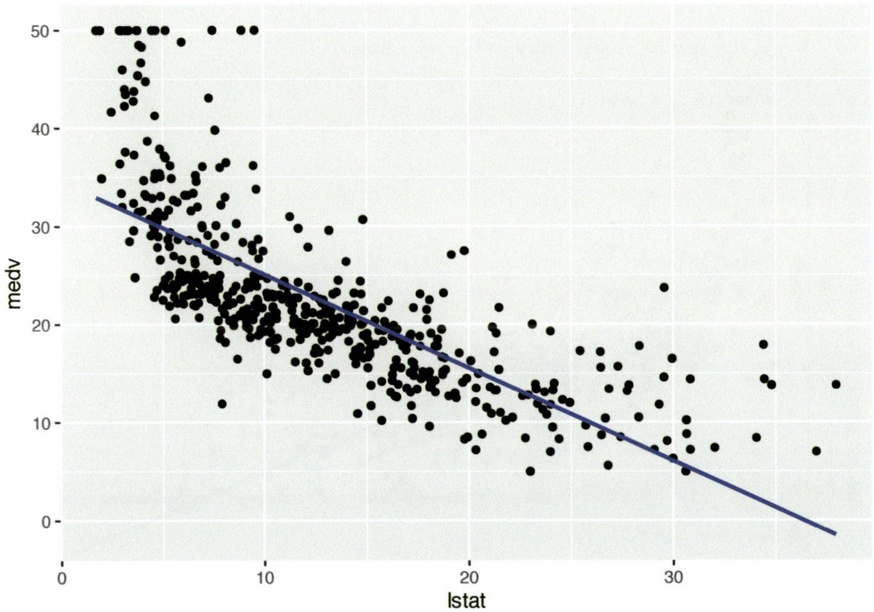

On the next layer, we put in the labels. We specify the title, x and y axes labels using the labs() function.

We also specify that we wish to have a classic theme for a clean graph using the theme_classic() function. Putting in all these commands lets us generate the graph. Following is the complete code to accomplish this:

```
data %>%
  ggplot(
    aes(
      x = lstat,
      y = medv
    )
  ) +
  geom_point() +
  geom_smooth(method = "lm", se = FALSE) +
  labs(
```

```
    title = "Population status vs. median home value",
    x = "Lower status of population (percent)",
    y = "Median property value in USD 1000s"
  ) +
  theme_classic()
```

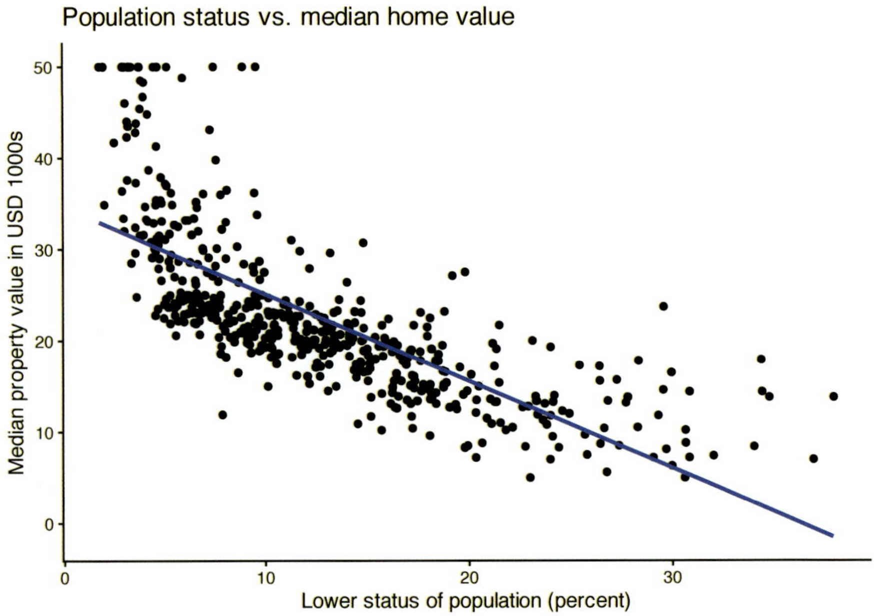

We observe that the straight line doesn't accurately represent the data, as many observations deviate significantly from it. In such instances, a polynomial line would be a more suitable fit to capture the trend of the data.

5.3 POLYNOMIAL REGRESSION

We've previously noticed that the `lm()` function in R is a reliable tool for fitting linear regression models. It can also be extended to perform polynomial regression by including polynomial terms in the model formula. The syntax for `lm()` in R is straightforward: `lm(formula, data)`.

Here, `formula` is a formula object that encapsulates the model structure to be fitted. Typically, the formula adheres to the format `response ~ predictors`, wherein `response` denotes the dependent variable, and `predictors` encompass the independent variables incorporated in the analysis. The `data` parameter signifies the data frame that houses the variables specified in the formula.

To incorporate polynomial terms, we simply replace `predictor` with `poly(predictor, degree, raw = TRUE)`. The `poly()` function signifies our intention to include polynomial terms, while the `degree` parameter specifies the degree of the polynomial. For instance, to include quadratic terms, we set `degree = 2`; to include cubic terms, we set `degree = 3`; and to include quartic terms, we set `degree = 4`. We also specify that we wish to keep raw polynomials by putting `raw=TRUE` in the options.

Let's explore this functionality using the `data` dataset. We store the outcomes of `lm()` in a variable named `model` using the following code:

```
model <- lm(
  medv ~ poly(lstat, degree = 2, raw = TRUE),
  data = data
)
```

The coefficients stored in the model can be accessed using the function `coefficients()`. We employ the following code:

```
model %>%
  coefficients() %>%
  enframe()
```

```
## # A tibble: 3 x 2
##   name                                  value
##   <chr>                                 <dbl>
## 1 (Intercept)                         42.9
## 2 poly(lstat, degree = 2, raw = TRUE)1 -2.33
## 3 poly(lstat, degree = 2, raw = TRUE)2  0.0435
```

Here `enframe()` is used to transform named atomic vectors or lists into one- or two-column data frames.

Let us now try with cubic terms:

```
model <- lm(
  medv ~ poly(lstat, degree = 3, raw = TRUE),
  data = data
)
```

As before, the coefficients stored in the model can be accessed using the following code:

```
model %>%
  coefficients() %>%
  enframe()
```

```
## # A tibble: 4 x 2
##   name                                       value
##   <chr>                                      <dbl>
## 1 (Intercept)                              48.6
## 2 poly(lstat, degree = 3, raw = TRUE)1 -3.87
## 3 poly(lstat, degree = 3, raw = TRUE)2  0.149
## 4 poly(lstat, degree = 3, raw = TRUE)3 -0.00200
```

We find that the model coefficients have changed a lot! Let us now try with quartic terms:

```
model <- lm(
  medv ~ poly(lstat, degree = 4, raw = TRUE),
  data = data
)

model %>%
  coefficients() %>%
  enframe()
```

```
## # A tibble: 5 x 2
##   name                                         value
##   <chr>                                        <dbl>
## 1 (Intercept)                              57.3
## 2 poly(lstat, degree = 4, raw = TRUE)1 -7.03
## 3 poly(lstat, degree = 4, raw = TRUE)2  0.495
## 4 poly(lstat, degree = 4, raw = TRUE)3 -0.0163
## 5 poly(lstat, degree = 4, raw = TRUE)4  0.000195
```

We notice that the model coefficients have changed significantly once more! This is one limitation of using polynomial regression—the results vary greatly depending on the degree of the polynomial used. Let's use `ggplot()` to plot the best-fit curves for different degrees of polynomial:

```
data %>%
  ggplot(
    aes(
      x = lstat,
      y = medv
    )
  ) +
  geom_point() +
  geom_smooth(
    method = "lm",
    se = FALSE,
    formula = y ~ poly(x,  1,  raw = TRUE),
    color = "blue",
    linetype = "dashed",
    aes(fill = "1")
  ) +
  geom_smooth(
    method = "lm",
    se = FALSE,
    formula = y ~ poly(x,  2,  raw = TRUE),
    color = "darkgreen",
    linetype = "solid",
    aes(fill = "2")
  ) +
  geom_smooth(
    method = "lm",
    se = FALSE,
    formula = y ~ poly(x,  3,  raw = TRUE),
    color = "orange",
    linetype = "twodash",
    aes(fill = "3")
  ) +
  geom_smooth(
    method = "lm",
    se = FALSE,
    formula = y ~ poly(x,  40,  raw = TRUE),
    color = "red",
    linetype = "dotdash",
    aes(fill = "40")
  ) +
  labs(
    title = "Population status vs. median value of homes",
    x = "Lower status of population (percent)",
```

```
    y = "Median property value in USD 1000s",
    fill = "Degree of polynomial"
  ) +
  theme_classic()
```

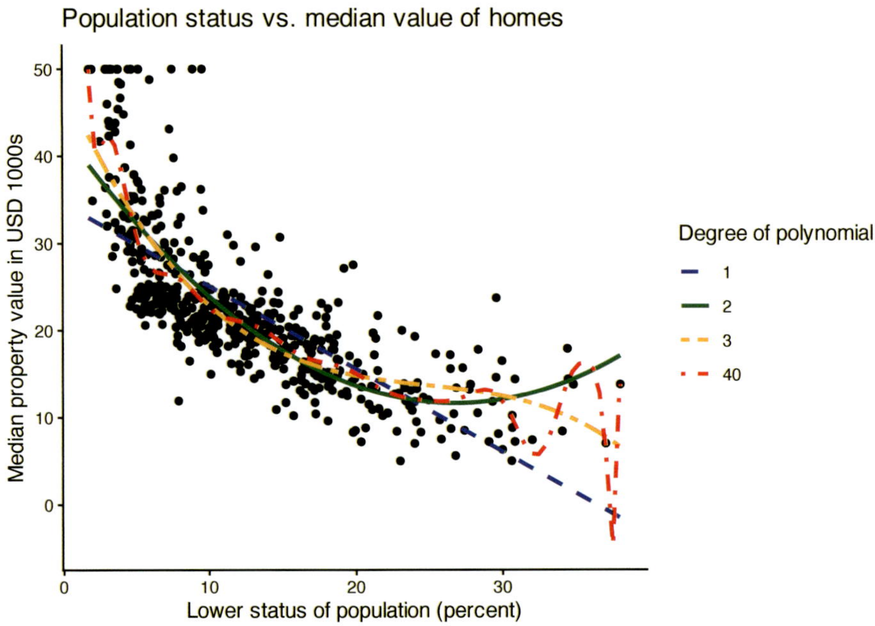

We notice that the curves vary greatly! Clearly, a polynomial of degree 1 (which gives a linear regression) is insufficient to capture the variations in the data. As the degree increases, the curves become more curved and better able to represent the dataset's variations. However, at extremely high polynomial degrees, the curves become so complex that they don't accurately represent the data, but rather resemble minor variations or even noise. Therefore, it's important to choose an appropriate polynomial degree—not too low so as to miss the dataset's variations and not too high so as to capture noise. In the next sections of this chapter, we'll discuss methods for determining the optimal polynomial degree.

When using the `lm()` function, it returns an object of class "lm" that contains important information about the fitted model. This object includes estimates of the **regression coefficients**, which show how much each independent variable affects the dependent variable. It also includes the **residuals**, which are the differences between the observed and predicted values. The **fitted values** are the predictions made by the model, and the **R-squared value** shows how well the model explains the variance in the dependent variable. The **p-values** tell us how likely it is that the coefficients are due to chance.

We can extract diagnostic information from the model object using various functions. For example, `summary(model)` provides a detailed summary that includes coefficients, R-squared, and p-values:

```
summary(model)
```

```
Call:
lm(formula = medv ~ poly(lstat, degree = 4, raw = TRUE),
 data = data)

Residuals:
    Min      1Q  Median      3Q     Max
-13.563  -3.180  -0.632   2.283  27.181

Coefficients:
                                      Estimate Std. Error
(Intercept)                          5.731e+01  2.280e+00
poly(lstat, degree = 4, raw = TRUE)1 -7.028e+00  7.308e-01
poly(lstat, degree = 4, raw = TRUE)2  4.955e-01  7.489e-02
poly(lstat, degree = 4, raw = TRUE)3 -1.631e-02  2.994e-03
poly(lstat, degree = 4, raw = TRUE)4  1.949e-04  4.043e-05
                                     t value Pr(>|t|)
(Intercept)                           25.134  < 2e-16 ***
poly(lstat, degree = 4, raw = TRUE)1  -9.618  < 2e-16 ***
poly(lstat, degree = 4, raw = TRUE)2   6.616 9.50e-11 ***
poly(lstat, degree = 4, raw = TRUE)3  -5.448 7.98e-08 ***
poly(lstat, degree = 4, raw = TRUE)4   4.820 1.90e-06 ***
---
Signif. codes:
0 '***' 0.001 '**' 0.01 '*' 0.05 '.' 0.1 ' ' 1

Residual standard error: 5.28 on 501 degrees of freedom
Multiple R-squared:  0.673, Adjusted R-squared:  0.6704
F-statistic: 257.8 on 4 and 501 DF,  p-value: < 2.2e-16
```

The `plot()` function helps generate diagnostic plots, such as residuals versus fitted values, to aid evaluation of the model, as discussed in the previous chapter:

```
plot(model)
```

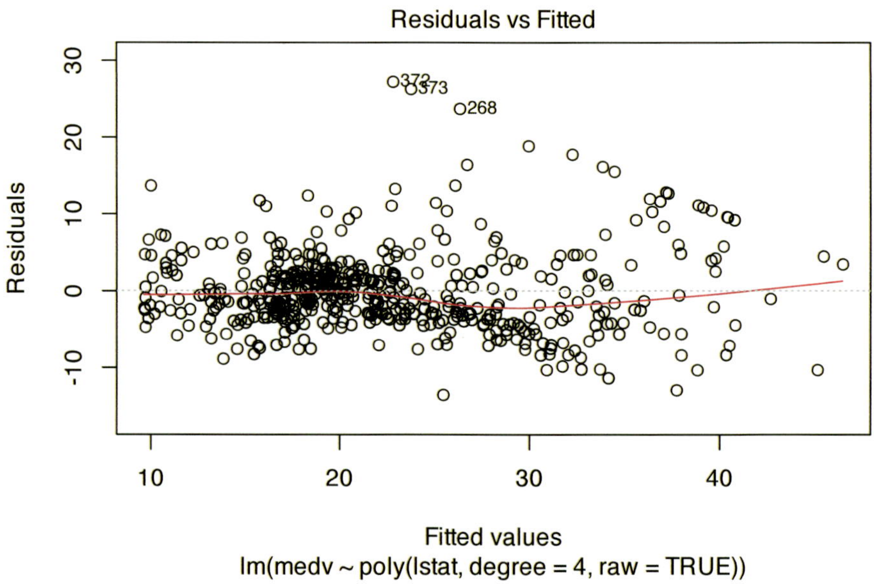
Residuals vs Fitted
Residuals
Fitted values
lm(medv ~ poly(lstat, degree = 4, raw = TRUE))
372
373
268

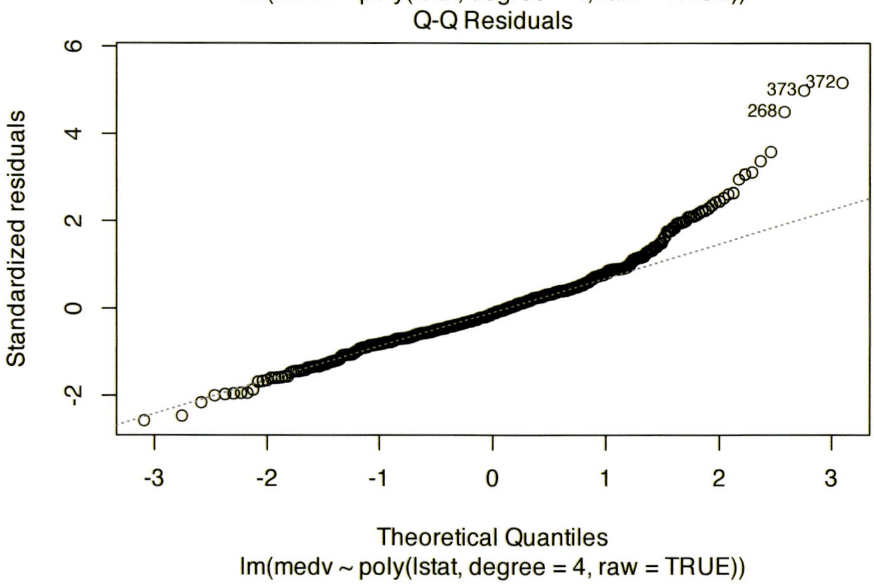
Q-Q Residuals
Standardized residuals
Theoretical Quantiles
lm(medv ~ poly(lstat, degree = 4, raw = TRUE))
372
373
268

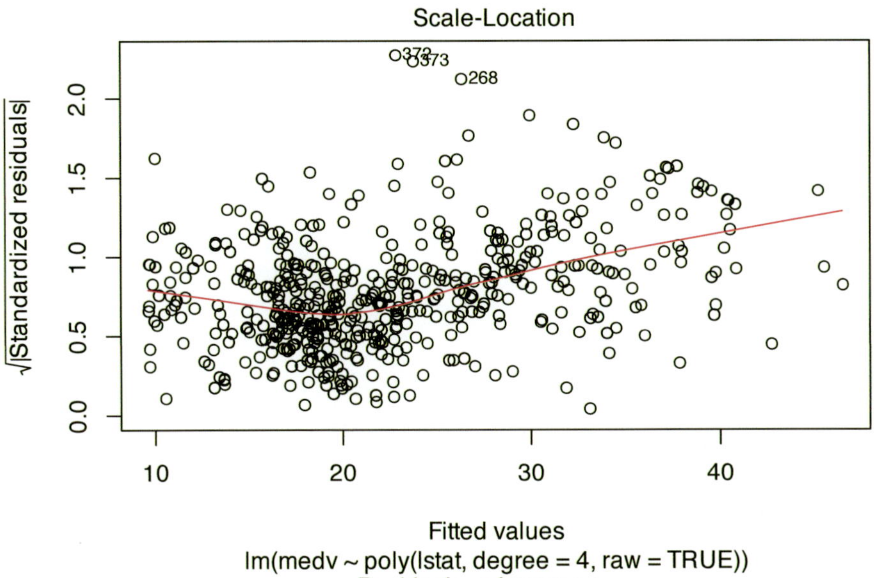

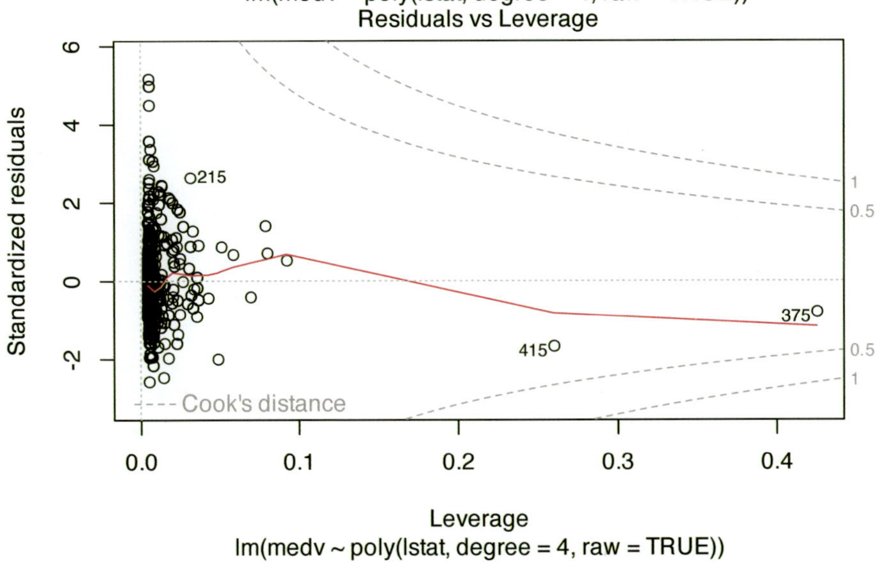

The `confint()` function applied on `model` calculates the confidence intervals for the model parameters:

```
confint(model)
```

```
##                                          2.5 %
## (Intercept)                      52.8301431590
```

```
## poly(lstat, degree = 4, raw = TRUE)1 -8.4642532976
## poly(lstat, degree = 4, raw = TRUE)2  0.3483392065
## poly(lstat, degree = 4, raw = TRUE)3 -0.0221917103
## poly(lstat, degree = 4, raw = TRUE)4  0.0001154414
##                                             97.5 %
## (Intercept)                           61.7897672779
## poly(lstat, degree = 4, raw = TRUE)1 -5.5926668730
## poly(lstat, degree = 4, raw = TRUE)2  0.6426230645
## poly(lstat, degree = 4, raw = TRUE)3 -0.0104286302
## poly(lstat, degree = 4, raw = TRUE)4  0.0002742941
```

In this way, the `lm()` function offers a straightforward way to analyze and interpret the relationships between variables. Its formula interface and comprehensive output make it a crucial tool in statistical modeling and data analysis.

5.4 TIDYMODELS EXAMPLE

We will now use `tidymodels` to perform polynomial regression from the beginning [Fig. 5.1]. The biggest benefit of using `tidymodels` is that once the steps have been coded, moving from one machine learning algorithm to another becomes a breeze. Throughout this book, you will find that we use very similar codes with various machine learning algorithms.

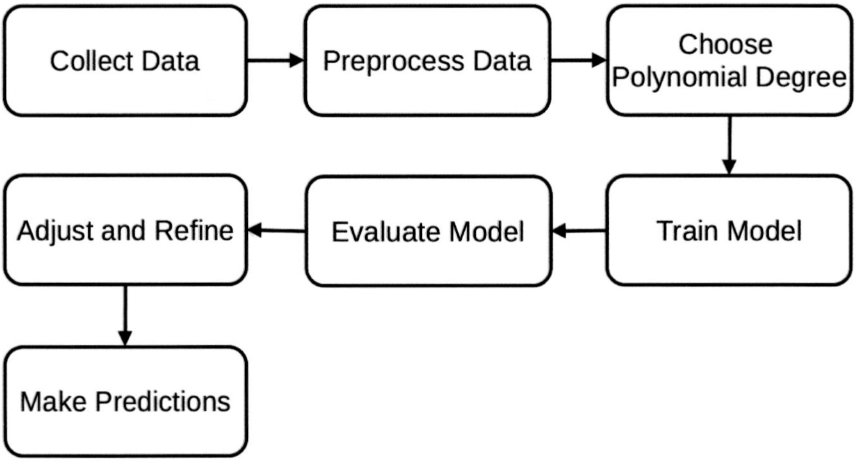

FIGURE 5.1: A schematic of polynomial regression process.

5.4.1 LOAD DATA

We start by loading the necessary information into a data file. This is also an opportune moment to convert the raw data into a tibble format using the as_tibble() function, which enhances visualization capabilities. Additionally, we remove any missing values (NA) from the dataset. Furthermore, we select only the required columns to reduce the dataset's size.

We employ the pipe operator %>% from tidyverse to keep our code intuitive and clean:

```
data <- Boston %>%
  na.omit() %>%
  as_tibble() %>%
  dplyr::select(lstat, medv)
```

Let us confirm the operations by examining the data variable. Since we've converted our data into a tibble, there is no longer a need to use the glimpse() function:

```
data
```

```
## # A tibble: 506 x 2
##    lstat  medv
##    <dbl> <dbl>
##  1  4.98  24
##  2  9.14  21.6
##  3  4.03  34.7
##  4  2.94  33.4
##  5  5.33  36.2
##  6  5.21  28.7
##  7 12.4   22.9
##  8 19.2   27.1
##  9 29.9   16.5
## 10 17.1   18.9
## # i 496 more rows
```

We observe that the number of rows remains unchanged, indicating the absence of any rows containing missing values.

5.4.2 SPLIT DATA INTO TRAINING AND TESTING

We partition the data into training and testing sets using the `initial_split()` function from the `rsample` package, a component of `tidymodels`. We retain 70% of the data for training and 30% for testing. These sets are then extracted using `training()` and `testing()` functions:

```
# Split data into training and testing sets
set.seed(2025)
split_7030 <- initial_split(data, prop = 0.7)
data_train <- training(split_7030)
data_test <- testing(split_7030)
```

In this exemplar case, we have set the seed to 2025. `set.seed` is a function used in programming to initialize the random number generator to a specific state, ensuring that the results of random operations are reproducible. By setting the seed, we ensure that the sequence of random numbers generated is consistent each time our code is executed. This consistency facilitates reliable and reproducible results in experiments and simulations, particularly in troubleshooting scenarios. In practice, this line will be removed or commented out upon code deployment, at which point it is not required.

5.4.3 CREATE A RECIPE

In the recipe, we input the formula as `medv ~ .` since we wish to predict the median value of the house using all the available variables in the dataset. Remember that we have already removed the unnecessary variables that we do not expect to have any relation with the final result. We further specify that this recipe should be used on `data_train`, i.e., the training data. We also create new columns that will be the basis expansions of the variable `lstat` using the `step_poly()` function. We specify that we wish expansions of the second degree (making for quadratic polynomials), and specify that we wish to keep raw polynomials by putting `raw=TRUE` in the options. The complete recipe code is provided below:

```
recipe_ml <- recipe(
  medv ~ .,
  data = data_train
) %>%
  step_poly(
    lstat,
```

```
    degree = 2,
    options=list(raw=TRUE)
  )
```

5.4.4 CREATE A MODEL DESIGN

We define a linear_reg model using the lm engine. In tidymodels, the lm engine is used for tidy and efficient linear and polynomial regression modeling. It integrates seamlessly with tidyverse tools like dplyr and ggplot2 for straightforward data manipulation, visualization, and model assessment. It follows a standard formula approach to specify variable relationships and provides robust methods for model validation, ensuring reliable statistical analysis in data projects. Since we wish to estimate the median value in numeric form using independent variables, we set the mode to regression. The code that follows demonstrates the creation of the model design:

```
model_lin_reg <- linear_reg() %>%
  set_engine("lm") %>%
  set_mode("regression")
```

5.4.5 CREATE A WORKFLOW

We add the recipe and model to the workflow() function, and train the model on data_train dataset using fit() function:

```
workflow_recipe_model <- workflow() %>%
  add_recipe(recipe_ml) %>%
  add_model(model_lin_reg) %>%
  fit(data_train)
```

5.4.6 ASSESS PREDICTION QUALITY WITH TEST DATA

We can now utilize our trained and fitted model to predict median value of homes using the test data set to assess model efficacy. To accomplish this, we employ the augment() function. The augment() function, originating from the broom package in R, appends predictions and residuals from a statistical model to the original dataset. This facilitates the evaluation of a model's fit by juxtaposing actual and predicted values. We use the following code:

```
data_test_with_prediction <- augment(
  workflow_recipe_model, data_test
)
```

Let us now have a glimpse of the augmented file, `data_test_with_prediction`. We select the columns that we are interested in, and consider the first few rows using the `head()` function:

```
data_test_with_prediction %>%
  dplyr::select(
    lstat,
    medv,
    .pred
  ) %>%
  head()
```

```
# A tibble: 6 x 3
  lstat  medv .pred
  <dbl> <dbl> <dbl>
1  4.98  24    32.3
2  4.03  34.7  34.0
3 29.9   16.5  11.4
4  6.58  23.1  29.4
5 11.3   18.2  22.3
6 12.8   18.4  20.4
```

We find that the predicted values (`.pred`) are close to the actual measured values (`medv`), albeit with some difference. Perhaps the quadratic polynomial is insufficient to explain all aspects of the dataset.

Additionally, we can compute the **accuracy metrics**. Accuracy metrics in machine learning assess a model's performance by comparing its predictions to the actual outcomes. We utilize the augmented `data_test_with_prediction` dataset to evaluate the estimated values against the actual truth using the `metrics()` function as follows:

```
metrics(
  data_test_with_prediction,
  truth = medv,
```

```
  estimate = .pred
)
```

```
## # A tibble: 3 x 3
##   .metric .estimator .estimate
##   <chr>   <chr>          <dbl>
## 1 rmse    standard       5.62
## 2 rsq     standard       0.632
## 3 mae     standard       3.99
```

We find that the root mean squared error is around 5.62 thousand dollars, the mean absolute error is around 3.99 thousand dollars, and we have been able to explain around 63.2% of the results (R squared = 0.632). The rest of the variations are probably due to other variables that we did not consider in this example (and perhaps even more variables that are absent from the `Boston` dataset), or because quadratic polynomial is insufficient to cover those.

5.5 HYPERPARAMETER TUNING

We selected 2 as the degree of the polynomial for our regression exercise. But the degree could as well have been 3, 4, or some higher value. As we increase the degree, the curve is a better fit to the data points. But this benefit also has a limit, after which a very high degree makes for a curvy line that veers and swerves to the tunes of the noise in the dataset, reducing the predictive powers of the model. But what should be the best value for `degree`? One way to find that is through hyperparameter tuning.

Hyperparameter tuning is the process of adjusting the settings of a machine learning model (called its **hyperparameters**) to improve its performance. In this case, we can create a grid with different values of `degree`, use them to fit models, and test those models. The value of `degree` that has the lowest error, or that is able to explain the majority of the data can then be taken as the best value of `degree`.

In this section, we'll employ a grid of hyperparameters and tune the workflow to compute the best value of `degree`. Most of the steps are similar to what we did before, with the addition of `tune()` being used to hold the value of `degree` in the model design, and `tune_grid()` function used to tune the workflow. We'll begin with the splitting step, and continue till the computation of accuracy metrics.

5.5.1 SPLIT DATA INTO TRAINING AND TESTING

As before, we split data into training and testing sets using `initial_split()` function from `rsample` package, a constituent of `tidymodels`. We keep 70% of data for training, and 30% for testing purposes, and extract these two sets using `training()` and `testing()` functions.

```
# Split data into training and testing sets
set.seed(2025)
split_7030 <- initial_split(data, prop = 0.7)
data_train <- training(split_7030)
data_test <- testing(split_7030)
```

As before, we have set the seed to 2025 to enable comparison with the previous results. The `set.seed` function in programming initializes the random number generator to a specific state, ensuring reproducibility of random operations. By setting the seed, we guarantee that the sequence of random numbers generated is consistent each time we run our code. This consistency is crucial for achieving repeatable results in experiments and simulations, particularly for troubleshooting. However, when we deploy the code, this line will be removed or commented out, since at that point it will not be required.

5.5.2 CREATE A RECIPE

As before, we input the formula as `medv ~ .` since we wish to predict the median value of the house using all the available variables in the dataset. Remember that we have already removed the unnecessary variables that we do not expect to have any relation with the final result. We further specify that this recipe should be used on `data_train`, i.e., the training data. We also create new columns that will be the basis expansions of the variable `lstat` using the `step_poly()` function. We specify the degree as `tune()`, which is a placeholder that will be replaced by the value(s) of degree that we provide for checking. We further specify that we wish to keep raw polynomials by putting `raw=TRUE` in the options. The complete recipe code is provided below:

```
recipe_ml <- recipe(
  medv ~ .,
  data = data_train
) %>%
  step_poly(
    lstat,
    degree = tune(),
    options=list(raw=TRUE)
  )
```

5.5.3 CREATE A MODEL DESIGN

We define a linear_reg model using the lm engine. In tidymodels, the lm engine is used for tidy and efficient linear and polynomial regression modeling in R. It integrates seamlessly with tidyverse tools like dplyr and ggplot2 for straightforward data manipulation, visualization, and model assessment. It follows a standard formula approach to specify variable relationships and provides robust methods for model validation, ensuring reliable statistical analysis in data projects. Since we wish to estimate the median value in numeric form using independent variables, we set the mode to regression. Here's the code for the model design:

```
model_lin_reg <- linear_reg() %>%
  set_engine("lm") %>%
  set_mode("regression")
```

5.5.4 CREATE A WORKFLOW

We add the recipe and model to the workflow() function, but leave training for later. Following is the code at this stage:

```
workflow_recipe_model <- workflow() %>%
  add_recipe(recipe_ml) %>%
  add_model(model_lin_reg)
```

5.5.5 CREATE A HYPER-PARAMETER GRID

We create a **hyper-parameter grid** of values to be used in degree using the data.frame() function to create a data frame of values. Here's the code we use:

```
hyper_parameter_grid <- data.frame(
  degree = c(1:10)
)
```

We check the values in the hyper_parameter_grid variable.

```
hyper_parameter_grid
```

```
##    degree
## 1       1
## 2       2
## 3       3
## 4       4
## 5       5
## 6       6
## 7       7
## 8       8
## 9       9
## 10     10
```

5.5.6 CREATE RESAMPLES FOR CROSS-VALIDATION

V-fold cross-validation, also known as **k-fold cross-validation**, is a method used to assess the performance of a model by dividing the data into V groups of roughly equal size, which are called "**folds**". In each iteration of the analysis, V-1 folds are used to train the model, while the remaining fold is used to test its performance. In its basic form, without any repeats, the number of resamples is equal to V, which provides a thorough evaluation of the model's capabilities.

The `vfold_cv()` function performs V-fold cross-validation on a data frame. It takes two arguments: `v`, which determines the number of partitions, and `repeats`, which specifies how many times the partitioning should be repeated. By default, `v` is set to 10 and `repeats` is set to 1, which corresponds to a standard cross-validation process. The function also supports **stratified sampling**, where a variable in the data frame is used to ensure that each resample maintains a similar distribution across different categories. This is achieved by using the `strata` argument.

The `vfold_cv()` function offers several customization options. The `data` parameter specifies the input data frame, while the `v` parameter determines the number of partitions for the data set. The `strata` parameter enables the use of a variable for stratified sampling, while the `breaks` parameter specifies the number of bins for numeric stratification variables. The `pool` parameter assesses the proportion of data used to evaluate if a group is too small and should be combined with another group. It is strongly advised to maintain the default value of 0.1 for this argument to avoid the risks of stratifying groups that are too small. The `dots` argument, ..., is reserved for future extensions and should remain empty when using the function.

In the `vfold_cv()` function, we specify that the strata should be generated using the `data_train` dataset. We want to create 10 such strata, and `medv` is the variable in the dataset that we'll use for stratified sampling. To accomplish this, we execute the following code:

```
set.seed(2025)
folds_for_tuning <- vfold_cv(
  data_train,
  v = 10,
  strata = medv
)
```

We can look at `folds_for_tuning` to observe the number of training and testing data in each of the cross-validation folds:

```
folds_for_tuning
```

```
## #  10-fold cross-validation using stratification
## # A tibble: 10 x 2
##    splits            id
##    <list>            <chr>
##  1 <split [317/37]> Fold01
##  2 <split [318/36]> Fold02
##  3 <split [318/36]> Fold03
##  4 <split [318/36]> Fold04
##  5 <split [318/36]> Fold05
##  6 <split [318/36]> Fold06
##  7 <split [319/35]> Fold07
##  8 <split [319/35]> Fold08
##  9 <split [320/34]> Fold09
## 10 <split [321/33]> Fold10
```

5.5.7 TUNE THE WORKFLOW

To find out the best value of `degree`, we tune the workflow with all the values of `degree` in the hyper-parameter grid. To do this, we utilize the `tune_grid()` function which tests different sets of model settings to see how well each one performs using various data samples, measuring metrics like accuracy or error.

We pass our workflow `workflow_recipe_model` to the `tune_grid()` function, along with `folds_for_tuning` as `resamples`, `hyper_parameter_grid` as `grid`, and specify the `metrics` on which to optimize using the `metric_set()` function. The following code emerges:

```
tune_results <- tune_grid(
  workflow_recipe_model,
  resamples = folds_for_tuning,
  grid = hyper_parameter_grid,
  metrics = metric_set(rmse, rsq, mae)
)
```

We save the results of tuning to tune_results.

5.5.8 EXTRACT THE BEST VALUE(S) OF HYPER-PARAMETER(S)

We can use the show_best() function from the tune package to extract the optimal hyperparameter values. For instance, to find the best values for RMSE, we can run the following code:

```
show_best(
  tune_results,
  metric = "rmse"
)
```

```
# A tibble: 5 x 7
  degree .metric .estimator  mean     n std_err .config
   <int> <chr>   <chr>      <dbl> <int>   <dbl> <chr>
1      5 rmse    standard    5.16    10   0.219 pre05_mod0_~
2      6 rmse    standard    5.18    10   0.210 pre06_mod0_~
3      4 rmse    standard    5.20    10   0.238 pre04_mod0_~
4      7 rmse    standard    5.22    10   0.213 pre07_mod0_~
5      9 rmse    standard    5.25    10   0.230 pre09_mod0_~
```

We discover that the optimal value for rmse is achieved when degree is set to 5, followed by 6, 4, and 7.

To identify the most suitable values for rsq, we can utilize the following code:

```
show_best(
  tune_results,
  metric = "rsq"
)
```

```
# A tibble: 5 x 7
  degree .metric .estimator  mean     n std_err .config
   <int> <chr>   <chr>      <dbl> <int>   <dbl> <chr>
1      5 rsq     standard   0.680    10  0.0363 pre05_mod0_~
2      6 rsq     standard   0.679    10  0.0349 pre06_mod0_~
3      4 rsq     standard   0.676    10  0.0378 pre04_mod0_~
4      7 rsq     standard   0.675    10  0.0353 pre07_mod0_~
5      9 rsq     standard   0.671    10  0.0384 pre09_mod0_~
```

We discover that the optimal value for rsq is achieved when `degree` is set to 5, followed by 6, 4, and 7.

To identify the most suitable values for mae, we can utilize the following code:

```
show_best(
  tune_results,
  metric = "mae"
)
```

```
# A tibble: 5 x 7
  degree .metric .estimator  mean     n std_err .config
   <int> <chr>   <chr>      <dbl> <int>   <dbl> <chr>
1      6 mae     standard    3.82    10   0.118 pre06_mod0_~
2      5 mae     standard    3.83    10   0.135 pre05_mod0_~
3      7 mae     standard    3.84    10   0.118 pre07_mod0_~
4      9 mae     standard    3.87    10   0.138 pre09_mod0_~
5      8 mae     standard    3.87    10   0.145 pre08_mod0_~
```

We discover that the optimal value for mae is achieved when `degree` is set to 6, followed by 5, 7, and 9.

We can use the `select_best()` function, a component of the `tune` package, to select the optimal hyperparameter value. For example, to find the best value of `degree` that yields the best rmse, we can execute the following code:

```
best_val <- select_best(
  tune_results,
  metric = "rmse"
)
best_val
```

```
## # A tibble: 1 x 2
##   degree .config
##    <int> <chr>
## 1      5 pre05_mod0_post0
```

5.5.9 FINALIZE AND TRAIN THE BEST MODEL

We can now finalize the workflow by using the `finalize_workflow()` function from the `tune` package. We pass the best value `best_val` as an argument to the function. Additionally, we train the model by fitting it to the `data_train` dataset using the following code:

```
best_model <- workflow_recipe_model %>%
  finalize_workflow(best_val) %>%
  fit(data_train)
```

5.5.10 ASSESS PREDICTION QUALITY WITH TEST DATA

We can now utilize our trained and fitted model to predict `medv` using the test data set to assess model efficacy. To accomplish this, we employ the `augment()` function. The `augment()` function, originating from the `broom` package in R, appends predictions and residuals from a statistical model to the original dataset. This facilitates the evaluation of a model's fit by juxtaposing actual and predicted values. We use the following code:

```
data_test_with_prediction <- augment(
  best_model, data_test
)
```

Let us now have a glimpse of the augmented file, `data_test_with_prediction`. We select the columns that we are interested in, and consider the first few rows using the `head()` function:

```
data_test_with_prediction %>%
  dplyr::select(
    lstat,
    medv,
    .pred
```

```
  ) %>%
  head()
```

```
# A tibble: 6 x 3
  lstat  medv .pred
  <dbl> <dbl> <dbl>
1  4.98  24    32.1
2  4.03  34.7  36.0
3 29.9   16.5  10.6
4  6.58  23.1  27.5
5 11.3   18.2  21.4
6 12.8   18.4  20.4
```

We find that the predicted values (`.pred`) are close to the actual measured values (`medv`), albeit with some difference.

We also compute the accuracy metrics. Accuracy metrics in machine learning assess a model's performance by comparing its predictions to the actual outcomes. We utilize the augmented `data_test_with_prediction` dataset to evaluate the estimated values against the actual truth using the `metrics()` function as follows:

```
metrics(
  data_test_with_prediction,
  truth = medv,
  estimate = .pred
)
```

```
## # A tibble: 3 x 3
##   .metric .estimator .estimate
##   <chr>   <chr>          <dbl>
## 1 rmse    standard       5.36
## 2 rsq     standard       0.665
## 3 mae     standard       3.70
```

We find that the root mean squared error has decreased from around 5.62 thousand dollars to around 5.36 thousand dollars, the mean absolute error has decreased from around 3.99 thousand dollars to around 3.7 thousand dollars, and we have been able to explain around 66.5% of the results (R squared = 0.665), greater than the previous value of 63.2% of the results (R squared = 0.632). The rest of the variations are probably due to other variables that we did not consider in this example (and perhaps even more variables that are absent from the `Boston` dataset).

5.6 WHEN TO USE POLYNOMIAL REGRESSION, AND WHEN NOT TO

Polynomial regression, a statistical analysis technique, models the relationship between independent and dependent variables using polynomial functions. It finds applications in various fields, but its limitations should be considered before use.

Polynomial regression excels when the relationship between variables is non-linear, exhibiting a curved pattern rather than a straight line. For instance, in modeling plant growth, where the growth rate fluctuates over time, polynomial regression effectively fits a curve capturing this dynamic pattern. It also proves beneficial in analyzing complex trends that a simple linear model cannot adequately represent. Additionally, polynomial regression aids in visualizing trends and estimating values within the data range.

However, polynomial regression may not be the ideal choice when overfitting is a concern, especially with high-degree polynomials and limited data. Overfitting occurs when the model captures noise instead of the underlying pattern. Furthermore, polynomial regression may yield unpredictable results when used for extrapolation beyond the training data range. Additionally, higher-degree polynomials can introduce multicollinearity issues, leading to unstable coefficient estimates. The increased complexity of higher-degree polynomials makes computations and interpretations challenging.

In conclusion, polynomial regression should be used judiciously when dealing with non-linear, complex relationships and when fitting smooth curves or interpolating within data range. However, it should be avoided in situations where overfitting is a risk, predictions are made outside the data range, multicollinearity is present, or model interpretability is paramount. The appropriate model selection depends on a thorough understanding of the data, the nature of variable relationships, and the specific analysis objectives.

5.7 CHAPTER SUMMARY

- Polynomial regression models non-linear relationships between variables by fitting a curved line to data points.
- Polynomial regression uses polynomial functions, such as quadratic or cubic equations, to capture complex patterns and trends in the data.
- We can fit polynomial regression models using the `lm()` function in R.
- Hyperparameter tuning is often employed to select an appropriate polynomial degree.
- The Boston Housing dataset, used for predicting home values, includes features like crime rate, land zoning, and median value. The dataset, originally from the 1970s, requires ethical considerations since it incorporates historical biases.

- Polynomial regression is a useful tool for modeling non-linear relationships and fitting curves. However, it should be used judiciously because it carries risks of overfitting, multicollinearity, and unpredictable results beyond the data range.

6 Logistic regression

Imagine yourself as an educator conducting a study to explore the relationship between study habits and exam outcomes among students. You've diligently gathered data from a diverse group of students, meticulously recording the number of hours they invested in studying and their subsequent exam results.

To determine the number of hours required to pass a course, we can plot the data on a graph. The horizontal axis represents the number of hours studied, while the vertical axis indicates whether the student passed or failed. While students who studied for a few hours may not have passed, those who studied more are more likely to have passed. However, we need to find a threshold number of hours that separates those who passed from those who failed.

Logistic regression aids in this regard. The curve exhibits an S-shape, indicating a gradual initial increase, a rapid rise as study hours accumulate, and a subsequent stabilization as the probability of passing approaches a peak, as illustrated in Figure 6.1.

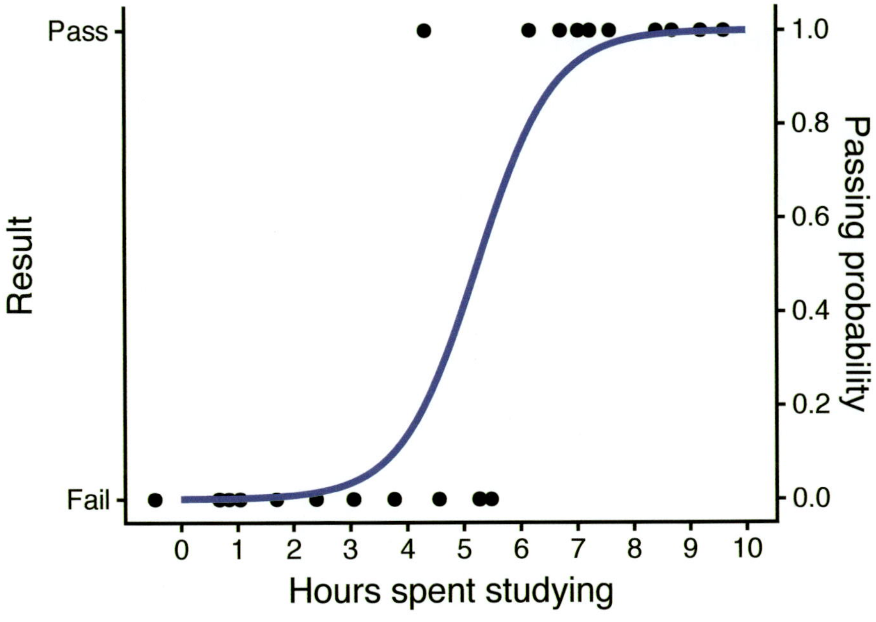

FIGURE 6.1: A plot of number of study hours versus the probability of passing.

Initially, when students study only a bit, their chances of passing are low, and their performance remains near the bottom of the curve. However, as they study more, the curve starts to rise, indicating that the probability of passing the exam increases with additional study hours. Nevertheless, after a certain point, the curve levels off, suggesting that further study doesn't significantly enhance the likelihood of passing (although it may improve the chances of passing with a higher percentage of marks). For now, we are only concerned with whether the student passed or failed.

When we have a new student who has studied for a specific number of hours, we can use this curve to predict their chances of passing. For instance, if a student studied for 4 hours, the curve might indicate only a 10% probability of passing. This suggests that based on the observed data, there's a 10% likelihood that this student will pass the exam. Conversely, for another student who studied for 6 hours, the curve might show an 80% probability of passing. This implies that, based on the observed data, there's an 80% likelihood that this student will pass the exam. For yet another student who studied for 5 hours, the curve might indicate a 40% probability of passing. This suggests that, based on the observed data, there's a 40% likelihood that this student will pass the exam.

In this way, logistic regression creates a curve that best fits the data to help understand and predict how study hours impact the likelihood of passing an exam. It provides a way to estimate probabilities based on the relationship between study time and exam results.

6.1 WHAT IS LOGISTIC REGRESSION?

Logistic regression is a statistical method employed for **binary classification**, which entails categorizing data into one of two distinct categories. In contrast to linear regression, which predicts continuous values, logistic regression is tailored for scenarios where the outcome is categorical, frequently represented as 0 or 1, yes or no, affirmative or negative, or success or failure. It is widely utilized in domains such as medicine, finance, and social sciences to ascertain the likelihood of an event's occurrence.

Logistic regression aims to model the probability of a given input belonging to a particular class. For instance, in the context of medical diagnosis, logistic regression can be used to estimate the likelihood of a patient having a particular disease, given the results of diagnostic tests and other relevant factors. This probability is then transformed into a **log-odds**, or **logit**, which is simply the natural logarithm of the odds ratio. The **odds ratio** represents the probability of an event occurring compared to the probability of it not occurring:

$$\text{Odds} = \frac{p}{1-p}$$

where p is the probability of the event occurring. Thus, we may write logit as:

$$\text{Logit} = \ln\left(\frac{p}{1-p}\right)$$

The **logistic function**, mathematically defined as $\sigma(z) = \frac{1}{1+e^{-z}}$, forms the cornerstone of logistic regression. This function, depicted in Figure 6.2, transforms any real-valued number into a value within the range of 0 and 1, rendering it suitable for binary probability estimation.

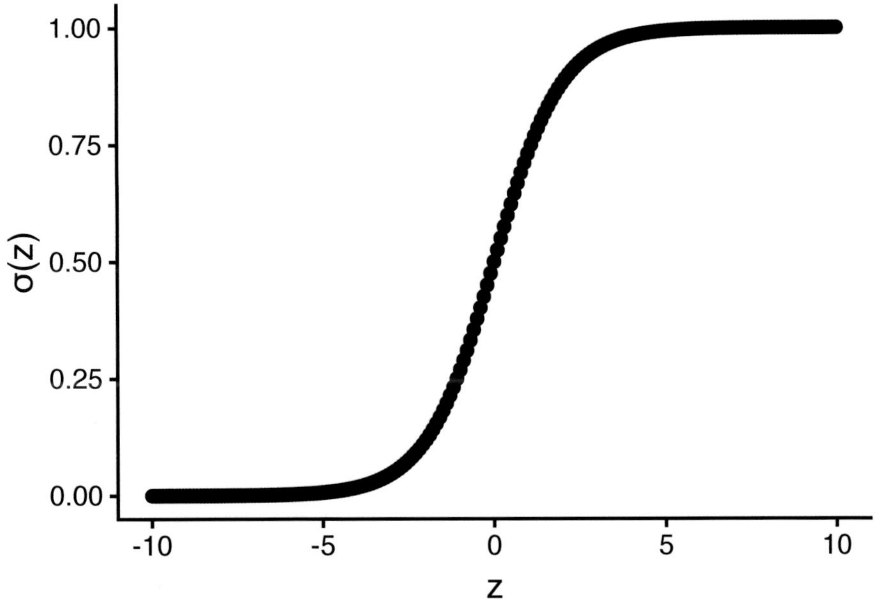

FIGURE 6.2: A plot of the logistic function.

In this scenario, z represents the linear combination of the input features, along with their corresponding weights. Logistic regression assumes a linear relationship between log-odds and the input variables. This means that the log-odds (logit) is modeled as a linear combination of the input features:

$$\text{Logit}(p) = \beta_0 + \beta_1 x_1 + \beta_2 x_2 + ... + \beta_n x_n$$

Here, β_0 is the intercept, and $\beta_1, \beta_2, \dots, \beta_n$ are coefficients for the input variables $x_1, x_2, \dots, x_n$.

Logistic regression aims to determine the optimal weights (coefficients) that effectively predict the probability of the outcome. In doing so, it tries to construct a curve that aligns with the observed data. This entails estimating parameters (weights) by employing techniques such as the **maximum likelihood estimation (MLE)**. MLE (MLE) is a statistical method employed to determine the parameters of a probability distribution by maximizing a likelihood function to ensure that under the presupposed statistical model, the observed data is most probable. The logistic function ensures that the predicted probabilities are confined to the range of 0 and 1. A threshold (usually set to 0.5) is employed for outcome classification. If the predicted probability surpasses or equals the threshold, the outcome is categorized into one category (e.g., pass); otherwise, it is categorized into the other category (e.g., fail).

Logistic regression, utilizing linear relationships for coefficients, provides high interpretability, enabling the analysis of how each feature influences the probability of an outcome. For example, a large positive coefficient for study hours indicates a strong positive association between increased study time and a higher chance of passing. Additionally, logistic regression can handle multiple features, allowing for the exploration of complex relationships between input data and the outcome.

However, logistic regression, while a powerful tool, has certain limitations. It assumes a linear relationship between input variables and the logarithm of the odds of the outcome. While it can handle some non-linearity through polynomial features or interaction terms, this may not be sufficient for complex relationships without significant modifications. Additionally, logistic regression is highly sensitive to the scale of input features, so **feature scaling** or **normalization** is often necessary to achieve optimal performance.

In practice, logistic regression often serves as a baseline model due to its simplicity and ease of interpretation. It is a fundamental tool for binary classification, offering a straightforward and powerful approach to prediction and classification. However, it is important to note that the assumptions and limitations of logistic regression need to be considered and justified carefully before employing the method in real-world applications.

6.2 THE DIABETES DATASET

The Pima Indians[1] Diabetes dataset is a well-established resource in machine learning and statistical analysis. It is used for predicting diabetes risk among Pima Indian women. It originates from a study conducted by the National Institute of Diabetes and Digestive and Kidney Diseases (NIDDK). The NIDDK is a component of the National Institutes of Health (NIH) in the United States, and is dedicated to advancing research on diseases and conditions related to diabetes, endocrinology, metabolic disorders, digestive diseases, liver diseases, and kidney diseases. The Pima Indians Diabetes dataset serves as a valuable resource for exploring binary classification problems and comprehending the correlations between various health indicators and diabetes risk.

This dataset, comprising 768 observations of individuals from the Pima Indian community, encompasses eight distinct health or demographic characteristics. Each feature quantifies a specific aspect of their well-being, including the number of pregnancies, plasma glucose levels after a two-hour glucose tolerance test, diastolic

[1]The Pima Indians are a Native American tribe from Arizona, known for their traditional farming skills and cultural heritage.

blood pressure, triceps skinfold thickness, serum insulin levels, body mass index (BMI), diabetes pedigree function (indicating genetic predisposition), and age. The dataset's outcome variable is binary, categorizing individuals into two groups: those with diabetes (pos) and those without diabetes (neg).

The selected features play a crucial role in predicting diabetes and offer valuable insights into various health aspects. For example, plasma glucose concentration and BMI are directly linked to metabolic health, while the diabetes pedigree function provides genetic risk information. These features collectively enable the development of an effective model for assessing diabetes risk.

The dataset is widely used for testing different machine learning algorithms, such as logistic regression, decision trees, random forests, and support vector machines. Its binary classification format makes it a practical choice for developing and improving predictive models. Additionally, it's commonly used in educational settings to teach students about classification problems, feature engineering, and model evaluation.

However, the dataset faces several challenges, particularly **class imbalance**, where the number of diabetes cases is significantly lower than the number of non-cases. This disparity can lead to biased models. To address this issue, techniques like resampling, class weights, or metrics tailored to class imbalance, such as the F1 score or the area under the Receiver operating characteristic (ROC) curve, are often employed.

Another concern is the presence of **missing values**. Certain features may contain missing or zero values, which could stem from data entry errors or unrecorded measurements. The appropriate handling of these missing values, such as imputation (filling in missing data with estimated values), removal, or other methods, is essential for building reliable models.

Feature scaling and **normalization** are crucial when using this dataset. The features have significant variations in scale and units, so normalization ensures that each feature contributes equally to the model's performance. This prevents any single feature from dominating the predictions and helps create a balanced and trustworthy model.

In summary, the Pima Indians Diabetes dataset is a valuable resource for developing and evaluating machine learning models in medical diagnosis and classification. Its well-defined features and binary outcome make it suitable for investigating various predictive techniques and understanding classification challenges. Despite class imbalance and missing values, the dataset remains a practical and insightful tool for practitioners and students in healthcare and machine learning.

6.2.1 USING THE DATASET

We begin by (installing and) loading all the required packages and libraries.

```
if (!require("tidymodels")){
  install.packages("tidymodels")
}
library(tidymodels)

if (!require("tidyverse")){
  install.packages("tidyverse")
}
library(tidyverse)

if (!require("janitor")){
  install.packages("janitor")
}
library(janitor)

if (!require("themis")){
  install.packages("themis")
}
library(themis)

if (!require("mlbench")){
  install.packages("mlbench")
}
library(mlbench)
```

The `themis` package in R (Hvitfeldt, 2025) is a valuable tool for handling imbalanced datasets, a prevalent challenge in data science and machine learning. It provides functions for over-sampling and under-sampling to rectify class distribution imbalances. `themis` incorporates methods such as SMOTE to enhance the performance of machine learning models on imbalanced data. It seamlessly integrates with the `tidymodels` ecosystem, facilitating the incorporation of sampling strategies into workflows. By ensuring adequate representation of minority classes during training, `themis` contributes to the creation of fairer and more accurate predictive models. We will delve into these aspects in more detail shortly.

The `mlbench` (Machine Learning Benchmark Problems) package (Blake and Merz, 1998) provides access to the Pima Indian Diabetes dataset. We load the dataset using the `data()` function:

```
data("PimaIndiansDiabetes")
```

We begin our analysis by looking at the top few values using the `glimpse()` function:

```
PimaIndiansDiabetes %>%
  glimpse(width = 50)
```

```
## Rows: 768
## Columns: 9
## $ pregnant <dbl> 6, 1, 8, 1, 0, 5, 3, 10, 2, 8, ~
## $ glucose  <dbl> 148, 85, 183, 89, 137, 116, 78,~
## $ pressure <dbl> 72, 66, 64, 66, 40, 74, 50, 0, ~
## $ triceps  <dbl> 35, 29, 0, 23, 35, 0, 32, 0, 45~
## $ insulin  <dbl> 0, 0, 0, 94, 168, 0, 88, 0, 543~
## $ mass     <dbl> 33.6, 26.6, 23.3, 28.1, 43.1, 2~
## $ pedigree <dbl> 0.627, 0.351, 0.672, 0.167, 2.2~
## $ age      <dbl> 50, 31, 32, 21, 33, 30, 26, 29,~
## $ diabetes <fct> pos, neg, pos, neg, pos, neg, p~
```

We find that this dataset contains 768 rows and 9 columns. The details of the columns as read from the help file using `?PimaIndiansDiabetes` are as follows:

1. `pregnant`: Number of times pregnant.
2. `glucose`: Plasma glucose concentration (glucose tolerance test).
3. `pressure`: Diastolic blood pressure (mm Hg).
4. `triceps`: Triceps skin fold thickness (mm).
5. `insulin`: 2-Hour serum insulin (mu U/ml).
6. `mass`: Body mass index.
7. `pedigree`: Diabetes pedigree function.
8. `age`: Age (years).
9. `diabetes`: Class variable (test for diabetes).

In this chapter, we will investigate whether diabetes (`diabetes`) can be modeled and predicted using other variables. To facilitate visualization, we convert the dataset to a tibble using the `as_tibble()` function from the `tibble` package. Subsequently, we remove any missing values (NA) from the dataset. In this instance, certain NA values are also represented as 0 values. We eliminate these values using the `filter()` function from the `dplyr` package. Finally, we store the results in a variable named `data`. The following is the code to perform these steps:

```
data <- PimaIndiansDiabetes %>%
  as_tibble() %>%
```

```
  na.omit() %>%
  filter(
    glucose != 0,
    pressure != 0,
    triceps != 0,
    insulin != 0
  )
```

Let us check if `data` is what we wanted it to be. Since we've converted our data into a tibble, there is no longer a need to use the `glimpse()` function. We use the following code:

```
data
```

```
# A tibble: 393 x 9
   pregnant glucose pressure triceps insulin  mass pedigree
      <dbl>   <dbl>    <dbl>   <dbl>   <dbl> <dbl>    <dbl>
 1        1      89       66      23      94  28.1    0.167
 2        0     137       40      35     168  43.1    2.29
 3        3      78       50      32      88  31      0.248
 4        2     197       70      45     543  30.5    0.158
 5        1     189       60      23     846  30.1    0.398
 6        5     166       72      19     175  25.8    0.587
 7        0     118       84      47     230  45.8    0.551
 8        1     103       30      38      83  43.3    0.183
 9        1     115       70      30      96  34.6    0.529
10        3     126       88      41     235  39.3    0.704
# i 383 more rows
# i 2 more variables: age <dbl>, diabetes <fct>
```

The values look good. We find that we are left with 393 rows (we had begun with 768 rows), depicting that there were several NA or 0 values.

We observe that the column `diabetes` has factor values. Let us check the levels of the factors using `levels()` function:

```
levels(data$diabetes)
```

```
## [1] "neg" "pos"
```

We find that `diabetes` has two levels: "neg" for negative, and "pos" for positive. The first value is referred to as the reference level. In the default case, the reference level is "neg". Later, when we plot the receiver operator characteristic curve, we will need "pos" as the reference level, since we are more interested in diabetes positive cases, or cases where patients have diabetes. To change the reference level of factors, we use `relevel()` function, setting the `ref` argument to "pos", and over-writing the `diabetes` column in `data`:

```
data$diabetes <- relevel(data$diabetes, ref = "pos")
```

Now when we check the levels, we find that the order is corrected:

```
levels(data$diabetes)
```

```
## [1] "pos" "neg"
```

The rest of the dataset is the same as before:

```
data
```

```
# A tibble: 393 x 9
   pregnant glucose pressure triceps insulin  mass pedigree
      <dbl>   <dbl>    <dbl>   <dbl>   <dbl> <dbl>    <dbl>
 1        1      89       66      23      94  28.1    0.167
 2        0     137       40      35     168  43.1    2.29
 3        3      78       50      32      88  31      0.248
 4        2     197       70      45     543  30.5    0.158
 5        1     189       60      23     846  30.1    0.398
 6        5     166       72      19     175  25.8    0.587
 7        0     118       84      47     230  45.8    0.551
 8        1     103       30      38      83  43.3    0.183
 9        1     115       70      30      96  34.6    0.529
10        3     126       88      41     235  39.3    0.704
# i 383 more rows
# i 2 more variables: age <dbl>, diabetes <fct>
```

6.3 LOGISTIC REGRESSION

Logistic regression is a statistical method for binary classification, and in R, the `glm()` (Generalized Linear Models) function is commonly used for this purpose. To perform logistic regression using `glm`, we need to follow a few key steps including data preprocessing, model training, making predictions, and evaluating the model [Fig. 6.3].

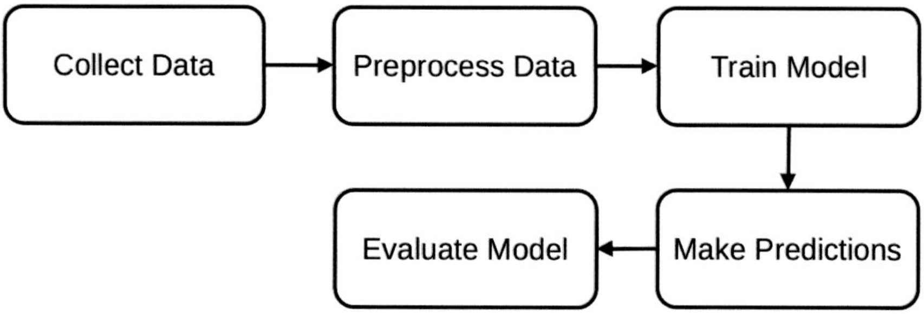

FIGURE 6.3: A schematic of the logistic regression process.

Before fitting the model, we must prepare the data to effectively manage the binary nature of the outcome variable.

To fit the logistic regression model, we use `glm(diabetes ~ ., data = data, family = "binomial")`. The formula `diabetes ~ .` indicates that `diabetes` is the response (dependent) variable and all other columns in the dataset are predictor (independent) variables. The `family = binomial` argument specifies logistic regression for binary outcomes. Let us store the result into a variable `logistic` and peruse a summary[2] of the same. The following code is used:

```
logistic <- glm(
  diabetes ~ .,
  data = data,
  family = "binomial"
)
summary(logistic)
```

[2]The summary() function in base R provides quick overviews of data based on their class. For vectors, data frames, or fitted models, it offers an instant snapshot of the data. For numeric data, it returns key statistics. For factors, it returns level frequencies. For linear models, it yields coefficients, residuals, significance levels, and goodness-of-fit. Essential for exploratory data analysis, it reveals data distribution and characteristics.

```
Call:
glm(formula = diabetes ~ ., family = "binomial", data = data)

Coefficients:
             Estimate Std. Error z value Pr(>|z|)
(Intercept) 10.0675428  1.2152353   8.284  < 2e-16 ***
pregnant    -0.0825268  0.0554279  -1.489  0.13651
glucose     -0.0382834  0.0057701  -6.635 3.25e-11 ***
pressure     0.0015612  0.0118184   0.132  0.89491
triceps     -0.0108243  0.0170224  -0.636  0.52485
insulin      0.0008296  0.0013068   0.635  0.52553
mass        -0.0718646  0.0268702  -2.675  0.00748 **
pedigree    -1.1293677  0.4249830  -2.657  0.00787 **
age         -0.0341502  0.0183691  -1.859  0.06301 .
---
Signif. codes:
0 '***' 0.001 '**' 0.01 '*' 0.05 '.' 0.1 ' ' 1

(Dispersion parameter for binomial family taken to be 1)

    Null deviance: 498.90  on 392  degrees of freedom
Residual deviance: 344.14  on 384  degrees of freedom
AIC: 362.14

Number of Fisher Scoring iterations: 5
```

We observe that the most significant variables, denoted by stars and dots, are glucose, mass, pedigree, and age. This finding is not surprising. glucose represents the plasma glucose concentration (glucose tolerance test). This test evaluates the body's ability to process glucose. Following a period of fasting, an individual consumes a glucose solution, and blood sugar levels are monitored at regular intervals to assess the rate and efficacy of glucose clearance from the bloodstream. It is widely utilized in diagnosing diabetes and other conditions that impact blood sugar regulation.

Similarly, mass represents the body mass index, or BMI. BMI, which assesses body mass in relation to height and weight, is associated with an increased risk of diabetes. A higher BMI typically indicates a higher body fat percentage, particularly around the abdominal region. This excess fat can impair the body's ability to utilize insulin effectively, leading to the development of type 2 diabetes.

pedigree represents the role of genetic factors. It is widely recognized that diabetes possesses a genetic predisposition, and both type 1 and type 2 diabetes exhibit familial patterns of inheritance.

age represents the age of person in years. It is well-established that the risk of developing diabetes increases with age. While type 1 diabetes can manifest at any age, type 2 diabetes is more prevalent among older individuals. As people age, their susceptibility to type 2 diabetes escalates due to various factors, including diminished insulin sensitivity, weight gain, and reduced physical activity.

We can utilize the `predict()` function to generate predictions with this model. However, in line with other chapters in this book, we will now transition to `tidymodels` for a more streamlined workflow.

6.4 TIDYMODELS EXAMPLE

We will now employ the `tidymodels` package to perform logistic regression. The primary advantage of utilizing `tidymodels` is that once the coding process has been completed, transitioning from one machine learning algorithm to another becomes an effortless task. Throughout this book, you will observe that we consistently employ comparable codes with various machine learning algorithms.

6.4.1 LOAD DATA

We begin by loading the required information into a data file. This is also an ideal time to convert the raw data into a tibble format. Furthermore, we remove any missing data values (represented by NA) from the dataset. We maintain a clean and intuitive code structure by using the pipe operator %>% from the `tidyverse` package. We also relevel the column `diabetes` as before. The following code emerges:

```
data <- PimaIndiansDiabetes %>%
  as_tibble() %>%
  na.omit() %>%
  filter(
    glucose != 0,
    pressure != 0,
    triceps != 0,
    insulin != 0
  )

data$diabetes <- relevel(data$diabetes, ref = "pos")
```

Let us confirm the operations by examining the `data` variable. Since we've converted our data into a tibble, there is no longer a need to use the `glimpse()` function:

```
data
```

```
## # A tibble: 393 x 9
##    pregnant glucose pressure triceps insulin  mass pedigree
##       <dbl>   <dbl>    <dbl>   <dbl>   <dbl> <dbl>    <dbl>
##  1        1      89       66      23      94  28.1    0.167
##  2        0     137       40      35     168  43.1    2.29
##  3        3      78       50      32      88  31      0.248
##  4        2     197       70      45     543  30.5    0.158
##  5        1     189       60      23     846  30.1    0.398
##  6        5     166       72      19     175  25.8    0.587
##  7        0     118       84      47     230  45.8    0.551
##  8        1     103       30      38      83  43.3    0.183
##  9        1     115       70      30      96  34.6    0.529
## 10        3     126       88      41     235  39.3    0.704
## # i 383 more rows
## # i 2 more variables: age <dbl>, diabetes <fct>
```

data looks good!

6.4.2 SPLIT DATA INTO TRAINING AND TESTING

We partition the data into training and testing sets using the `initial_split()` function from the `rsample` package, a component of `tidymodels`. We retain 70% of the data for training and 30% for testing. These sets are then extracted using `training()` and `testing()` functions:

```
# Split data into training and testing sets
set.seed(2025)
split_7030 <- initial_split(data, prop = 0.7)
data_train <- training(split_7030)
data_test <- testing(split_7030)
```

In this exemplar case, we have set the seed to 2025. `set.seed` is a function used in programming to initialize the random number generator to a specific state, ensuring that the results of random operations are reproducible. By setting the seed, we ensure that the sequence of random numbers generated is consistent each time our code is executed. This consistency facilitates reliable and reproducible results in experiments and simulations, particularly in troubleshooting scenarios. In practice, this line will be removed or commented out upon code deployment, at which point it is not required.

6.4.3 CREATE A RECIPE

We create a recipe using the `recipe()` function. In the recipe, we input the formula as `diabetes ~ .` since we wish to predict diabetes using all the available variables in the dataset. We further specify that this recipe should be applied to `data_train`, which corresponds to the training data. We also normalize all predictors using the `step_normalize()` function. The complete recipe code is depicted below:

```
recipe_ml <- recipe(
  diabetes ~ .,
  data = data_train
) %>%
  step_normalize(all_predictors())
```

6.4.4 CREATE A MODEL DESIGN

Given our interest in logistic regression, we utilize the `logistic_reg()` function to construct a model design. We further specify the engine as generalized linear model (`glm`). Since our objective is to predict whether an individual has diabetes or not based on the predictor variables, we set the mode to `classification`. The code that follows demonstrates the creation of the model design:

```
model_log_reg <- logistic_reg() %>%
  set_engine("glm") %>%
  set_mode("classification")
```

6.4.5 CREATE A WORKFLOW

We add the recipe and model to the `workflow()` function, and train the model on `data_train` dataset using `fit()` function:

```
workflow_recipe_model <- workflow() %>%
  add_recipe(recipe_ml) %>%
  add_model(model_log_reg) %>%
  fit(data_train)
```

6.4.6 ASSESS PREDICTION QUALITY WITH TEST DATA

Once the model has been trained on the training data, we can use it to make predictions on the testing data, data_test. We make use of augment() function. The augment() function, originating from the broom package in R, appends predictions and residuals from a statistical model to the original dataset. This facilitates the evaluation of a model's fit by juxtaposing actual and predicted values. The function makes predictions for data_test based on the trained model in workflow_recipe_model, and adds the predicted classes and class probabilities for negative and positive predictions in columns .pred_class, .pred_neg, and .pred_pos, respectively. We use the following code:

```
data_test_with_prediction <- augment(
  workflow_recipe_model, data_test
)
```

To examine the predictions, we select the diabetes column and the prediction columns using the select() function:

```
data_test_with_prediction %>%
  select(
  diabetes, .pred_class,
  .pred_neg, .pred_pos
)
```

```
## # A tibble: 118 x 4
##    diabetes .pred_class .pred_neg .pred_pos
##    <fct>    <fct>           <dbl>     <dbl>
##  1 neg      neg            0.982     0.0180
##  2 pos      neg            0.970     0.0299
##  3 pos      neg            0.610     0.390
##  4 pos      neg            0.804     0.196
##  5 pos      pos            0.304     0.696
##  6 neg      neg            0.506     0.494
##  7 neg      neg            0.834     0.166
##  8 pos      pos            0.0556    0.944
##  9 neg      neg            0.984     0.0155
## 10 neg      neg            0.624     0.376
## # i 108 more rows
```

We observe that the first data row had a neg class for diabetes, and it was correctly predicted to be neg, as indicated in the .pred_class column. Furthermore, we note that the independent predictor variables for the first data row resulted in a strong negative prediction. The class probability for .pred_neg is 0.982, while the corresponding class probability for .pred_pos is $1 - .pred\_neg = 0.0180$.

However, the second, third, and fourth data rows are incorrectly classified. They should have been classified as pos, but they were incorrectly classified as neg.

To have a fuller understanding of what is going on, we can create a confusion matrix. A **confusion matrix** is a tabular representation employed to assess the efficacy of a classification model by quantifying the occurrences of true positives, true negatives, false positives, and false negatives. We can create a confusion matrix using the conf_mat() function, giving it three arguments—data, truth, and estimate. In this case, data comprises the augmented data_test_with_prediction, the truth value is in the column diabetes, and the estimated value is in the column .pred_class. Thus, we execute the following code:

```
conf_mat(
  data = data_test_with_prediction,
  truth = diabetes,
  estimate = .pred_class
)
```

```
##            Truth
## Prediction pos neg
##        pos  17   9
##        neg  20  72
```

We observe from the confusion matrix that 72 true negative values were correctly predicted to be negative, and 17 true positive values were correctly predicted to be positive. However, 20 true positive values were incorrectly predicted to be negative, and 9 true negative values were incorrectly predicted to be positive.

We can also compute various metrics discussed in Section 3.11. To this end, we calculate the number of true positive, true negative, false positive, and false negative values:

```
TP <- sum(
  data_test_with_prediction$.pred_class == "pos" &
    data_test_with_prediction$diabetes == "pos"
)
```

```
TN <- sum(
  data_test_with_prediction$.pred_class == "neg" &
    data_test_with_prediction$diabetes == "neg"
)
FP <- sum(
  data_test_with_prediction$.pred_class == "pos" &
    data_test_with_prediction$diabetes == "neg"
)
FN <- sum(
  data_test_with_prediction$.pred_class == "neg" &
    data_test_with_prediction$diabetes == "pos"
)
```

We also compute the total number of positive and negative values:

```
P <- TP + FP
N <- TN + FN
```

Next, we code in the formulae for various metrics:

```
Accuracy <- (TP + TN) / (P + N)
Accuracy
```

```
## [1] 0.7542373
```

The value of accuracy is found to be 0.7542373.

```
Sensitivity <- TPR <- TP / (TP + FN)
Sensitivity
```

```
## [1] 0.4594595
```

The value of sensitivity or true positive rate is found to be 0.4594595.

```
Specificity <- TNR <- TN / (TN + FP)
Specificity
```

```
## [1] 0.8888889
```

The value of specificity or true negative rate is found to be 0.8888889.

```
Precision <- PPV <- TP / (TP + FP)
Precision
```

```
## [1] 0.6538462
```

The value of precision or positive predictive value is found to be 0.6538462.

```
F_score <- 2 * PPV * TPR / (PPV + TPR)
F_score
```

```
## [1] 0.5396825
```

The value of F score is found to be 0.5396825.

To facilitate our task, R also has functions to compute these metrics. For instance, we may directly compute the accuracy metric using the `accuracy()` function in `yardstick` package, a component of `tidymodels`. To this end, we may write the following code:

```
accuracy(
  data = data_test_with_prediction,
  truth = diabetes,
  estimate = .pred_class
)
```

```
## # A tibble: 1 x 3
##   .metric  .estimator .estimate
##   <chr>    <chr>          <dbl>
## 1 accuracy binary         0.754
```

We get the same estimate for accuracy as before.

In place of computing various metrics separately, we may also club them in a metric set using the `metric_set()` function in the `yardstick` package, and compute the metrics values in the same way that we had computed the value of accuracy:

```
metrics_set <- metric_set(
  accuracy,
  sensitivity,
  specificity,
  precision,
  f_meas,
  kap,
  mcc
)
metrics_set(
  data = data_test_with_prediction,
  truth = diabetes,
  estimate = .pred_class
)
```

```
## # A tibble: 7 x 3
##   .metric     .estimator .estimate
##   <chr>       <chr>          <dbl>
## 1 accuracy    binary         0.754
## 2 sensitivity binary         0.459
## 3 specificity binary         0.889
## 4 precision   binary         0.654
## 5 f_meas      binary         0.540
## 6 kap         binary         0.379
## 7 mcc         binary         0.390
```

We notice that the metrics have relatively low values. This is mainly because the current dataset is imbalanced, with a much higher number of negative values than positive ones. This imbalance makes the model more likely to predict the class that is more common in the training data. So, it becomes important to balance the dataset before training. We discuss this in the next section.

6.5 BALANCING DATASET

Balancing dataset is a pivotal step in preparing data for machine learning, particularly in scenarios involving imbalanced classes. **Imbalanced classes** arise when one class is substantially more prevalent than another. For instance, in medical diagnostics, there may be a substantial number of healthy patients compared to a small number of patients with a rare disease. This imbalance can result in models that excel at predicting the majority class but exhibit diminished effectiveness in predicting the minority class. For instance, an algorithm optimizing accuracy may be biased to predict that a subject is healthy, disregarding other pertinent data, solely because an individual is more likely to be healthy than diseased. This approach would diminish the model's capability to accurately identify a diseased individual.

Balancing a dataset addresses the issue of class imbalance by ensuring that the machine learning model receives a more equitable representation of each class. Various techniques can be employed to achieve this, each addressing the imbalance in its own unique manner.

One common approach to address class imbalance is through **resampling** the data. This involves modifying the dataset to create a more balanced distribution of classes. **Oversampling** is a technique where we increase the number of instances of the minority class. This can be achieved by duplicating existing minority class examples or generating new examples through methods such as the Synthetic Minority Over-sampling Technique (SMOTE). **SMOTE** works by creating synthetic samples through interpolation between existing data points of the minority class, thereby enriching the dataset without simply duplicating existing examples.

Conversely, **undersampling** entails reducing the frequency of instances belonging to the majority class until it matches the frequency of instances belonging to the minority class. Although this approach facilitates class distribution balancing, it can result in the discarding of valuable information, as some instances from the majority class are eliminated.

In addition to resampling techniques, **algorithmic approaches** can be employed to address imbalanced datasets. One such approach involves adjusting the **class weights** within the model. Numerous machine learning algorithms permit the assignment of varying weights to different classes. By assigning higher weights to the minority class, the algorithm becomes more attuned to the minority class, thereby enhancing its predictive performance.

Another algorithmic approach involves employing **ensemble methods**. Techniques such as boosting and bagging can be adapted to address class imbalance. **Boosting** prioritizes instances that are challenging to classify, thereby allocating greater attention to the minority class. Conversely, **bagging** combines multiple models to augment overall prediction performance. These methods can enhance the model's ability to generalize effectively to both classes.

When assessing the performance of models on imbalanced datasets, it is crucial to employ suitable evaluation metrics. A **confusion matrix** offers a comprehensive analysis of predictions, elucidating the counts of true positives, true negatives, false positives, and false negatives. This analysis facilitates the identification of areas where the model encounters errors.

Metrics such as precision, recall, and F1 score provide a more comprehensive evaluation of model performance, particularly in imbalanced classification settings. **Precision** measures the proportion of predicted positive cases that are truly positive, while **recall** measures the proportion of actual positive cases correctly identified. The **F1 score** combines precision and recall into a single harmonic mean, balancing their trade-offs. These metrics are valuable because they reveal how well the model handles the positive—and often minority—class, which overall accuracy may obscure.

In essence, achieving data balance is paramount for developing machine learning models that yield optimal performance across all classes. This ensures that the model is not biased toward the majority class and enhances its capacity to make accurate predictions for both prevalent and uncommon classes. Through the application of techniques such as resampling, adjusting class weights, and selecting suitable evaluation metrics, we can construct models that exhibit greater resilience and fairness, resulting in more reliable and effective predictions.

6.5.1 CHECK BALANCE

To ascertain the balance of the dataset, we can utilize the `count()` function from the `dplyr` package, which is part of the `tidyverse`. We pass the training dataset, `data_train`, to `count()` and specify the argument as `diabetes`, indicating the column whose entries need to be counted. The resulting code is as follows:

```
data_train %>% count(diabetes)
```

```
## # A tibble: 2 x 2
##   diabetes     n
##   <fct>    <int>
## 1 pos         93
## 2 neg        182
```

We observe that the `neg` label is significantly overrepresented, with `neg` values approximately twice as likely as `pos` values. Consequently, it is unsurprising that the model exhibited an overprediction of `neg` values.

6.5.2 DOWNSAMPLE

Downsampling, or **undersampling**, is a technique employed to mitigate class imbalance in a dataset by randomly selecting a subset of instances from the majority class to align with the number of instances in the minority class. This approach facilitates class distribution balancing, thereby preventing the model from becoming biased toward the majority class. Nevertheless, it entails the discarding of valuable information from the majority class, potentially resulting in a loss of pertinent data.

When performing downsampling using the tidymodels framework in R, we leverage the `recipes` package, which is specifically designed for preprocessing and feature engineering. To initiate the process, we construct a preprocessing recipe that includes a step for downsampling the training data to address class imbalance. We utilize the `step_downsample()` function from the `themis` package, which is integrated into `tidymodels`, to accomplish this. This function facilitates class balancing by reducing the number of instances in the majority class to align with the minority class. Following is the code that we can use for downsampling:

```
recipe_ml <- recipe(
  diabetes ~ .,
  data = data_train
) %>%
  step_downsample(diabetes)
```

6.5.3 PREP AND JUICE

In the `tidymodels` framework, the `prep()` function plays a crucial role in preparing data for machine learning models. This function is used to finalize the preprocessing steps defined in a recipe object. The recipe specifies the transformations and preprocessing operations to be applied to our data. By using `prep()`, we fit these preprocessing steps to the training data. This also enables us to apply the same preprocessing consistently to both training and new data.

To use `prep()`, we first define a recipe object that outlines the preprocessing steps we want to apply. Once the recipe is prepared, the `juice()` function can be used to extract the transformed training data from a prepared recipe object. After preparing the recipe with `prep()`, `juice()` provides the fully transformed version of the data that is used during training. This function is particularly useful for accessing and analyzing the data after transformation.

Conversely, we can apply the transformations to new data using the `bake()` function. This step can be used to apply the fitted recipe to both training and new datasets, ensuring that all data is processed consistently. While `juice()` retrieves the transformed data from the original dataset, `bake()` is used to apply the same preprocessing steps to new datasets.

In the current context, we apply the `prep()` and `juice()` functions through the following code and save the transformed training data to `data_train_updated`:

```
data_train_updated <- recipe_ml %>%
  prep() %>%
  juice()
```

We can now analyze the impact of downsampling using the count() function as previously demonstrated:

```
count(data_train_updated, diabetes)
```

```
## # A tibble: 2 x 2
##   diabetes     n
##   <fct>    <int>
## 1 pos         93
## 2 neg         93
```

We observe that the dataset is now balanced, with an equal number of negative and positive values. Furthermore, we note that during the downsampling process, the underrepresented class has remained unchanged, while the observations in the overrepresented class have been reduced to match the number of observations in the underrepresented class.

6.5.4 UPSAMPLE

Upsampling is a technique employed to mitigate class imbalance in datasets by augmenting the number of instances belonging to the minority class. This augmentation facilitates the creation of a more balanced dataset, thereby enhancing the performance and ensuring fairness of machine learning models.

The primary objective of upsampling is to achieve class balance. When one class is underrepresented relative to another, upsampling augments the number of instances of the minority class to align with the count of instances in the majority class. This balanced dataset facilitates improved model learning and mitigates the likelihood of bias favoring the majority class. Consequently, a more balanced dataset often yields more reliable and equitable predictions, particularly for the minority class.

This is done by increasing the representation of the minority class. This can be achieved through replicating existing instances or generating synthetic samples. Techniques such as the SMOTE create synthetic examples by combining existing minority class samples, thereby augmenting the data.

To perform upsampling in tidymodels, we utilize the themis package, which integrates seamlessly with the recipes package for preprocessing tasks. Following data splitting, we define a recipe for preprocessing that incorporates upsampling. The step_upsample() function from the themis package is employed to specify this step. The resulting code is as follows:

```
recipe_ml <- recipe(
  diabetes ~ .,
  data = data_train
) %>%
  step_upsample(diabetes)
```

Once the recipe has been defined, we use the prep() function to fit it to the training data. This prepares the recipe by applying the upsampling step. We then use the juice() function to extract the upsampled training data, which includes all the transformations applied by the recipe.

In the current context, we apply the prep() and juice() functions through the following code and save the transformed training data to data_train_updated:

```
data_train_updated <- recipe_ml %>%
  prep() %>%
  juice()
```

We can now analyze the impact of upsampling using the count() function as previously demonstrated:

```
count(data_train_updated, diabetes)
```

```
## # A tibble: 2 x 2
##   diabetes      n
##   <fct>     <int>
## 1 pos         182
## 2 neg         182
```

We observe that the dataset is now balanced, with an equal number of negative and positive values. Furthermore, we note that during the upsampling process, the overrepresented class has remained unchanged, while the observations in the underrepresented class have been increased to match the number of the overrepresented class.

With the upsampled data, we can proceed to define and fit a model.

6.5.5 SMOTE

SMOTE is a widely employed method to address class imbalance in machine learning datasets. It generates synthetic instances of the minority class to augment the dataset, thereby creating a more balanced representation that can improve the performance of classification models.

SMOTE achieves this by selecting a data point from the minority class, identifying its nearest neighbors, and generating new examples along the line segments connecting the selected point and its neighbors. For each selected point in the minority class, SMOTE calculates the differences between the feature vectors of the point and its neighbors, scales these differences by a random factor, and adds the scaled difference to the original point to create synthetic examples.

To implement SMOTE in R, we utilize the `themis` package, which integrates seamlessly with the `recipes` package for preprocessing tasks. Following the dataset's partitioning into training and testing sets, we construct a recipe that incorporates SMOTE to augment the minority class. The `step_smote()` function is employed to specify this step, as exemplified below:

```
recipe_ml <- recipe(
  diabetes ~ .,
  data = data_train
) %>%
  step_smote(diabetes)
```

After defining the recipe, we apply the SMOTE transformation using the `prep()` function to fit the recipe to the training data. Subsequently, we utilize the `juice()` function to extract the transformed training data, which comprises both the original and synthetic minority class instances.

In the current context, we use the following code and save the transformed training data to `data_train_updated`:

```
data_train_updated <- recipe_ml %>%
  prep() %>%
  juice()
```

We can now analyze the impact of upsampling using the `count()` function as previously demonstrated:

```
count(data_train_updated, diabetes)
```

```
## # A tibble: 2 x 2
##   diabetes     n
##   <fct>    <int>
## 1 pos        182
## 2 neg        182
```

We observe that the dataset is now balanced, with an equal number of negative and positive values. Furthermore, we note that during the SMOTE process, the overrepresented class has remained unchanged, while the observations in the underrepresented class have been increased to match the number of observations in the overrepresented class.

With the transformed data, we can proceed to define and fit a model.

6.5.6 COMBINATION USING RATIOS

Within the `tidymodels` framework, the `under_ratio` and `over_ratio` parameters are often utilized within the `themis` package's resampling functions for undersampling and oversampling techniques. These parameters serve to control the impact of resampling on class distribution by specifying the ratio of instances between classes.

The `under_ratio` parameter is employed in undersampling techniques to regulate the proportion of majority class instances relative to the minority class instances following resampling. This parameter specifies the number of majority class instances that will be retained in comparison to the minority class. For instance, if `under_ratio` is set to 1.0, the count of majority class instances will be equal to the count of minority class instances. Conversely, if `under_ratio` is set to 2.0, the majority class will have twice as many instances as the minority class.

Similarly, the `over_ratio` parameter plays a pivotal role in oversampling algorithms, determining the proportion of minority class instances to majority class instances after resampling. In the context of oversampling techniques such as SMOTE, the `over_ratio` parameter specifies the number of synthetic examples generated for the minority class to achieve a balanced dataset with the majority class. For instance, setting `over_ratio` to 1.0 ensures that the number of minority class instances equals the number of majority class instances. Conversely, setting `over_ratio` to 0.5 results in a dataset where the minority class is half the size of the majority class.

Proper use of these parameters can improve model performance and ensure fairness in class distribution. An exemplar code is depicted below:

```
recipe_ml <- recipe(
  diabetes ~ .,
  data = data_train
) %>%
  step_smote(diabetes, over_ratio = 0.9) %>%
  step_downsample(diabetes, under_ratio = 1)
```

After defining the recipe, we apply the transformation using the `prep()` function to fit the recipe to the training data. Subsequently, we utilize the `juice()` function to extract the transformed training data. In the current context, we use the following code and save the transformed training data to `data_train_updated`:

```
data_train_updated <- recipe_ml %>%
  prep() %>%
  juice()
```

We can now analyze the impact of transformation using the `count()` function as previously demonstrated:

```
count(data_train_updated, diabetes)
```

```
## # A tibble: 2 x 2
##   diabetes     n
##   <fct>    <int>
## 1 pos        163
## 2 neg        163
```

We observe that the dataset is now balanced, with an equal number of negative and positive values. With the set `over_ratio` and `under_ratio` parameters, both the overrepresented and underrepresented classes have been changed. With the transformed data, we can proceed to define and fit a model. We will use the current `data_train_updated` in the subsequent steps.

6.5.7 CREATE A MODEL DESIGN

We utilize the `logistic_reg()` function to construct a model design, as before. We specify the engine as generalized linear model (`glm`). Since our objective is to predict whether an individual has diabetes or not based on the predictor variables, we set the mode to classification. The code that follows demonstrates the creation of the model design:

```
model_log_reg <- logistic_reg() %>%
  set_engine("glm") %>%
  set_mode("classification")
```

6.5.8 CREATE A WORKFLOW

We construct a workflow using the workflow() function and add the recipe and the model we generated in the preceding steps. Subsequently, we train the model on the training data (data_train_updated) using the fit() function. The complete workflow creation code is provided below:

```
workflow_recipe_model <- workflow() %>%
  add_recipe(recipe_ml) %>%
  add_model(model_log_reg) %>%
  fit(data_train_updated)
```

6.5.9 ASSESS PREDICTION QUALITY WITH TEST DATA

Once the model has been trained on the training data, we can use it to make predictions on the testing data, data_test. We make use of augment() function. The augment() function, originating from the broom package in R, appends predictions and residuals from a statistical model to the original dataset. This facilitates the evaluation of a model's fit by juxtaposing actual and predicted values. The function makes predictions for data_test based on the trained model in workflow_recipe_model, and adds the predicted classes and class probabilities for negative and positive predictions in columns .pred_class, .pred_neg, and .pred_pos, respectively. We use the following code:

```
data_test_with_prediction <- augment(
  workflow_recipe_model, data_test
)
```

To examine the predictions, we select the diabetes column and the prediction columns using the select() function:

```
data_test_with_prediction %>%
  select(
  diabetes, .pred_class,
  .pred_neg, .pred_pos
)
```

```
## # A tibble: 118 x 4
##    diabetes .pred_class .pred_neg .pred_pos
##    <fct>    <fct>           <dbl>     <dbl>
##  1 neg      neg            0.974     0.0258
##  2 pos      neg            0.960     0.0402
##  3 pos      pos            0.380     0.620
##  4 pos      neg            0.715     0.285
##  5 pos      pos            0.183     0.817
##  6 neg      pos            0.417     0.583
##  7 neg      neg            0.689     0.311
##  8 pos      pos            0.0165    0.983
##  9 neg      neg            0.981     0.0194
## 10 neg      pos            0.500     0.500
## # i 108 more rows
```

We observe that the first data row had a neg class for `diabetes`, and it was correctly predicted to be neg, as indicated in the `.pred_class` column. Furthermore, we note that the independent predictor variables for the first data row resulted in a strong negative prediction. The class probability for `.pred_neg` is 0.974, while the corresponding class probability for `.pred_pos` is $1 - .pred\_neg = 0.0258$.

The second data row is incorrectly classified, as before. However, the third data row is now correctly classified. Thus, there appears to be some improvement from before.

To fully comprehend the scenario, we create the confusion matrix. As we have observed, a confusion matrix is a tabular representation employed to assess the efficacy of a classification model by quantifying the occurrences of true positives, true negatives, false positives, and false negatives. We create a confusion matrix using the `conf_mat()` function as before, giving it three arguments—data, truth, and estimate. In this case, data comprises the augmented `data_test_with_prediction`, the truth value is in the column `diabetes`, and the estimated value is in the column `.pred_class`. Thus, we execute the following code:

```
conf_mat(
  data = data_test_with_prediction,
  truth = diabetes,
  estimate = .pred_class
)
```

```
##            Truth
## Prediction pos neg
##        pos  24  19
##        neg  13  62
```

We observe from the confusion matrix that 62 true negative values were correctly predicted to be negative, and 24 true positive values were correctly predicted to be positive. However, 13 true positive values were incorrectly predicted to be negative, and 19 true negative values were incorrectly predicted to be positive.

We also compute various metrics using the `metric_set()` function in the `yardstick` package as before:

```
metrics_set <- metric_set(
  accuracy,
  sensitivity,
  specificity,
  precision,
  f_meas,
  kap,
  mcc
)
metrics_set(
  data = data_test_with_prediction,
  truth = diabetes,
  estimate = .pred_class
)
```

```
## # A tibble: 7 x 3
##   .metric     .estimator .estimate
##   <chr>       <chr>          <dbl>
## 1 accuracy    binary         0.729
## 2 sensitivity binary         0.649
## 3 specificity binary         0.765
## 4 precision   binary         0.558
```

```
## 5 f_meas        binary         0.6
## 6 kap           binary         0.397
## 7 mcc           binary         0.399
```

We observe that the metrics have changed from before. There has been a slight expected reduction in accuracy from 0.754 to 0.729, specificity from 0.889 to 0.765, and precision from 0.654 to 0.558. However, there is an increase in sensitivity from 0.459 to 0.649, kappa from 0.379 to 0.397, mcc from 0.390 to 0.399, and f_meas from 0.540 to 0.6, indicating that the model can now better identify true positives and negatives. Thus, balancing the dataset has improved certain metrics, while degrading certain others, even in a small dataset like the `diabetes` dataset.

In practice, the selection of appropriate metrics is contingent upon the specific application at hand. Balancing a dataset is frequently employed to enhance certain metrics, particularly in scenarios involving imbalanced training datasets.

6.6 RECEIVER OPERATING CHARACTERISTIC

The ROC curve is a fundamental graphical representation employed in statistical analysis and machine learning to evaluate the performance of a classification model. It visually depicts the model's capacity to distinguish between positive and negative classes.

The ROC curve plots the true positive rate (TPR), also known as sensitivity or recall, against the false positive rate (FPR), which is defined as one minus specificity. TPR quantifies the model's ability to correctly identify actual positives, while FPR indicates the proportion of actual negatives that are erroneously classified as positives.

To construct an ROC curve, we calculate the TPR and FPR at various threshold levels. A threshold is the cutoff value used to convert a model's predicted probability (or score) into a binary class prediction (positive vs. negative). These thresholds dictate how the model's predicted probabilities are converted into binary outcomes. By adjusting the threshold, we can plot different points on the ROC curve, reflecting the model's performance at each threshold.

The ROC curve demonstrates the trade-off between sensitivity and specificity. An ideal model would have an ROC curve close to the top-left corner of the plot, signifying high sensitivity and a low FPR. In contrast, a model with no discriminative ability would show a diagonal line from the bottom-left to the top-right, representing random guessing.

The area under the ROC curve (AUC) is a summary metric that measures the overall performance of the model. An AUC of 1.0 indicates a perfect model, while an AUC of 0.5 suggests no ability to distinguish between classes, akin to random guessing. The area under the curve can be computed using the following code:

```
roc_auc(data_test_with_prediction,
        truth = diabetes,
        .pred_pos)
```

```
## # A tibble: 1 x 3
##   .metric .estimator .estimate
##   <chr>   <chr>          <dbl>
## 1 roc_auc binary         0.825
```

We find that the value is above 0.8, meaning that the model has a good ROC value. We can plot the ROC curve using the following code:

```
data_test_with_prediction %>%
  roc_curve(truth = diabetes, .pred_pos) %>%
  autoplot()
```

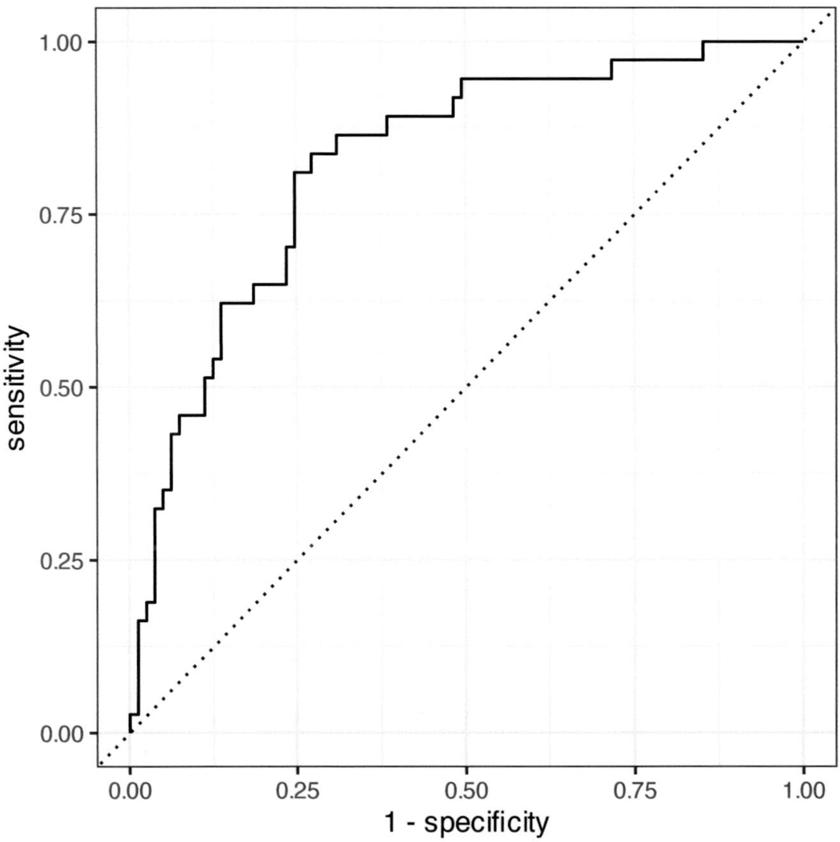

In practice, we use the ROC curve and AUC to compare different models and select the one that best distinguishes between positive and negative classes across various thresholds.

6.7 WHEN TO USE LOGISTIC REGRESSION, AND WHEN NOT TO

Logistic regression, a widely used statistical model, excels in binary classification tasks. Its primary objective is to predict one of two distinct outcomes. This model finds particular utility when the predicted outcome is categorical, possessing only two values. For instance, it can be employed to determine whether an email is spam or not, or to identify a patient's likelihood of having a specific disease based on a set of health indicators.

Logistic regression is a powerful tool when the relationship between predictor variables and the outcome can be approximated by a linear model in the log-odds scale. This implies that the model assumes that changes in the predictors result in proportional changes in the odds of the outcome. Additionally, logistic regression is valuable when we need to model the probability of an event occurring. It provides a prediction that falls between 0 and 1, offering a clear indication of the likelihood of the event.

One of the strengths of logistic regression is that it is relatively easy to interpret. Each coefficient in the model represents how a one-unit change in a predictor affects the log-odds of the outcome, which can be converted into odds ratios for more intuitive understanding. Due to its simplicity and efficiency, logistic regression often serves as a strong starting point or baseline model, especially when working with small datasets.

However, there are situations where logistic regression may not be the best choice. It is primarily designed for binary outcomes, so if we are dealing with problems involving more than two categories, logistic regression may not be ideal. While there are extensions like multinomial logistic regression for multiclass problems, they may not always handle complex classification tasks effectively.

If the relationship between predictors and the outcome isn't linear on the log-odds scale, logistic regression may struggle to capture the underlying data patterns. In cases with a large number of features compared to observations, logistic regression may face challenges and potentially overfit the data. Additionally, when the data involves complex interactions between predictors or intricate patterns, logistic regression's linear assumptions may hinder its effectiveness. While logistic regression can be adjusted to handle imbalanced datasets, other methods specifically designed to address class imbalance may yield better results. For extremely large datasets with complex patterns, advanced models like deep learning algorithms may offer superior performance.

In summary, logistic regression is a robust and interpretable model suitable for binary classification tasks with a linear relationship in log-odds. It offers valuable probabilistic predictions and serves as a suitable baseline model. However, it may not be the optimal choice for multiclass problems, non-linear relationships, high-dimensional data, intricate interactions, or highly imbalanced datasets. In such scenarios, alternative modeling techniques or advanced methodologies may be more appropriate.

6.8 CHAPTER SUMMARY

- Logistic regression, a statistical model for binary classification with a linear relationship in log-odds, offers interpretable results and is often used as a baseline, but may not be best for multiclass or non-linear problems.
- The Pima Indians Diabetes dataset, from an NIDDK study, is a model used in predicting diabetes risk. It contains health and demographic info of Pima Indian women, including pregnancies, glucose levels, blood pres-

sure, and BMI. Despite class imbalance and missing values, it is a valuable resource for testing machine learning algorithms.

- Techniques like resampling and class weight adjustment are used to address class imbalance. Appropriate evaluation metrics like precision, recall, and F1 score are crucial for imbalanced datasets.
- The `tidymodels` framework in R uses the `recipes` package for preprocessing and feature engineering, especially for class imbalance. Techniques like downsampling, upsampling, and SMOTE balance the dataset using the `themis` package. Parameters like `under_ratio` and `over_ratio` help create a balanced dataset for improved performance.
- The ROC curve with AUC summarizes a classification model's ability to distinguish between positive and negative classes.

7 K-nearest neighbors

Picture yourself with a group of friends, and you have a good idea about the pets they like. Some friends are fond of dogs, others like cats, and a few might even be interested in birds or hamsters. Now, imagine a new child coming in your town, and you have no clue about their favorite pet.

To try and figure it out, you observe the new child's best friends. Let's say they all adore dogs. If a majority of their best friends like dogs, you might assume that the new child will probably like dogs as well. After all, birds of a feather flock together!

K-nearest neighbors (KNN) operates in a similar manner: it examines the nearest examples (similar to the new child's friends) to determine what category something unfamiliar (the new child) might belong to. If most of the nearest examples (nearest neighbors) are in one category, KNN makes a prediction that the new item will also be in that category!

7.1 WHAT IS KNN?

KNN is a versatile and intuitive machine learning algorithm primarily used for classification and regression tasks. Its simplicity and effectiveness make it a popular choice for various predictive modeling scenarios. KNN is based on a straightforward concept: making predictions based on the proximity of data points within a dataset.

At its core, KNN operates on the **principle of similarity**. To predict or classify a new data point, KNN examines the "K" closest known data points from the existing dataset and determines the most probable outcome based on these neighbors. This approach is often compared to seeking opinions from a group of friends on a matter and then making a decision based on the majority vote. For instance, to determine whether a new email is spam, KNN would examine the "K" closest emails in the dataset, identify which of these are labeled as spam, and then classify the new email based on the most frequent label among these neighbors.

The process commences with the selection of a value for "**K**", which represents the number of neighbors to be considered in the algorithm. This choice holds significant importance, as it directly impacts the algorithm's performance. A smaller "K" may render the algorithm susceptible to noise and outliers, resulting in less precise predictions. Conversely, a larger "K" might smooth out the distinctions between categories, potentially leading to overly generalized outcomes. Therefore, determining an optimal "K" often requires experimentation.

Once the value of "K" is determined, the algorithm computes the distance between the new data point and every point in the training dataset. Various distance metrics can be employed, with **Euclidean distance** being one of the most prevalent. Euclidean distance quantifies the straight-line distance between two points in a multidimensional space. Other metrics, such as **Manhattan distance** or **Minkowski distance**, may be selected based on the characteristics of the data and the specific problem being addressed.

After calculating distances, KNN identifies the "K" closest neighbors. For classification tasks, the algorithm analyzes the classes of these neighbors and assigns the most prevalent class to the new data point. This process is similar to a voting system where the majority vote determines the class label. In regression tasks, KNN calculates an average—sometimes a weighted average—of the values of the "K" closest neighbors to predict the value for the new data point.

One of the key advantages of KNN is its simplicity. Unlike other machine learning algorithms, KNN doesn't require an explicit training phase. Instead, it uses the entire dataset to make predictions, which can be particularly advantageous when dealing with datasets that may be computationally expensive or infeasible to train. Additionally, KNN is adaptive, meaning it can easily adjust to new data points without the need for a complete retraining of the model.

However, KNN is not without its drawbacks. The algorithm can be computationally intensive, especially for large datasets. As the number of data points increases, calculating distances and identifying neighbors becomes significantly slower. This can result in slower prediction times, particularly in high-dimensional spaces where the computational cost becomes prohibitive. Additionally, KNN's performance heavily relies on the selection of distance metric and the value of "K", making **hyperparameter tuning** a crucial and often critical step in the modeling process.

Another factor to consider is the **curse of dimensionality**. As the number of features or dimensions in a dataset increases, the concept of "closeness" loses its significance, and the distances between points tend to become more uniform. This phenomenon can negatively impact the performance of KNN, as distinguishing between neighboring points in high-dimensional spaces becomes increasingly difficult.

In summary, KNN is a simple yet effective algorithm for classification and regression tasks. It relies on the concept of similarity to make predictions. KNN's simplicity and lack of a formal training phase make it appealing for various applications. However, practitioners should be aware of its high computational requirements and the impact of dimensionality on its performance. By carefully selecting the value of "K", choosing suitable distance metrics, and considering the dataset's characteristics, KNN can be a valuable tool in a machine learning toolkit.

7.2 THE IRIS DATASET

Iris flowers [Fig. 7.1], renowned for their beauty and longevity, boast three petals standing upright and three sepals hanging down, collectively referred to as the six "tepals". Their vibrant colors, including purple, blue, white, and yellow, evoke associations with royalty and wisdom. Throughout history and mythology, irises have been imbued with symbolic meanings, representing faith, hope, and courage. Gardeners and florists alike are captivated by their graceful and bold appearance.

FIGURE 7.1: A schematic of variations in iris flowers.

The **Iris dataset**, based on these beautiful flowers, is a highly esteemed and extensively used dataset in the fields of machine learning and statistics. Introduced by the renowned British biologist and statistician Ronald A. Fisher in 1936, its simplicity and well-defined structure have made it an indispensable resource for teaching and benchmarking machine learning algorithms.

The Iris dataset comprises data about three distinct species of iris flowers: *Iris setosa*, *Iris versicolor*, and *Iris virginica*. Each species is represented by 50 samples, resulting in a total of 150 observations in the dataset. The primary objective of this dataset is to serve as a classification challenge, where the task is to differentiate between these three species based on their physical attributes.

The dataset includes four key features measured for each iris flower: sepal length, sepal width, petal length, and petal width. These features are recorded in centimeters and provide essential information for distinguishing the species. The sepal is the part of the flower that encloses the petals, and the petal is often the more colorful part of the flower situated above the sepals. Measurements of these parts are used to identify and classify the species.

The dataset presents these four features for each flower in a tabular format, where each row represents a distinct observation. The columns include sepal length, sepal width, petal length, and petal width. The final column indicates the species of the flower, which is one of the three iris species. The goal is to predict the species of a flower based on these characteristics.

The Iris dataset's popularity stems from its simplicity and ease of comprehension and application. Its relatively small size makes it ideal for initial experimentation with machine learning algorithms. The straightforward nature of the dataset facilitates quick visualization and interpretation of the data. For example, plotting the data using pairwise comparisons of features can reveal how effectively different species can be distinguished based on measurements.

The Iris dataset is frequently used to evaluate and compare the performance of various machine learning algorithms. Its well-defined structure and clear separation between species make it an ideal benchmark for testing classification methods like KNN, Decision Trees, Support Vector Machines, and others. Additionally, the dataset is commonly used to teach fundamental concepts in data analysis, machine learning, and statistical classification.

The Iris dataset, despite its simplicity, provides valuable insights into classification problems and the application of various techniques. It serves as a clear example of how data can be utilized to make predictions and understand the relationships between different features. The dataset's small size and distinct separations between classes make it particularly useful for illustrating concepts such as dimensionality reduction, model evaluation, and feature selection.

The Iris dataset holds historical significance too. Ronald Fisher employed it to illustrate his **linear discriminant analysis** (LDA) technique,[1] a method employed to identify the linear combinations of features that effectively distinguish distinct classes. Fisher's contributions with the Iris dataset formed the groundwork for numerous contemporary statistical and machine learning approaches.

In summary, the Iris dataset is a fundamental resource in the field of machine learning and statistics. The dataset's simplicity and clear structure make it an exceptional tool for teaching, learning, and benchmarking. By presenting a straightforward classification problem with quantifiable features, the dataset enables practitioners and researchers to delve into and comprehend various machine learning techniques and concepts. Its historical significance and enduring relevance guarantee that the Iris dataset continues to serve as a cornerstone of data science education and practice.

7.2.1 USING THE DATASET

We begin by (installing and) loading all the required packages and libraries:

---

[1]Imagine data points grouped into red and blue dots on a graph. LDA helps draw a line (or plane in higher dimensions) that best separates these groups. It projects data onto a new axis to maximize group distance while minimizing group spread. This aids classification and is used in pattern and face recognition.

```r
if (!require("tidymodels")){
  install.packages("tidymodels")
}
library(tidymodels)

if (!require("tidyverse")){
  install.packages("tidyverse")
}
library(tidyverse)

if (!require("janitor")){
  install.packages("janitor")
}
library(janitor)

if (!require("GGally")){
  install.packages("GGally")
}
library(GGally)

if (!require("dslabs")){
  install.packages("dslabs")
}
library(dslabs)

if (!require("kknn")){
  install.packages("kknn")
}
library(kknn)

if (!require("kableExtra")){
  install.packages("kableExtra")
}
library(kableExtra)
```

The `iris` dataset, a part of base R's `datasets` package (R Core Team, 2022), is readily accessible via the prompt: `iris`. Let's delve into this dataset. To begin, we'll examine the first few values using the `head()` function:

```r
iris %>%
  head()
```

```
##   Sepal.Length Sepal.Width Petal.Length Petal.Width Species
## 1          5.1         3.5          1.4         0.2  setosa
## 2          4.9         3.0          1.4         0.2  setosa
## 3          4.7         3.2          1.3         0.2  setosa
## 4          4.6         3.1          1.5         0.2  setosa
## 5          5.0         3.6          1.4         0.2  setosa
## 6          5.4         3.9          1.7         0.4  setosa
```

The dataset comprises five columns:

1. `Sepal.Length`: the length of the sepal.
2. `Sepal.Width`: the width of the sepal.
3. `Petal.Length`: the length of the petal.
4. `Petal.Width`: the width of the petal.
5. `Species`: the name of the species.

The lengths and widths are provided in centimeters. To obtain the dimensions of the dataset, we can utilize the `dim()` function:

```
iris %>%
  dim()
```

```
## [1] 150   5
```

We notice that the dataset contains 150 observations and 5 columns. To obtain a concise summary of the results, we can utilize the `summary()` function in base R. This function provides a quick overview of data based on its class. For vectors, data frames, or fitted models, it offers an instantaneous snapshot of the data. For numerical data, it returns essential statistics. For factors, it provides level frequencies. For linear models, it yields coefficients, residuals, significance levels, and goodness-of-fit. This function is particularly useful for exploratory data analysis (EDA), as it reveals the data distribution and characteristics. To apply the `summary()` function to the `iris` dataset, we use the following code:

```
iris %>%
  summary()
```

```
##   Sepal.Length    Sepal.Width     Petal.Length
##  Min.   :4.300   Min.   :2.000   Min.   :1.000
```

```
##  1st Qu.:5.100   1st Qu.:2.800   1st Qu.:1.600
##  Median :5.800   Median :3.000   Median :4.350
##  Mean   :5.843   Mean   :3.057   Mean   :3.758
##  3rd Qu.:6.400   3rd Qu.:3.300   3rd Qu.:5.100
##  Max.   :7.900   Max.   :4.400   Max.   :6.900
##   Petal.Width          Species
##  Min.   :0.100   setosa    :50
##  1st Qu.:0.300   versicolor:50
##  Median :1.300   virginica :50
##  Mean   :1.199
##  3rd Qu.:1.800
##  Max.   :2.500
```

For the four independent variables, we obtain statistical summaries including the minimum, first quartile, median, mean, third quartile, and maximum values. For the dependent variable, we have 50 entries for each of the three species: setosa, versicolor, and virginica, demonstrating that the dataset is balanced.

7.3 PLOTTING THE DATASET

`ggpairs()` (ggplot2 generalized pairs) is a function from the `GGally` package (Schloerke et al., 2024) in R, which is an extension of the popular `ggplot2` package (Wickham, 2016). `ggpairs()` is used to create a matrix of scatterplots to visualize the relationships between multiple variables in a dataset. A **scatterplot**, also known as a **pairs plot** or **pairwise plot**, is a statistical graph used to visualize and compare two variables. It presents a set of points plotted on a Cartesian coordinate system, where each point represents the values of two variables. One variable is assigned to the x-axis, while the other is assigned to the y-axis. Scatterplots are particularly valuable in identifying relationships or correlations between variables, such as trends, clusters, or outliers. By visually examining the distribution and pattern of the points, one can infer whether a positive, negative, or no correlation exists between the variables, which aids in data analysis and interpretation.

`ggpairs()` generates scatterplots for every pair of variables in a dataset. By visualizing these relationships, it becomes easy to identify patterns such as linear relationships, clusters, or outliers. The diagonal of the matrix frequently contains histograms or density plots of the individual variables, illustrating their distribution and aiding in understanding their unique characteristics, such as their range and distribution. Additionally, `ggpairs()` can display **correlation coefficients** in the upper or lower triangle of the matrix, offering numerical insights into the intensity and direction of relationships between variables.

The plots generated by `ggpairs()` are highly customizable, enabling users to modify plot types, incorporate regression lines, and adjust themes using additional ggplot2 functions. Moreover, users can enhance their plots by adding other layers or elements, such as color-coding by groups, or customizing plot aesthetics to improve the visual appeal.

`ggpairs()` is a valuable tool for **EDA** because it provides a simple and comprehensive method for visualizing and understanding the relationships between multiple variables. It can help identify potential issues in the data, such as **multicollinearity**,[2] and guide the selection of features for further analysis or modeling.

We can create a `ggpairs()` plot for the `iris` dataset using the following code:

```
ggpairs(
  data = iris,
  columns = 1:4,
  aes(color = Species, alpha = 0.5)
)
```

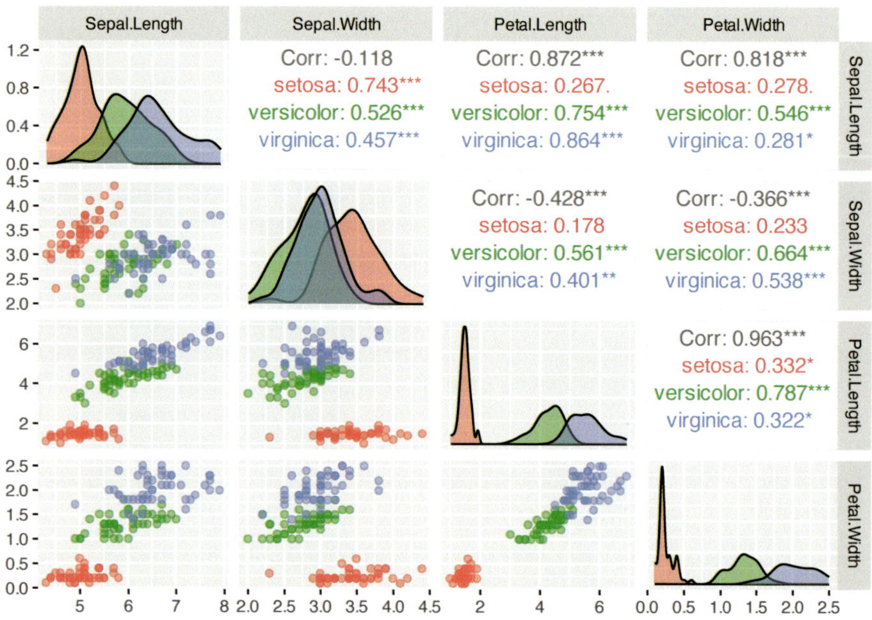

The plots reveal that the flower dimensions of the three species differ significantly from one another. Consequently, it is feasible to use flower dimensions as a means of identifying the species.

[2] Multicollinearity is a statistical phenomenon where highly correlated predictor variables in a multiple regression model can be accurately predicted from each other. This condition inflates standard errors, making coefficients unreliable and conclusions misleading. Multicollinearity also makes it difficult to determine the individual effect of each predictor on the dependent variable, complicating model interpretation in machine learning.

To better understand the distribution of flower dimensions, we can create box plots. A **box plot** consists of three main components: the box, which represents the **interquartile range** (**IQR**), the distance between the first and third quartiles; the median line, which signifies the median (the 50th percentile); and whiskers extending from the box to the smallest and largest values that fall within 1.5 times the IQR from the lower and upper quartiles, respectively. Any data points that extend beyond the whiskers are considered potential outliers. An **outlier** is a data point that significantly differs from the rest of the dataset. It can be much higher or lower than the majority, indicating measurement variability, errors, or unusual occurrences. Outliers can skew the results, so their identification and analysis are crucial. Statistical tests and graphical methods, such as box plots and scatter plots, can help identify outliers.

Box plots are particularly useful for comprehending data spread, skewness,[3] and identifying outliers. They provide a concise summary of central tendency, dispersion, and skewness, making them ideal for comparing distributions across different categories. However, it's crucial to remember that box plots don't offer information about the underlying distribution, such as bimodality or multimodality. To gain a more comprehensive understanding, box plots may need to be complemented with other plots like histograms or density plots.

`geom_boxplot()`, a function in the `ggplot2` package for R, enables the creation of box plots. To achieve this, we employ the following code, where we modify the variable in y to generate various box plots:

```
iris %>%
  ggplot(aes(x = Species, y = Petal.Length)) +
  geom_boxplot() +
  theme_classic(base_size = 18)
```

[3]Skewness measures how unevenly data spreads around the center, showing whether it extends more to one side than the other.

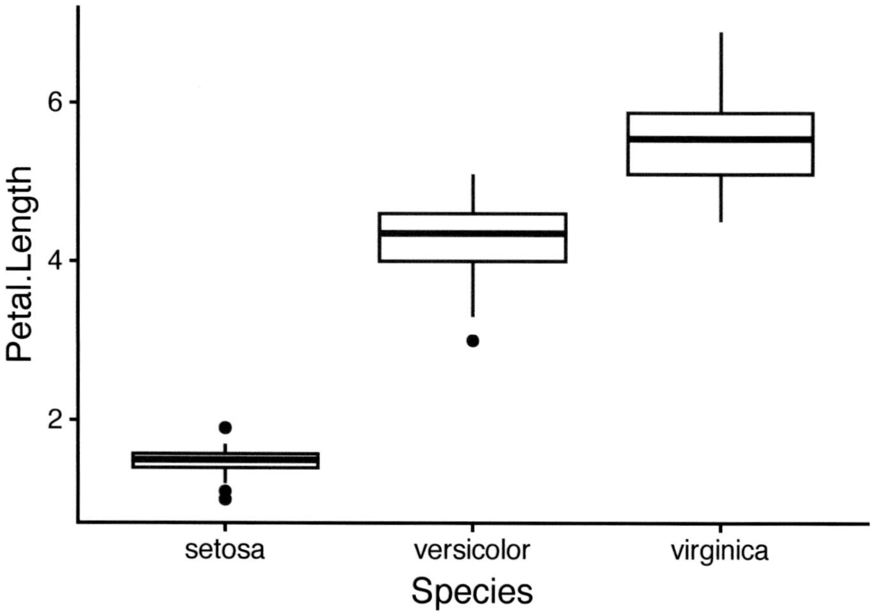

We notice that the values of `Petal.Length` for *setosa* are significantly lower compared to those of the *versicolor* and *virginica* species. Considering the substantial differences in these values, it becomes evident that these measurements can be effectively utilized to differentiate between the species.

We also observe the variations in the values of other independent variables among the three species. These variations can be used to identify the most similar (nearest) neighbors.

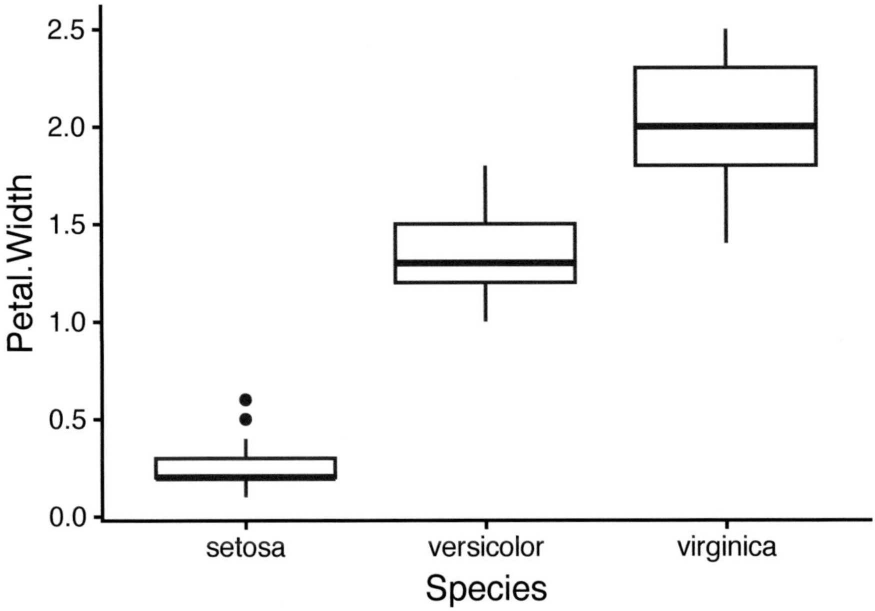
2.5
2.0
1.5
1.0
0.5
0.0
Petal.Width
setosa
versicolor
virginica
Species

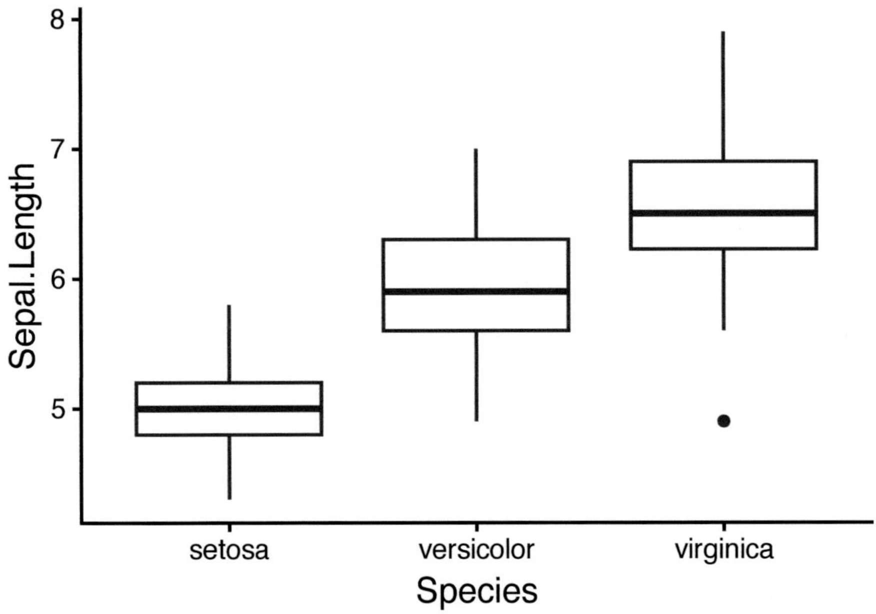
8
7
6
5
Sepal.Length
setosa
versicolor
virginica
Species

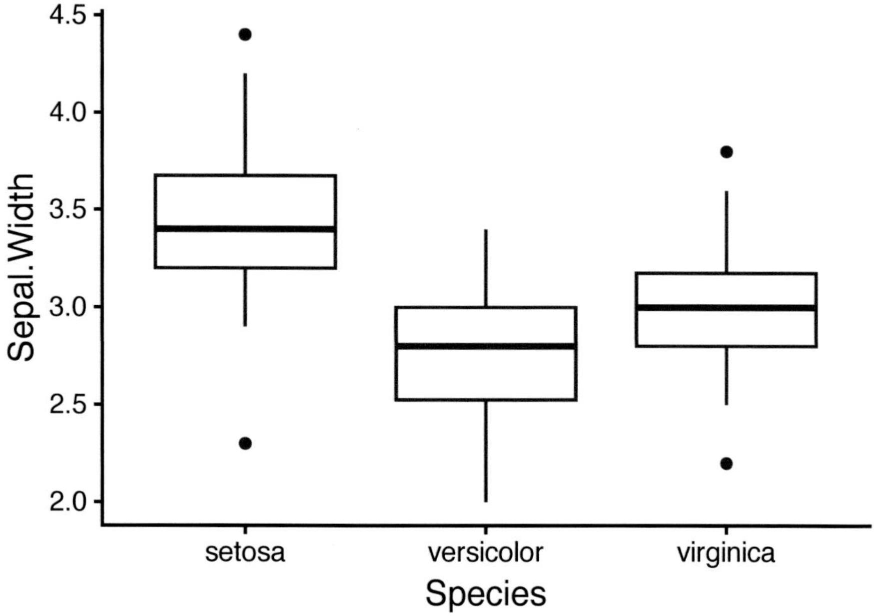

We will now delve into the application of KNN using the `tidymodels` package.

7.4 TIDYMODELS EXAMPLE

We will now employ the `tidymodels` package to perform KNN classification. The primary advantage of utilizing `tidymodels` is that once the coding process has been completed, transitioning from one machine learning algorithm to another becomes an effortless task. Throughout this book, you will observe that we consistently employ comparable codes with various machine learning algorithms.

7.4.1 LOAD DATA

We begin by loading the required information into a data file. This is also an ideal time to convert the raw data into a tibble format for enhanced visualization and to eliminate any missing values (NA) that might be present in the dataset.

We also convert variable names to snake case. This optional step is highly recommended for better comprehension in subsequent steps and for enhanced visualization in tables and charts.

As before, we use the pipe operator `%>%` from the `tidyverse` package to maintain a clean and intuitive code structure. We employ the following code and store the final results in the `data` variable:

```
data <- iris %>%
  # Convert to tibble
  as_tibble() %>%
  # Remove NA values
  na.omit() %>%
  # Convert to snake case
  clean_names("snake")
```

Here # is used to insert comments—lines used to make the code more human-readable and understandable, but which are not processed by R.

Let us confirm the operations by examining the `data` variable. Since we've converted our data into a tibble, there is no need to use `glimpse()` or `head()` functions:

```
data
```

```
## # A tibble: 150 x 5
##    sepal_length sepal_width petal_length petal_width species
##           <dbl>       <dbl>        <dbl>       <dbl> <fct>
##  1          5.1         3.5          1.4         0.2 setosa
##  2          4.9         3            1.4         0.2 setosa
##  3          4.7         3.2          1.3         0.2 setosa
##  4          4.6         3.1          1.5         0.2 setosa
##  5          5           3.6          1.4         0.2 setosa
##  6          5.4         3.9          1.7         0.4 setosa
##  7          4.6         3.4          1.4         0.3 setosa
##  8          5           3.4          1.5         0.2 setosa
##  9          4.4         2.9          1.4         0.2 setosa
## 10          4.9         3.1          1.5         0.1 setosa
## # i 140 more rows
```

We find that there is no change in the number of rows, indicating that there are no rows with NA values. `data` looks good!

7.4.2 SPLIT DATA INTO TRAINING AND TESTING

We partition the data into training and testing sets using the `initial_split()` function from the `rsample` package, a component of `tidymodels`. We retain 70% of the data for training and 30% for testing. These sets are then extracted using `training()` and `testing()` functions:

```
# Split data into training and testing sets
set.seed(2025)
split_7030 <- initial_split(data, prop = 0.7)
data_train <- training(split_7030)
data_test <- testing(split_7030)
```

In this exemplar case, we have set the seed to 2025. `set.seed` is a function used in programming to initialize the random number generator to a specific state, ensuring that the results of random operations are reproducible. By setting the seed, we ensure that the sequence of random numbers generated is consistent each time our code is executed. This consistency facilitates reliable and reproducible results in experiments and simulations, particularly in troubleshooting scenarios. In practice, this line will be removed or commented out upon code deployment, at which point it is not required.

7.4.3 CREATE A RECIPE

We create a recipe using the `recipe()` function. In the recipe, we input the formula `species ~ .` to predict the species using all the available variables in the dataset. We further specify that this recipe should be applied to `data_train`, which is the training data. We also normalize all the predictor variables using the `step_normalize()` function. This step normalizes numeric data to have a standard deviation of one and a mean of zero. We normalize data for KNN to ensure that all features contribute equally to the distance calculation, avoiding bias toward features with larger scales. The complete recipe code is provided below:

```
recipe_ml <- recipe(
  species ~ .,
  data = data_train
) %>%
  step_normalize(all_predictors())
```

7.4.4 CREATE A MODEL DESIGN

We define a `nearest_neighbor` model using the `nearest_neighbor()` function from `parsnip` package and the `kknn` engine (Schliep and Hechenbichler, 2025). KNN is versatile and can be used for both classification and regression tasks. In this particular scenario, we intend to classify data into three categories: `setosa`, `versicolor`, and `virginica`. Therefore, we set the `mode` parameter to `classification`.

In the model design, we define neighbors to be 1 (meaning K = 1) and weight_func to be "rectangular." **Weighted K-Nearest Neighbors (WKNN)** considers the distance of each neighboring point to the query point, typically giving more weight to closer neighbors. By setting weight_func to "rectangular," we indicate that we don't want to give more weight to more nearby points; instead, we want to apply a uniform weight regardless of the distances between points. In essence, we're converting the default weighted KNN to a classic, unweighted KNN. Here's the code for the model design:

```
model_kknn <- nearest_neighbor(
  neighbors = 1,
  weight_func = "rectangular"
) %>%
  set_engine("kknn") %>%
  set_mode("classification")
```

7.4.5 CREATE A WORKFLOW

We add the recipe and model to the workflow() function, and train the model on data_train dataset using fit() function:

```
workflow_recipe_model <- workflow() %>%
  add_recipe(recipe_ml) %>%
  add_model(model_kknn) %>%
  fit(data_train)
```

7.4.6 ASSESS PREDICTION QUALITY WITH TEST DATA

We can now evaluate the performance of our trained and fitted model by using the test data set, data_test. We make use of augment() function. The augment() function, originating from the broom package in R, appends predictions and residuals from a statistical model to the original dataset. This facilitates the evaluation of a model's fit by juxtaposing actual and predicted values. The function makes predictions for data_test based on the trained model in workflow_recipe_model, and adds the predicted classes and class probabilities for predictions. We use the following code:

```
data_test_with_prediction <- augment(
  workflow_recipe_model, data_test
)
```

Let's take a look at the augmented file, data_test_with_prediction.

```
data_test_with_prediction %>%
  glimpse(width = 50)
```

```
## Rows: 45
## Columns: 9
## $ .pred_class      <fct> setosa, setosa, setosa,~
## $ .pred_setosa     <dbl> 1, 1, 1, 1, 1, 1, 1, 1,~
## $ .pred_versicolor <dbl> 0, 0, 0, 0, 0, 0, 0, 0,~
## $ .pred_virginica  <dbl> 0, 0, 0, 0, 0, 0, 0, 0,~
## $ sepal_length     <dbl> 5.1, 5.4, 4.4, 5.1, 5.7~
## $ sepal_width      <dbl> 3.5, 3.9, 2.9, 3.5, 3.8~
## $ petal_length     <dbl> 1.4, 1.7, 1.4, 1.4, 1.7~
## $ petal_width      <dbl> 0.2, 0.4, 0.2, 0.3, 0.3~
## $ species          <fct> setosa, setosa, setosa,~
```

We notice that four new columns have been added to the dataset:

1. The .pred_class column contains the predictions made by the model. For the first three rows, it correctly classifies the species as setosa, which aligns with the values in the species column.
2. The .pred_setosa column provides the binary identification of a flower being setosa, calculated by the model.
3. The .pred_versicolor column provides the binary identification of a flower being versicolor, calculated by the model.
4. The .pred_virginica column provides the binary identification of a flower being virginica, calculated by the model.

The three species predictions sum up to 1. To have a fuller understanding of what is going on, we can create a confusion matrix. A **confusion matrix** is a tabular representation employed to assess the efficacy of a classification model by quantifying the occurrences of true positives, true negatives, false positives, and false negatives. We can create a confusion matrix using the conf_mat() function, giving it three arguments—data, truth, and estimate. In this case, data comprises the augmented data_test_with_prediction, the truth value is in the column species, and the estimated value is in the column `.pred_class`. Thus, we execute the following code:

```
conf_mat(
  data_test_with_prediction,
  truth = species,
  estimate = .pred_class
)
```

```
            Truth
Prediction   setosa versicolor virginica
  setosa         14          0         0
  versicolor      1         16         0
  virginica       0          3        11
```

We discover that all 14 flowers predicted by the model to be setosa were indeed setosa. Similarly, out of the 17 flowers predicted to be versicolor (1 + 16), 16 were versicolor, and 1 was setosa. Lastly, out of the 14 flowers predicted to be virginica (3 + 11), 11 were virginica, and 3 were versicolor. It seems that a majority of the predictions were indeed accurate!

We can also calculate the accuracy metrics. In machine learning, **accuracy metrics** evaluate a model's performance by comparing its predictions to the actual outcomes. We define the desired metrics in a metric set using the `metric_set()` function in the `yardstick` package, and use the augmented `data_test_with_prediction` dataset to compare the predicted values to the actual results. To achieve this, we employ the following code:

```
metrics_set <- metric_set(
  accuracy,
  sensitivity,
  specificity,
  precision,
  f_meas,
  kap,
  mcc
)
metrics_set(data_test_with_prediction,
               truth = species,
               estimate = .pred_class)
```

```
## # A tibble: 7 x 3
##   .metric     .estimator .estimate
##   <chr>       <chr>          <dbl>
## 1 accuracy    multiclass     0.911
## 2 sensitivity macro          0.925
## 3 specificity macro          0.958
## 4 precision   macro          0.909
## 5 f_meas      macro          0.911
## 6 kap         multiclass     0.865
## 7 mcc         multiclass     0.870
```

We discover that the model's accuracy, sensitivity, specificity, and precision scores exceed 90%, suggesting that it is indeed quite effective in classifying data!

7.5 HYPERPARAMETER TUNING

So far, we've set the number of nearest neighbors to be 1 (K = 1). However, this might not be the most suitable value. In KNN, the algorithm attributes the characteristics of the closest neighbors to a data point. When K = 1, the algorithm considers only the properties of a single nearest neighbor, which could potentially lead to overfitting. Conversely, when K is very large, points that are not closely related to the data point in question may also be included in the neighborhood, negating the purpose of KNN. Therefore, determining the optimal value of K is crucial.

This is typically achieved through **hyperparameter tuning**. By systematically exploring a range of K values, we can evaluate the model's accuracy for each K using techniques such as grid search or cross-validation. The objective is to find a balance where the model generalizes effectively without overfitting or underfitting. A smaller K makes the model more susceptible to noise, while a larger K smooths out predictions but may miss local patterns.

In this section, we'll employ a grid of hyperparameters and tune the workflow to compute the best value of `K`. Most of the steps are similar to what we did before, with the addition of `tune()` being used to hold the value of `neighbors` in the model design, and `tune_grid()` function used to tune the workflow. We'll begin with the splitting step, and continue till the computation of accuracy metrics.

7.5.1 SPLIT DATA INTO TRAINING AND TESTING

We divide our data into training and testing sets using the `initial_split()` function from the `rsample` package, which is part of the `tidymodels` library. As before, we allocate 70% of the data for training and 30% for testing. Subsequently, we extract these two sets using the `training()` and `testing()` functions.

```
# Split data into training and testing sets
set.seed(2025)
split_7030 <- initial_split(data, prop = 0.7)
data_train <- training(split_7030)
data_test <- testing(split_7030)
```

As before, we have set the seed to 2025 to enable comparison with the previous results. The `set.seed` function in programming initializes the random number generator to a specific state, ensuring reproducibility of random operations. By setting the seed, we guarantee that the sequence of random numbers generated is consistent each time we run our code. This consistency is crucial for achieving repeatable results in experiments and simulations, particularly for troubleshooting. However, when we deploy the code, this line will be removed or commented out, since at that point it will not be required.

7.5.2 CREATE A RECIPE

In the recipe, we input the formula `species ~ .` to predict the species using all the available variables in the dataset. We further specify that this recipe should be applied to `data_train`, which is the training data. We also normalize all the predictor variables using the `step_normalize()` function. This step normalizes numeric data to have a standard deviation of one and a mean of zero. We normalize data for KNN to ensure that all features contribute equally to the distance calculation, avoiding bias toward features with larger scales. The complete recipe code is provided below:

```
recipe_ml <- recipe(
  species ~ .,
  data = data_train
) %>%
  step_normalize(all_predictors())
```

7.5.3 CREATE A MODEL DESIGN

Just as before, we define a `nearest_neighbor` model using the `kknn` engine. KNN is versatile and can be used for both classification and regression tasks. In this particular scenario, we intend to classify data into three categories: `setosa`, `versicolor`, and `virginica`. Therefore, we set the `mode` parameter to `classification`.

In the model design, we define neighbors to be tune() and weight_func to be "rectangular." tune() is a placeholder for the value that is computed during hyperparameter tuning. By setting weight_func to "rectangular," we indicate that we don't want to give more weight to more nearby points; instead, we want to apply a uniform weight regardless of the distances between points. In essence, we're converting the default weighted KNN to a classic, unweighted KNN. Here's the code for the model design:

```
model_kknn <- nearest_neighbor(
  neighbors = tune(),
  weight_func = "rectangular"
) %>%
  set_engine("kknn") %>%
  set_mode("classification")
```

7.5.4 CREATE A WORKFLOW

We add the recipe and model to the workflow() function, but leave training for later. Following is the code at this stage:

```
workflow_recipe_model <- workflow() %>%
  add_recipe(recipe_ml) %>%
  add_model(model_kknn)
```

7.5.5 CREATE A HYPERPARAMETER GRID

Let's examine all values of K from 1 to 10. We construct a hyperparameter grid using the crossing() function. The crossing() function, a component of the tidyr package, generates a data frame encompassing all possible combinations of the specified vectors or factors. Here's the code we use:

```
hyper_parameter_grid <- crossing(
  neighbors = c(1:10)
)
```

7.5.6 CREATE RESAMPLES FOR CROSS-VALIDATION

V-fold cross-validation, also known as **k-fold cross-validation**, is a method used to assess the performance of a model by dividing the data into V groups of roughly equal size, which are called "**folds**." In each iteration of the analysis, V-1 folds are used to train the model, while the remaining fold is used to test its performance. In its basic form, without any repeats, the number of resamples is equal to V, which provides a thorough evaluation of the model's capabilities.

The `vfold_cv()` function performs V-fold cross-validation on a data frame. It takes two arguments: `v`, which determines the number of partitions, and `repeats`, which specifies how many times the partitioning should be repeated. By default, `v` is set to 10 and `repeats` is set to 1, which corresponds to a standard cross-validation process. The function also supports stratified sampling, where a variable in the data frame is used to ensure that each resample maintains a similar distribution across different categories. This is achieved by using the `strata` argument.

The `vfold_cv()` function offers several customization options. The `data` parameter specifies the input data frame, while the `v` parameter determines the number of partitions for the data set. The `strata` parameter enables the use of a variable for stratified sampling, while the `breaks` parameter specifies the number of bins for numeric stratification variables. The `pool` parameter assesses the proportion of data used to evaluate if a group is too small and should be combined with another group. It is strongly advised to maintain the default value of 0.1 for this argument to avoid the risks of stratifying groups that are too small. The `dots` argument, ..., is reserved for future extensions and should remain empty when using the function.

In our specific scenario, we use the `vfold_cv()` function on the `data_train` dataset. We set the number of partitions to 10 and stratify the data based on the `species` column to ensure that each fold contains an equal proportion of species. To achieve this, we execute the following code:

```
set.seed(2025)
folds_for_tuning <- vfold_cv(
  data = data_train,
  v = 10,
  strata = species
)
```

We can use the following code to observe the number of training and testing data in each of the cross-validation folds:

```
folds_for_tuning
```

```
## #  10-fold cross-validation using stratification
## # A tibble: 10 x 2
##    splits          id
##    <list>          <chr>
##  1 <split [93/12]> Fold01
##  2 <split [94/11]> Fold02
##  3 <split [94/11]> Fold03
##  4 <split [94/11]> Fold04
##  5 <split [94/11]> Fold05
##  6 <split [95/10]> Fold06
##  7 <split [95/10]> Fold07
##  8 <split [95/10]> Fold08
##  9 <split [95/10]> Fold09
## 10 <split [96/9]>  Fold10
```

7.5.7 TUNE THE WORKFLOW

The tune_grid() function in R is used for hyperparameter tuning in machine learning models. It helps find the best combination of parameters for a model by trying out different sets of values (tuning grid) and evaluating their performance using cross-validation. This way, it helps in optimizing the model and improving its predictions.

We pass our workflow workflow_recipe_model to the tune_grid() function, along with folds_for_tuning as resamples, hyper_parameter_grid as grid, and specify the metrics on which to optimize using the following code:

```
tune_results <- tune_grid(
  workflow_recipe_model,
  resamples = folds_for_tuning,
  grid = hyper_parameter_grid,
  metrics = metric_set(accuracy, sensitivity, specificity)
)
```

We save the results of tuning to tune_results.

7.5.8 EXTRACT THE BEST VALUE(S) OF HYPERPARAMETER(S)

We can use the show_best() function from the tune package to extract the optimal hyperparameter values. For instance, to find the best values for accuracy, we can run the following code:

```
show_best(
  tune_results,
  metric = "accuracy"
)
```

```
## # A tibble: 5 x 7
##   neighbors .metric  .estimator  mean     n std_err .config
##       <int> <chr>    <chr>      <dbl> <int>   <dbl> <chr>
## 1         7 accuracy multiclass 0.959    10  0.0225 pre0_mo~
## 2         8 accuracy multiclass 0.959    10  0.0225 pre0_mo~
## 3         3 accuracy multiclass 0.950    10  0.0225 pre0_mo~
## 4         4 accuracy multiclass 0.950    10  0.0225 pre0_mo~
## 5         5 accuracy multiclass 0.940    10  0.0306 pre0_mo~
```

We discover that the optimal value for accuracy is achieved when K is set to 7 or 8, followed by 3, 4, and 5. Interestingly, the first two values (K = 7 and 8) yield the same mean value of accuracy.

To identify the most suitable values for sensitivity, we can utilize the following code:

```
show_best(
  tune_results,
  metric = "sensitivity"
)
```

```
## # A tibble: 5 x 7
##   neighbors .metric   .estimator  mean     n std_err .config
##       <int> <chr>     <chr>      <dbl> <int>   <dbl> <chr>
## 1         7 sensitiv~ macro      0.961    10  0.0216 pre0_m~
## 2         8 sensitiv~ macro      0.961    10  0.0216 pre0_m~
## 3         3 sensitiv~ macro      0.95     10  0.0222 pre0_m~
## 4         4 sensitiv~ macro      0.95     10  0.0222 pre0_m~
## 5         1 sensitiv~ macro      0.939    10  0.0237 pre0_m~
```

We discover that the optimal value for sensitivity is achieved when K is set to 7 or 8, followed by 3, 4, and 1. Interestingly, the first two values (K = 7 and 8) yield the same mean value of sensitivity.

To identify the most suitable values for specificity, we can utilize the following code:

```
show_best(
  tune_results,
  metric = "specificity"
)
```

```
## # A tibble: 5 x 7
##   neighbors .metric     .estimator  mean     n std_err .config
##       <int> <chr>       <chr>      <dbl> <int>   <dbl> <chr>
## 1         7 specific~ macro        0.979    10  0.0114 pre0_m~
## 2         8 specific~ macro        0.979    10  0.0114 pre0_m~
## 3         3 specific~ macro        0.975    10  0.0115 pre0_m~
## 4         4 specific~ macro        0.975    10  0.0115 pre0_m~
## 5         1 specific~ macro        0.969    10  0.0119 pre0_m~
```

We discover that the optimal value for specificity is achieved when K is set to 7 or 8, followed by 3, 4, and 1. Interestingly, the first two values (K = 7 and 8) yield the same mean value of specificity.

We can use the `select_best()` function, a component of the `tune` package, to select the optimal hyperparameter value. For example, to find the best value of K that yields the highest accuracy, we can execute the following code:

```
best_val <- select_best(
  tune_results,
  metric = "accuracy"
)
best_val
```

```
## # A tibble: 1 x 2
##   neighbors .config
##       <int> <chr>
## 1         7 pre0_mod07_post0
```

7.5.9 FINALIZE AND TRAIN THE BEST MODEL

We can now finalize the workflow by using the finalize_workflow() function from the tune package. We pass the best value best_val as an argument to the function. Additionally, we train the model by fitting it to the data_train dataset using the following code:

```
best_model <- workflow_recipe_model %>%
  finalize_workflow(best_val) %>%
  fit(data_train)
```

7.5.10 ASSESS PREDICTION QUALITY WITH TEST DATA

We can now evaluate the performance of our trained and fitted model by using the test data set. To accomplish this, we employ the augment() function. The augment() function, originating from the broom package in R, appends predictions and residuals from a statistical model to the original dataset. This facilitates the evaluation of a model's fit by juxtaposing actual and predicted values. We use the following code:

```
data_test_with_prediction <- augment(
  best_model, data_test
)
```

Let us now have a glimpse of the augmented file, data_test_with_prediction:

```
data_test_with_prediction %>%
  glimpse(width = 50)
```

```
## Rows: 45
## Columns: 9
## $ .pred_class      <fct> setosa, setosa, setosa,~
## $ .pred_setosa     <dbl> 1.0000000, 1.0000000, 1~
## $ .pred_versicolor <dbl> 0.0000000, 0.0000000, 0~
## $ .pred_virginica  <dbl> 0.0000000, 0.0000000, 0~
## $ sepal_length     <dbl> 5.1, 5.4, 4.4, 5.1, 5.7~
## $ sepal_width      <dbl> 3.5, 3.9, 2.9, 3.5, 3.8~
## $ petal_length     <dbl> 1.4, 1.7, 1.4, 1.4, 1.7~
```

```
## $ petal_width      <dbl> 0.2, 0.4, 0.2, 0.3, 0.3~
## $ species          <fct> setosa, setosa, setosa,~
```

We notice that as before, four new columns have been added to the dataset:

1. The `.pred_class` column contains the predictions made by the model. For the first three rows, it correctly classifies the species as `setosa`, which aligns with the values in the `species` column.
2. The `.pred_setosa` column provides the binary identification of a flower being setosa, calculated by the model.
3. The `.pred_versicolor` column provides the binary identification of a flower being versicolor, calculated by the model.
4. The `.pred_virginica` column provides the binary identification of a flower being virginica, calculated by the model.

The three species predictions sum up to 1. Next, we construct a confusion matrix to understand how many flowers were correctly and incorrectly classified. To achieve this, we utilize the `conf_mat()` function. This function, part of the `yardstick` package in R, generates a confusion matrix from model predictions. This matrix aids in evaluating the performance of a classification model by presenting the counts of correct and incorrect predictions. This enables us to assess metrics such as accuracy, precision, recall, and F1 score. We employ the following code:

```
conf_mat(
  data_test_with_prediction,
  truth = species,
  estimate = .pred_class
)
```

```
##             Truth
## Prediction   setosa versicolor virginica
##   setosa         15          0         0
##   versicolor      0         17         0
##   virginica       0          2        11
```

We discover that all 15 flowers predicted to be setosa by the model were indeed setosa. Similarly, all 17 flowers predicted to be versicolor were indeed versicolor. Lastly, out of the 13 flowers predicted to be virginica (2 + 11), 11 were virginica, and 2 were versicolor. It appears that a majority of the predictions were indeed accurate! Moreover, we can observe that the accuracy is superior to what we had achieved earlier with K = 1, demonstrating the effectiveness of hyperparameter tuning.

We can also calculate the accuracy metrics. In machine learning, **accuracy metrics** evaluate a model's performance by comparing its predictions to the actual outcomes. We define the desired metrics in a metric set using the `metric_set()` function in the `yardstick` package, and use the augmented `data_test_with_prediction` dataset to compare the predicted values to the actual results. To achieve this, we employ the following code:

```
metrics_set <- metric_set(
  accuracy,
  sensitivity,
  specificity,
  precision,
  f_meas,
  kap,
  mcc
)
metrics_set(data_test_with_prediction,
               truth = species,
               estimate = .pred_class)
```

```
## # A tibble: 7 x 3
##   .metric     .estimator .estimate
##   <chr>       <chr>          <dbl>
## 1 accuracy    multiclass     0.956
## 2 sensitivity macro          0.965
## 3 specificity macro          0.980
## 4 precision   macro          0.949
## 5 f_meas      macro          0.954
## 6 kap         multiclass     0.933
## 7 mcc         multiclass     0.935
```

We discover that the model's accuracy, sensitivity, specificity, and precision scores exceed 90%, suggesting that it is indeed quite effective in classifying data! And these metrics are superior to what we had achieved earlier with $K = 1$, further establishing the effectiveness of hyperparameter tuning.

7.6 EXAMPLE: IDENTIFYING HANDWRITING

Processing of human handwriting is crucial in various applications. For instance, it is essential in mail processing to sort mail based on ZIP or PIN codes. Additionally, it is vital in automated form processing where optical character recognition is utilized. Training a machine to recognize handwriting involves teaching it to understand the variations in handwriting among different individuals. The MNIST dataset is a widely used dataset commonly employed for such training.

The **MNIST dataset**, a short form of the Modified National Institute of Standards and Technology database, is a foundational resource in the field of machine learning and computer vision. It is extensively used for training and evaluating image processing systems, particularly those focused on handwritten digit recognition. MNIST serves as a benchmark for various machine learning algorithms, providing a standardized platform for testing and comparing models.

The dataset consists of a collection of grayscale images, each representing a handwritten digit from 0 to 9. Each image is 28x28 pixels in size, resulting in a total of 784 features for each image. Each feature has certain value of brightness dependent on the quantity of ink that got applied to that pixel when writing the digit. To ensure uniformity and facilitate processing, the images are centered, normalized, and anti-aliased. **Anti-aliasing**, a technique in digital image and graphics processing, smooths jagged edges by blending colors at the edges, creating a smoother transition and a more natural appearance. The dataset is divided into two main subsets: a training set and a test set.

The training set comprises 60,000 images, each meticulously paired with its corresponding label, indicating the digit depicted in the image. This extensive dataset serves as a robust foundation for the development and refinement of machine learning models. The ample variety of handwriting styles and variations captured in the large dataset provides ample examples of each digit, enabling the training of highly accurate models. The test set, comprising 10,000 images, plays a crucial role in evaluating the performance of these trained models. The consistent format and well-defined classes of MNIST make it an ideal starting point for developers and experimenters in the field of image classification algorithms.

One of the key features of MNIST is its accessibility and simplicity. The dataset was meticulously designed to be user-friendly and straightforward to load, which has contributed to its widespread popularity in educational settings and as a benchmark in the machine learning community. It is frequently used in introductory machine learning courses to illustrate fundamental concepts like image processing, classification, and neural networks.

In addition to its educational value, MNIST has played a pivotal role in the development of cutting-edge machine learning techniques. It has been instrumental in evaluating a wide range of algorithms, from traditional methods like KNN and support vector machines to modern approaches such as deep learning, including convolutional neural networks (CNNs). The dataset's simplicity allows researchers and practitioners to focus on developing and testing novel methodologies without the added complexity that might arise from more intricate datasets.

Despite its widespread use and significance, MNIST also has certain limitations. While it serves as a valuable resource for introductory purposes, it may not fully capture the complexities of real-world handwriting or the diverse styles of digits encountered in various applications. Consequently, researchers have constructed more challenging datasets and tasks, such as the **EMNIST (Extended MNIST) dataset**, which expands MNIST by including a wider range of handwriting samples, including letters, and the **Fashion MNIST dataset**, which involves classifying images of fashion items from 10 categories instead of digits.

We can access MNIST dataset using the `read_mnist()` function from the `dslabs` package (Irizarry and Gill, 2024) as follows:

```
mnist <- read_mnist()
```

We can explore the structure of the dataset using the `str()` function:

```
str(mnist)
```

```
## List of 2
##  $ train:List of 2
##   ..$ images: int [1:60000, 1:784] 0 0 0 0 0 0 0 0 0 0 ...
##   ..$ labels: int [1:60000] 5 0 4 1 9 2 1 3 1 4 ...
##  $ test :List of 2
##   ..$ images: int [1:10000, 1:784] 0 0 0 0 0 0 0 0 0 0 ...
##   ..$ labels: int [1:10000] 7 2 1 0 4 1 4 9 5 9 ...
```

We observe that the dataset is divided into two sub-datasets. The training set comprises 60,000 images and their corresponding 60,000 labels. The test set contains 10,000 images and their corresponding 10,000 labels.

The images are stored as **digital values** for each pixel. These can be converted to 28×28 matrix for interpretation [Figure 7.2].

The values can also be converted to an image using the `image()` function using the following code (for the first image in the training dataset):

```
image(matrix(mnist$train$images[1,], 28, 28, byrow = TRUE))
```

0	0	0	0	0	0	0	0	0	0	0	0	0	0	0	0	0	0	0	0	0	0	0	0	0	0	0	0
0	0	0	0	0	0	0	0	0	0	0	0	0	0	0	0	0	0	0	0	0	0	0	0	0	0	0	0
0	0	0	0	0	0	0	0	0	0	0	0	0	0	0	0	0	0	0	0	0	0	0	0	0	0	0	0
0	0	0	0	0	0	0	0	0	0	0	0	0	0	0	0	0	0	0	0	0	0	0	0	0	0	0	0
0	0	0	0	0	0	0	0	0	0	0	0	0	0	0	0	0	0	0	0	0	0	0	0	0	0	0	0
0	0	0	0	0	0	0	0	0	0	0	0	3	18	18	18	126	136	175	26	166	255	247	127	0	0	0	0
0	0	0	0	0	0	0	0	30	36	94	154	170	253	253	253	253	253	225	172	253	242	195	64	0	0	0	0
0	0	0	0	0	0	0	49	238	253	253	253	253	253	253	253	253	251	93	82	82	56	39	0	0	0	0	0
0	0	0	0	0	0	0	18	219	253	253	253	253	253	198	182	247	241	0	0	0	0	0	0	0	0	0	0
0	0	0	0	0	0	0	0	80	156	107	253	253	205	11	0	43	154	0	0	0	0	0	0	0	0	0	0
0	0	0	0	0	0	0	0	0	14	1	154	253	90	0	0	0	0	0	0	0	0	0	0	0	0	0	0
0	0	0	0	0	0	0	0	0	0	0	139	253	190	2	0	0	0	0	0	0	0	0	0	0	0	0	0
0	0	0	0	0	0	0	0	0	0	0	11	190	253	70	0	0	0	0	0	0	0	0	0	0	0	0	0
0	0	0	0	0	0	0	0	0	0	0	0	35	241	225	160	108	1	0	0	0	0	0	0	0	0	0	0
0	0	0	0	0	0	0	0	0	0	0	0	0	81	240	253	253	119	25	0	0	0	0	0	0	0	0	0
0	0	0	0	0	0	0	0	0	0	0	0	0	0	45	186	253	253	150	27	0	0	0	0	0	0	0	0
0	0	0	0	0	0	0	0	0	0	0	0	0	0	0	16	93	252	253	187	0	0	0	0	0	0	0	0
0	0	0	0	0	0	0	0	0	0	0	0	0	0	0	0	0	249	253	249	64	0	0	0	0	0	0	0
0	0	0	0	0	0	0	0	0	0	0	0	0	0	46	130	183	253	253	207	2	0	0	0	0	0	0	0
0	0	0	0	0	0	0	0	0	0	0	0	39	148	229	253	253	253	250	182	0	0	0	0	0	0	0	0
0	0	0	0	0	0	0	0	0	0	24	114	221	253	253	253	253	201	78	0	0	0	0	0	0	0	0	0
0	0	0	0	0	0	0	0	23	66	213	253	253	253	253	198	81	2	0	0	0	0	0	0	0	0	0	0
0	0	0	0	0	0	18	171	219	253	253	253	253	195	80	9	0	0	0	0	0	0	0	0	0	0	0	0
0	0	0	0	55	172	226	253	253	253	253	244	133	11	0	0	0	0	0	0	0	0	0	0	0	0	0	0
0	0	0	0	136	253	253	253	212	135	132	16	0	0	0	0	0	0	0	0	0	0	0	0	0	0	0	0
0	0	0	0	0	0	0	0	0	0	0	0	0	0	0	0	0	0	0	0	0	0	0	0	0	0	0	0
0	0	0	0	0	0	0	0	0	0	0	0	0	0	0	0	0	0	0	0	0	0	0	0	0	0	0	0
0	0	0	0	0	0	0	0	0	0	0	0	0	0	0	0	0	0	0	0	0	0	0	0	0	0	0	0

FIGURE 7.2: Pixel values for the digit 5, taken from a sample in the MNIST dataset.

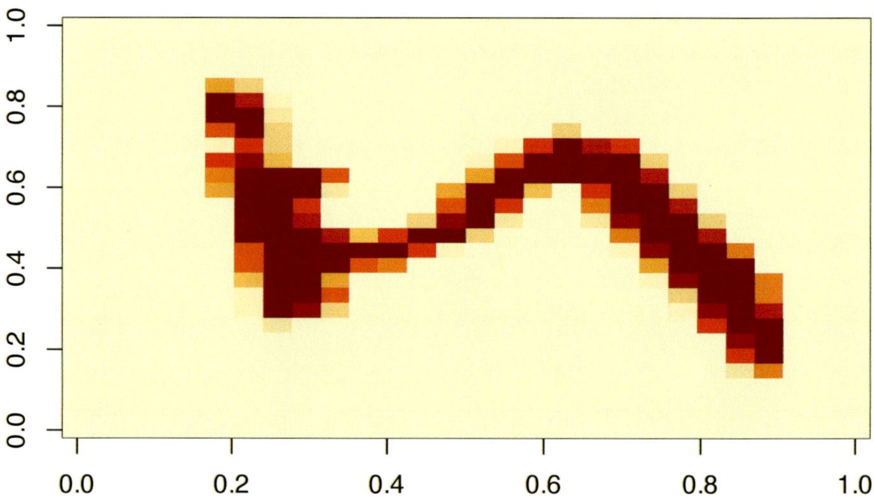

Since the image appears rotated counter-clockwise, we can rotate the same using a `rotate()` function as follows:

```
rotate <- function(x) t(apply(x, 2, rev))
image(
  rotate(matrix(mnist$train$images[1,], 28, 28, byrow = TRUE))
)
```

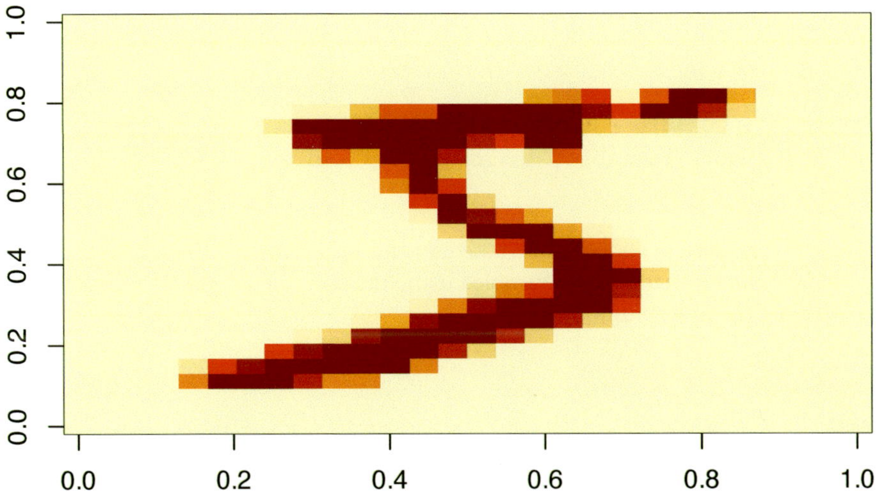

Since the MNIST dataset is quite extensive, while KNN is a computationally intensive algorithm, we will apply KNN to a small subset of MNIST, comprising 5000 entries from `mnist$train`. To do this, we convert the dataset to tibble using `as_tibble()` function, and sample 5000 values using the `sample_n()` function. We set the seed to 2025 for reproducibility.

```
set.seed(2025)
data <- mnist$train %>%
  as_tibble() %>%
  sample_n(5000)
```

Let us explore `data`:

```
data
```

```
## # A tibble: 5,000 x 2
##    images[,1]  [,2]  [,3]  [,4]  [,5]  [,6]  [,7] labels
##         <int> <int> <int> <int> <int> <int> <int>  <int>
```

```
## 1            0      0      0      0      0      0      0      1
## 2            0      0      0      0      0      0      0      3
## 3            0      0      0      0      0      0      0      5
## 4            0      0      0      0      0      0      0      8
## 5            0      0      0      0      0      0      0      8
## 6            0      0      0      0      0      0      0      6
## 7            0      0      0      0      0      0      0      6
## 8            0      0      0      0      0      0      0      6
## 9            0      0      0      0      0      0      0      0
## 10           0      0      0      0      0      0      0      6
## # i 4,990 more rows
## # i 1 more variable: images[8:784] <int>
```

We find that data is a tibble with 5000 rows and 2 columns. The 5000 rows comprise the 5000 entries that we have sampled using the sample_n(5000) command. Let us explore the first column:

```
data[,1]
```

```
## # A tibble: 5,000 x 1
##    images[,1]  [,2]  [,3]  [,4]  [,5]  [,6]  [,7]  [,8]
##         <int> <int> <int> <int> <int> <int> <int> <int>
## 1           0     0     0     0     0     0     0     0
## 2           0     0     0     0     0     0     0     0
## 3           0     0     0     0     0     0     0     0
## 4           0     0     0     0     0     0     0     0
## 5           0     0     0     0     0     0     0     0
## 6           0     0     0     0     0     0     0     0
## 7           0     0     0     0     0     0     0     0
## 8           0     0     0     0     0     0     0     0
## 9           0     0     0     0     0     0     0     0
## 10          0     0     0     0     0     0     0     0
## # i 4,990 more rows
## # i 1 more variable: images[9:784] <int>
```

We find that the first column itself is a tibble comprising 784 variables named images[,1], images[,2], ...images[,784]. These are the 784 (28×28) cells of the matrix. Similarly, we can explore the second column:

```
data[,2]
```

```
## # A tibble: 5,000 x 1
##    labels
##     <int>
##  1      1
##  2      3
##  3      5
##  4      8
##  5      8
##  6      6
##  7      6
##  8      6
##  9      0
## 10      6
## # i 4,990 more rows
```

We find that the second column is also a tibble containing the labels of the 5000 entries. We observe that the first entry is 1, followed by 3, 5, 8, and 8.

We can simplify the structure of data by converting its columns into tibbles, and binding them together using the cbind() function of base R to bind columns:

```
im <- as_tibble(data$images)
lb <- as_tibble(data$labels)
data <- cbind(im, lb)
```

Let us explore data again. We begin by looking at the dimensions of data:

```
dim(data)
```

```
## [1] 5000  785
```

We find that data now has 5000 rows and 785 columns. Let us look at the first few column names using the colnames() and head() functions:

```
colnames(data) %>% head()
```

```
## [1] "V1" "V2" "V3" "V4" "V5" "V6"
```

Let us now look at the last few column names using the `colnames()` and `tail()` functions:

```
colnames(data) %>% tail()
```

```
## [1] "V780"   "V781"   "V782"   "V783"   "V784"   "value"
```

We find that the columns are named "V1," "V2," ... "V784," and "value." Columns "V1" to "V784" contain the values in cells 1 to 784 of the matrix. Column "value" contains the label. For our purposes, columns "V1" to "V784" are the independent (predictor) variables, and the column "value" is the dependent (predicted) variable. We convert the digit values to factors for classification purposes using the `as_factor()` function:

```
data$value <- as_factor(data$value)
```

Now our `data` is fully prepared for employing KNN.

7.6.1 SPLIT DATA INTO TRAINING AND TESTING

We divide our data into training and testing sets using the `initial_split()` function from the `rsample` package, which is part of the `tidymodels` library. As before, we allocate 70% of the data for training and 30% for testing. Subsequently, we extract these two sets using the `training()` and `testing()` functions.

```
# Split data into training and testing sets
set.seed(2025)
split_7030 <- initial_split(data, prop = 0.7, strata = value)
data_train <- training(split_7030)
data_test <- testing(split_7030)
```

The set.seed function in programming initializes the random number generator to a specific state, ensuring reproducibility of random operations. By setting the seed, we guarantee that the sequence of random numbers generated is consistent each time we run our code. This consistency is crucial for achieving repeatable results in experiments and simulations, particularly for troubleshooting. However, when we deploy the code, this line will be removed or commented out, since at that point it will not be required.

7.6.2 CREATE A RECIPE

In the recipe, we input the formula value ~ . to predict the value using all the available variables in the dataset. We further specify that this recipe should be applied to data_train, which is the training data. We also use step_zv() function from recipes package to remove variables that contain only a single value (zv = zero variance). Since the data is already normalized, we do not need to add an additional normalization step. The complete recipe code is provided below:

```
recipe_ml <- recipe(
  value ~ .,
  data = data_train
) %>%
  step_zv()
```

7.6.3 CREATE A MODEL DESIGN

Just as before, we define a nearest_neighbor model using the kknn engine. KNN is versatile and can be used for both classification and regression tasks. In this particular scenario, we intend to classify data into 10 categories, the 10 digits. Therefore, we set the mode parameter to classification.

In the model design, we define neighbors to be tune() and weight_func to be "rectangular." tune() is a placeholder for the value that is computed during hyperparameter tuning. WKNN considers the distance of each neighboring point to the query point, typically giving more weight to closer neighbors. By setting weight_func to "rectangular," we indicate that we don't want to give more weight to more nearby points; instead, we want to apply a uniform weight regardless of the distances between points. In essence, we're converting the default weighted KNN to a classic, unweighted KNN. Here's the code for the model design:

```
model_kknn <- nearest_neighbor(
  neighbors = tune(),
  weight_func = "rectangular"
) %>%
  set_engine("kknn") %>%
  set_mode("classification")
```

7.6.4 CREATE A WORKFLOW

We add the recipe and model to the `workflow()` function, but leave training for later. Following is the code at this stage:

```
workflow_recipe_model <- workflow() %>%
  add_recipe(recipe_ml) %>%
  add_model(model_kknn)
```

7.6.5 CREATE A HYPERPARAMETER GRID

We'll examine all values of K from 1 to 10. We construct a hyperparameter grid using the `crossing()` function. The `crossing()` function, a component of the `tidyr` package, generates a data frame encompassing all possible combinations of the specified vectors or factors. This function proves particularly useful in expanding data to include every possible combination of specific variables, which is often necessary in data analysis tasks. Here's the code we use:

```
hyper_parameter_grid <- crossing(
  neighbors = c(1:10)
)
```

7.6.6 CREATE RESAMPLES FOR CROSS-VALIDATION

We use the `vfold_cv()` function on the `data_train` dataset. We set the number of partitions to 10 and stratify the data based on the `value` column to ensure that each fold contains an equal proportion of digits. To achieve this, we execute the following code:

```
set.seed(2025)
folds_for_tuning <- vfold_cv(
  data_train,
  v = 10,
  strata = value
)
```

7.6.7 TUNE THE WORKFLOW

The tune_grid() function in R is used for hyperparameter tuning in machine learning models. It helps find the best combination of parameters for a model by trying out different sets of values (tuning grid) and evaluating their performance using cross-validation. This way, it helps in optimizing the model to improve its predictions.

We pass our workflow workflow_recipe_model to the tune_grid() function, along with folds_for_tuning as resamples, hyper_parameter_grid as grid, and specify the metrics on which to optimize using the following code:

```
tune_results <- tune_grid(
  workflow_recipe_model,
  resamples = folds_for_tuning,
  grid = hyper_parameter_grid,
  metrics = metric_set(accuracy, specificity)
)
```

We save the results of tuning to tune_results.

7.6.8 EXTRACT THE BEST VALUE(S) OF HYPER-PARAMETER(S)

We can extract the optimal hyperparameter values using the show_best() function, which is part of the tune package. For example, to determine the best values for accuracy, we can use the following code:

```
show_best(
  tune_results,
  metric = "accuracy"
)
```

```
## # A tibble: 5 x 7
##   neighbors .metric  .estimator  mean     n std_err .config
##       <int> <chr>    <chr>      <dbl> <int>   <dbl> <chr>
## 1         4 accuracy multiclass 0.889    10 0.00420 pre0_mo~
## 2         5 accuracy multiclass 0.887    10 0.00434 pre0_mo~
## 3         6 accuracy multiclass 0.886    10 0.00359 pre0_mo~
## 4         3 accuracy multiclass 0.886    10 0.00526 pre0_mo~
## 5         7 accuracy multiclass 0.885    10 0.00396 pre0_mo~
```

We discover that the optimal value for accuracy is achieved when K is set to 4, followed by 5, 6, and 3.

To identify the most suitable values for specificity, we can utilize the following code:

```
show_best(
  tune_results,
  metric = "specificity"
)
```

```
## # A tibble: 5 x 7
##   neighbors .metric   .estimator  mean     n std_err .config
##       <int> <chr>     <chr>      <dbl> <int>   <dbl> <chr>
## 1         4 specific~ macro      0.988    10 4.64e-4 pre0_m~
## 2         5 specific~ macro      0.987    10 4.79e-4 pre0_m~
## 3         6 specific~ macro      0.987    10 3.98e-4 pre0_m~
## 4         3 specific~ macro      0.987    10 5.88e-4 pre0_m~
## 5         7 specific~ macro      0.987    10 4.37e-4 pre0_m~
```

We discover that the optimal value for specificity is achieved when K is set to 4, followed by 5, 6, and 3.

We use the `select_best()` function, a component of the `tune` package, to select the optimal hyperparameter value. For example, to find the best value of K that yields the highest accuracy, we execute the following code:

```
best_val <- select_best(
  tune_results,
  metric = "accuracy"
)
best_val
```

```
## # A tibble: 1 x 2
##   neighbors .config
##       <int> <chr>
## 1         4 pre0_mod04_post0
```

7.6.9 FINALIZE AND TRAIN THE BEST MODEL

We now finalize the workflow by using the `finalize_workflow()` function from the `tune` package. We pass the best value `best_val` as an argument to the function. Additionally, we train the model by fitting it to the `data_train` dataset using the following code:

```
best_model <- workflow_recipe_model %>%
  finalize_workflow(best_val) %>%
  fit(data_train)
```

7.6.10 ASSESS PREDICTION QUALITY WITH TEST DATA

We evaluate the performance of our trained and fitted model by using the test data set. To achieve this, we employ the `augment()` function from the `broom` package in R. The `augment()` function extends the original dataset by including predictions and residuals generated by a statistical model. This visualization approach provides a direct comparison between the actual values and the predicted values, enabling us to assess the model's accuracy. We use the following code:

```
data_test_with_prediction <- augment(
  best_model, data_test
)
```

Next, we construct a confusion matrix to understand how many digits were correctly and incorrectly classified. To achieve this, we utilize the `conf_mat()` function. This function, part of the `yardstick` package in R, generates a confusion matrix from model predictions. This matrix aids in evaluating the performance of a classification model by presenting the counts of correct and incorrect predictions. This enables us to assess metrics such as accuracy, precision, recall, and F1 score. We employ the following code:

```
conf_mat(
  data_test_with_prediction,
  truth = value,
  estimate = .pred_class,
  dnn = c("Pred", "Truth")
)
```

```
##       Truth
## Pred   0   1   2   3   4   5   6   7   8   9
##    0 152   0   2   0   0   0   5   0   2   1
##    1   0 170   3   2   5   3   0   9   2   2
##    2   0   3 121   2   4   0   0   0   0   1
##    3   0   0   5 133   2   8   0   2   3   1
##    4   0   1   4   2 128   2   0   2   2   8
##    5   0   0   0   3   2 117   2   0   6   0
##    6   2   0   3   0   0   1 133   0   2   0
##    7   0   0   3   2   1   1   0 130   1   7
##    8   0   1   2   5   1   7   0   0 127   1
##    9   0   0   0   3   7   7   0  12   2 120
```

We discover that a majority of the digits are correctly predicted. Next, we calculate the accuracy metrics. In machine learning, accuracy metrics evaluate a model's performance by comparing its predictions to the actual outcomes. We define the desired metrics in a metrics set and use the augmented `data_test_with_prediction` dataset to compare the predicted values to the actual results. To achieve this, we employ the following code:

```
metrics_set <- metric_set(
  accuracy,
  sensitivity,
  specificity,
  precision,
  f_meas,
  kap,
  mcc
)
metrics_set(data_test_with_prediction,
            truth = value,
            estimate = .pred_class)
```

```
## # A tibble: 7 x 3
```

```
##   .metric     .estimator .estimate
##   <chr>       <chr>          <dbl>
## 1 accuracy    multiclass     0.886
## 2 sensitivity macro          0.884
## 3 specificity macro          0.987
## 4 precision   macro          0.887
## 5 f_meas      macro          0.884
## 6 kap         multiclass     0.873
## 7 mcc         multiclass     0.873
```

We discover that the model's accuracy, sensitivity, specificity, and precision scores are close to, or exceed 90%, suggesting that it is indeed quite effective in classifying data!

7.6.11 FINAL ASSESSMENT WITH ALL OF MNIST TEST DATA

Let us now deploy our trained model to all of mnist test data. We create `mnist_test` dataset by combining the columns as before:

```
im <- as_tibble(mnist$test$images)
lb <- as_tibble(mnist$test$labels)
mnist_test <- cbind(im, lb)
mnist_test$value <- as_factor(mnist_test$value)
```

We evaluate the performance of our trained and fitted model by using this `mnist_test` data set. To achieve this, we employ the `augment()` function from the `broom` package in R. We have seen that the `augment()` function extends the original dataset by including predictions and residuals generated by a statistical model. This visualization approach provides a direct comparison between the actual values and the predicted values, enabling us to assess the model's accuracy. We use the following code:

```
data_test_with_prediction <- augment(
  best_model, mnist_test
)
```

Next, we construct a confusion matrix to understand how many digits were correctly and incorrectly classified. To achieve this, we utilize the `conf_mat()` function. This function, part of the `yardstick` package in R, generates a confusion matrix from model predictions. This matrix aids in evaluating the performance of a classification model by presenting the counts of correct and incorrect predictions. This enables us to assess metrics such as accuracy, precision, recall, and F1 score. We employ the following code:

```
conf_mat(
  data_test_with_prediction,
  truth = value,
  estimate = .pred_class,
  dnn = c("Pred", "Truth")
)
```

```
##      Truth
## Pred    0    1    2    3    4    5    6    7    8    9
##    0  956    0   26    1    2   11   18    0   24   12
##    1    1 1127   30   12   21   12    9   37    9   10
##    2    4    2  861   12    7    2    7    8    8    3
##    3    2    1   36  902    1   41    1    3   18   11
##    4    1    2   20    5  861    8   11   12   10   37
##    5    8    0    0   24    5  756   18    2   54    4
##    6    6    3   15    0    9   17  887    0    5    0
##    7    1    0   15   16    9    9    2  900   11   56
##    8    1    0   27   28    2   24    4    1  816    4
##    9    0    0    2   10   65   12    1   65   19  872
```

We discover that a majority of the digits are correctly predicted. Next, we calculate the accuracy metrics. We define the desired metrics in a metrics set and use the augmented `data_test_with_prediction` dataset to compare the predicted values to the actual results. To achieve this, we employ the following code:

```
metrics_set <- metric_set(
  accuracy,
  sensitivity,
  specificity,
  precision,
  f_meas,
  kap,
  mcc
)
metrics_set(data_test_with_prediction,
```

```
              truth = value,
              estimate = .pred_class)
```

```
## # A tibble: 7 x 3
##   .metric     .estimator .estimate
##   <chr>       <chr>          <dbl>
## 1 accuracy    multiclass     0.894
## 2 sensitivity macro          0.892
## 3 specificity macro          0.988
## 4 precision   macro          0.895
## 5 f_meas      macro          0.893
## 6 kap         multiclass     0.882
## 7 mcc         multiclass     0.882
```

We discover that the model's accuracy, sensitivity, specificity, and precision scores are close to, or exceed 90%, suggesting that it is indeed quite effective in classifying data! Since the test portion of the MNIST dataset was never seen by the model during the entire training process, we can consider it to be a good representative of real-world data that will be seen by the model once it is deployed. Thus, upon deployment, we can expect the deployment metrics to come close to these values. The scores indicate that there is a scope for improvement using better model(s), and a need to keep alternative arrangements—such as human oversight—for the inaccuracies of the model.

7.7 WHEN TO USE K-NEAREST NEIGHBORS, AND WHEN NOT TO

KNN is a versatile and straightforward machine learning algorithm employed for both classification and regression tasks. It operates by identifying the "K" most similar data points to a test instance and making predictions based on the majority class (for classification) or the average value (for regression) of these neighboring points. Although KNN is a powerful and intuitive algorithm, its effectiveness depends on various factors.

KNN performs well on small to medium-sized datasets because the algorithm requires computing distances between the test instance and all training instances. With smaller datasets, this computational cost is manageable. However, as the dataset grows, the time complexity of KNN increases, making it less practical for very large datasets.

KNN is effective when the data is not too complex and has a relatively low number of features. In low-dimensional spaces, the concept of "nearness" is more meaningful and easier to interpret, leading to better performance.

KNN is capable of capturing non-linear decision boundaries due to its reliance on local data. This makes it useful for problems where the decision boundary between classes is not a straight line or hyperplane. For example, KNN can effectively classify complex shapes or clusters in feature space that other algorithms might struggle with.

KNN is a non-parametric algorithm, meaning it does not make assumptions about the underlying data distribution. This is advantageous when we have little prior knowledge about the data and when the relationship between features and the target variable is not well-understood.

KNN allows the use of various distance metrics (such as Euclidean, Manhattan, or Minkowski distances), making it adaptable to different types of data and problem domains. By choosing an appropriate distance metric, we can tailor KNN to the specific characteristics of our dataset.

On the other hand, KNN is computationally expensive because it necessitates calculating distances between the test instance and all training instances for each prediction. This leads to significant memory and time costs, particularly with large datasets. As the dataset size increases, KNN becomes impractical due to its linear time complexity in relation to the number of training samples. This observation can be seen even in the MNIST example presented in this chapter, where the KNN takes a considerable amount of time to process.

In high-dimensional spaces, the concept of distance loses its significance due to the "**curse of dimensionality**." As the number of features increases, the distance between points becomes less discriminative, leading to a decline in KNN's performance. **Dimensionality reduction** techniques, such as **Principal Component Analysis** (**PCA**), can help mitigate this issue, but high-dimensional data generally presents a challenge for KNN.

KNN can struggle with imbalanced datasets, where one class is significantly underrepresented compared to others. In such cases, the majority class may dominate the decision-making process, resulting in biased predictions. To address this issue, techniques like resampling, weighting, or using algorithms specifically designed for imbalanced data may be necessary, adding complexity to the analysis.

KNN is highly sensitive to noisy data and outliers because it makes predictions based on the local neighborhood of each instance. Consequently, noisy or irrelevant data points can negatively affect the classification or regression results. Therefore, **preprocessing steps** like data cleaning and outlier detection become essential when using KNN on noisy datasets.

KNN may not be the most suitable choice for real-time or online applications where quick predictions are crucial. This is because its high computational cost during the prediction phase becomes a significant concern. Each prediction necessitates distance calculations for all training instances, which can be a time-consuming process.

In essence, KNN is a valuable algorithm for classification and regression, especially when dealing with small to medium-sized datasets with simple feature spaces, and non-linear decision boundaries. Its flexibility and ease of implementation make it a popular choice in various scenarios. However, KNN may not be the best option for large or high-dimensional datasets, imbalanced data, or situations requiring quick, real-time predictions. Understanding the strengths and limitations of KNN helps in selecting the appropriate algorithm for a given problem and in making informed decisions about preprocessing and data handling.

7.8 CHAPTER SUMMARY

- KNN is a machine learning algorithm used for classification and regression tasks. It identifies the "K" most similar data points to a test instance and makes predictions based on the majority class or average value of these neighboring points.
- KNN performs well on small to medium-sized datasets with low-dimensional data.
- KNN is computationally expensive and impractical for large datasets due to its linear time complexity. It struggles with high-dimensional spaces, imbalanced datasets, and noisy data, potentially leading to biased predictions and poor performance.
- Hyperparameter tuning is crucial in KNN to determine the optimal value of K, for a generalized model that avoids overfitting.
- The Iris dataset, featuring measurements of iris flowers, is a popular dataset for benchmarking and comparing machine learning algorithms.
- The MNIST dataset is a valuable resource for training and testing image classification algorithms.

8 Support vector machines

Imagine that you're tasked with categorizing animals into sea and land creatures. You could meticulously examine their physical characteristics, such as the presence of fins or legs, or whether they possess a streamlined body. Alternatively, you could employ a simple line drawn on the beach to accomplish the same task.

Imagine standing on a beach, with the sea to your left and the land to your right. A line separates the sea from the land. Any animal to the left of this line is a sea animal, while any animal to the right is a land animal.

Support vector machine (SVM) operates similarly. Instead of relying on a natural line, however, SVM endeavors to construct the optimal "line" that divides the two groups. Even better, it looks at all the animals and figures out where to place the line so that there's the most space between the sea animals and the land animals, thereby achieving the clearest separation possible.

Once the line is drawn, sea animals occupy one side, and land animals occupy the other. By observing which side a new animal falls on, SVM can now easily determine whether it's a sea animal or a land animal!

8.1 WHAT IS SVM?

SVM is a supervised machine learning algorithm employed for classification and regression tasks. Its primary objective is to identify the optimal decision boundary that effectively separates objects distributed across a vector space.

A **decision boundary** or **decision surface** is a hypersurface that divides the underlying vector space into two distinct sets, each representing a specific class. The classifier categorizes all points on one side of the decision boundary as belonging to one class, while those on the other side are classified as belonging to the other class. When the decision surface is a flat hypersurface of dimension one less than the ambient space, we refer to the classification problem as linear and the classes as **linearly separable**.

SVMs are designed to identify the optimal boundary that effectively separates different classes of data.

For simplicity, let us consider a two-dimensional space where we want to separate data into two classes [Fig. 8.1]. There can be numerous lines and curves drawn between the two groups of points depicted in red and blue. Any of these lines or curves can separate the points into two groups and serve as a decision boundary. SVM aims to find the line that creates the greatest separation between the two groups of points.

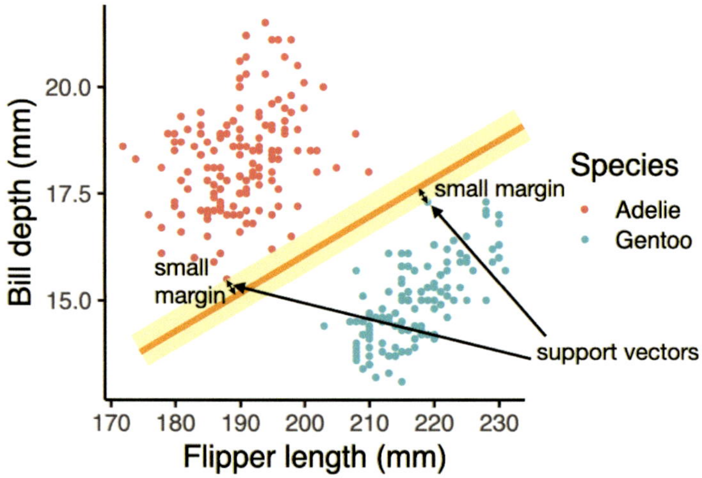

(a) A hyperplane separating two species of penguins based on their characteristics, with a small margin. The support vectors are marked.

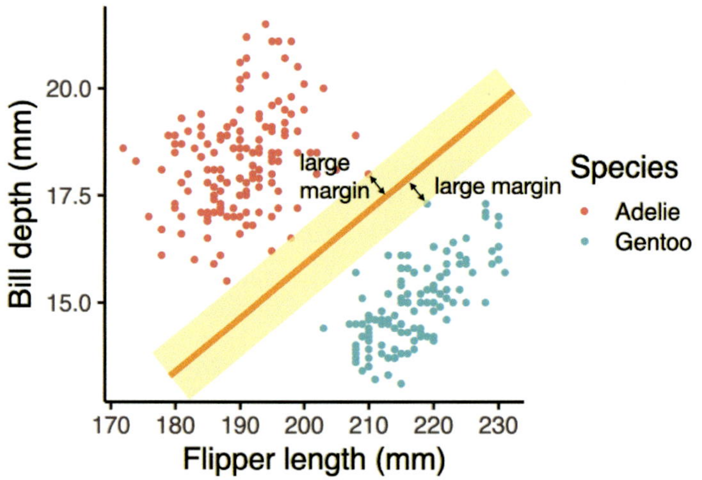

(b) A hyperplane separating the two species of penguins based on their characteristics, with a large margin.

FIGURE 8.1: A schematic representation of the workings of a support vector machine.

SVMs can be described as a method that aims to separate two classes with the maximum margin. **Margin** refers to the distance between the decision boundary and the closest data point from either class. The data points within each group that are closest to the line are known as **support vectors**. Consequently, we can also define the margin as the distance between the support vectors and the decision boundary.

In the current instance, while both the decision boundaries in Figure 8.1 are able to separate the red and blue points, we observe that the margin in Figure 8.1a is small, while the margin in Figure 8.1b is larger. By rotating and moving the decision boundary, we can alter the value of the margin.

SVMs identify the decision boundary with the largest margin among all the available decision boundaries. This is why SVMs are also referred to as "**maximum margin models**."

SVMs can also be applied when the data is not linearly separable. In such cases, SVMs utilize a technique called the "**kernel trick**." This trick transforms the input data into a higher-dimensional space, enabling the creation of a hyperplane that effectively separates the data. Common kernels employed in this context include the **polynomial kernel** and the **radial basis function** (RBF) kernel.

SVMs are particularly effective in high-dimensional spaces. They find applications in various fields, including image classification, bioinformatics, and text categorization. SVMs are robust to overfitting, especially in high-dimensional spaces, and they can handle non-linear data using kernel functions.

8.2 THE PALMER PENGUIN DATASET

The `penguins` dataset, accessible through the `palmerpenguins` package in R, offers size measurements for adult foraging penguins situated near Palmer Station in Antarctica. This dataset is a subset of the larger `penguins_raw` collection. It includes comprehensive details, such as penguin species, the island within the Palmer Archipelago where they reside, various size measurements like flipper length, body mass, and bill dimensions, and their sex. The data spans the study years of 2007, 2008, and 2009.

The dataset, presented as a tibble, comprises 344 rows and 8 variables. The `species` variable categorizes penguins into three types: Adélie, Chinstrap, and Gentoo. Similarly, the `island` variable identifies islands in the Palmer Archipelago as Biscoe, Dream, or Torgersen. The `bill_length_mm` and `bill_depth_mm` variables represent bill length and depth in millimeters, respectively. The `flipper_length_mm` and the `body_mass_g` variables are integers indicating measurements in millimeters and grams, respectively. The `sex` variable categorizes penguins based on their sex, and the `year` variable specifies the study year as an integer.

8.2.1 USING THE DATASET

We begin by (installing and) loading all the required packages and libraries.

```
if (!require("tidymodels")){
  install.packages("tidymodels")
}
library(tidymodels)

if (!require("tidyverse")){
  install.packages("tidyverse")
}
library(tidyverse)

if (!require("janitor")){
  install.packages("janitor")
}
library(janitor)

if (!require("kernlab")){
  install.packages("kernlab")
}
library(kernlab)

if (!require("GGally")){
  install.packages("GGally")
}
library(GGally)

if (!require("palmerpenguins")){
  install.packages("palmerpenguins")
}
library(palmerpenguins)
```

The `kernlab` package (Karatzoglou et al., 2004) in R is a versatile toolkit for kernel-based[1] machine learning methods. It offers a wide range of algorithms, including SVM, Kernel Principal Component Analysis (KPCA), and other kernel methods. These methods are useful for classification, regression, clustering, and dimensionality reduction. The package is renowned for its flexibility and efficiency, allowing

[1]In machine learning, a kernel function calculates the similarity between two data points in a transformed feature space without explicitly mapping the data. This technique is widely used in algorithms like SVMs and PCA.

users to customize kernels and fine-tune model parameters to achieve optimal performance.

The palmerpenguins package (Horst et al., 2020) offers access to the penguins dataset. Since this dataset is a tibble, accessing it directly will automatically display the initial few rows, eliminating the need to overtly use the head() function. We employ the following code:

```
penguins
```

```
## # A tibble: 344 x 8
##    species island    bill_length_mm bill_depth_mm
##    <fct>   <fct>              <dbl>         <dbl>
##  1 Adelie  Torgersen           39.1          18.7
##  2 Adelie  Torgersen           39.5          17.4
##  3 Adelie  Torgersen           40.3          18
##  4 Adelie  Torgersen           NA            NA
##  5 Adelie  Torgersen           36.7          19.3
##  6 Adelie  Torgersen           39.3          20.6
##  7 Adelie  Torgersen           38.9          17.8
##  8 Adelie  Torgersen           39.2          19.6
##  9 Adelie  Torgersen           34.1          18.1
## 10 Adelie  Torgersen           42            20.2
## # i 334 more rows
## # i 4 more variables: flipper_length_mm <int>,
## #   body_mass_g <int>, sex <fct>, year <int>
```

We find that this dataset contains 344 rows and 8 columns. We observe several missing (NA) values. Let us remove them using the na.omit() function, and save the result as data. Following is the code to perform these steps:

```
data <- penguins %>%
  na.omit()
```

Let us check if data is what we wanted it to be.

```
data
```

```
## # A tibble: 333 x 8
##    species island    bill_length_mm bill_depth_mm
##    <fct>   <fct>              <dbl>         <dbl>
##  1 Adelie  Torgersen           39.1          18.7
##  2 Adelie  Torgersen           39.5          17.4
##  3 Adelie  Torgersen           40.3          18
##  4 Adelie  Torgersen           36.7          19.3
##  5 Adelie  Torgersen           39.3          20.6
##  6 Adelie  Torgersen           38.9          17.8
##  7 Adelie  Torgersen           39.2          19.6
##  8 Adelie  Torgersen           41.1          17.6
##  9 Adelie  Torgersen           38.6          21.2
## 10 Adelie  Torgersen           34.6          21.1
## # i 323 more rows
## # i 4 more variables: flipper_length_mm <int>,
## #   body_mass_g <int>, sex <fct>, year <int>
```

The values appear to be good; there are no missing values indicated by NA. Upon further inspection, we discover that we are left with 333 rows. Since we had initially started with 344 rows, there were in total 11 rows that had missing values.

Let us now look at a summary of the dataset using the summary() function. The summary() function in R provides quick overviews of data based on their class. For vectors, data frames, or fitted models, it offers an instant snapshot of the data. In case of numeric data, it returns key statistics. For factors, it returns level frequencies. For linear models, it yields coefficients, residuals, significance levels, and goodness-of-fit. We pass data to the summary() function using the pipe operator %>%:

```
data %>%
  summary()
```

```
##       species          island    bill_length_mm
##  Adelie    :146   Biscoe   :163   Min.   :32.10
##  Chinstrap: 68   Dream    :123   1st Qu.:39.50
##  Gentoo    :119   Torgersen: 47   Median :44.50
##                                   Mean   :43.99
##                                   3rd Qu.:48.60
##                                   Max.   :59.60
##  bill_depth_mm   flipper_length_mm  body_mass_g
##  Min.   :13.10   Min.   :172       Min.   :2700
##  1st Qu.:15.60   1st Qu.:190       1st Qu.:3550
##  Median :17.30   Median :197       Median :4050
```

```
##  Mean   :17.16   Mean   :201        Mean   :4207
##  3rd Qu.:18.70   3rd Qu.:213        3rd Qu.:4775
##  Max.   :21.50   Max.   :231        Max.   :6300
##      sex             year
##  female:165   Min.   :2007
##  male  :168   1st Qu.:2007
##               Median :2008
##               Mean   :2008
##               3rd Qu.:2009
##               Max.   :2009
```

We can also create `ggpairs()` plot. `ggpairs()` (ggplot2 generalized pairs) is a function from the `GGally` package (Schloerke et al., 2024), which is an extension of the popular `ggplot2` package. `ggpairs()` is used to create a matrix of scatterplots to visualize the relationships between multiple variables in a dataset. A **scatterplot**, also known as a **pairs plot** or **pairwise plot**, is a statistical graph used to visualize and compare two variables. It presents a set of points plotted on a Cartesian coordinate system, where each point represents the values of two variables. One variable is assigned to the x-axis, while the other is assigned to the y-axis. Scatterplots are particularly valuable in identifying relationships or correlations between variables, such as trends, clusters, or outliers. By visually examining the distribution and pattern of the points, one can infer whether a positive, negative, or no correlation exists between the variables, which aids in data analysis and interpretation.

`ggpairs()` generates scatterplots for every pair of variables in a dataset. By visualizing these relationships, it becomes easy to identify patterns such as linear relationships, clusters, or outliers. The diagonal of a matrix frequently contains histograms or density plots of the individual variables, illustrating their distribution and aiding in understanding their unique characteristics, such as their range and distribution. Additionally, `ggpairs()` can display **correlation coefficients** in the upper or lower triangle of the matrix, offering numerical insights into the intensity and direction of relationships between variables.

The plots generated by `ggpairs()` are highly customizable, enabling users to modify plot types, incorporate regression lines, and adjust themes using additional ggplot2 functions. Moreover, users can enhance their plots by adding other layers or elements, such as color-coding by groups, or customizing plot aesthetics to improve the visual appeal.

`ggpairs()` is a valuable tool for **exploratory data analysis** (**EDA**) because it provides a simple and comprehensive method for visualizing and understanding the relationships between multiple variables. It can help identify potential issues in the data, such as **multicollinearity**,[2] and guide the selection of features for further analysis or modeling.

[2] Multicollinearity is a statistical phenomenon where highly correlated predictor variables in a multi-

We will use the physical features `bill_length_mm`, `bill_depth_mm`, `flipper_length_mm`, and `body_mass_g` in columns 3 to 6 in `data`, and color on the basis of species. Let us also set a transparency (`alpha`) of 50%. Following is the code at this stage:

```
ggpairs(
  data,
  columns = 3:6,
  aes(color = species, alpha = 0.5)
)
```

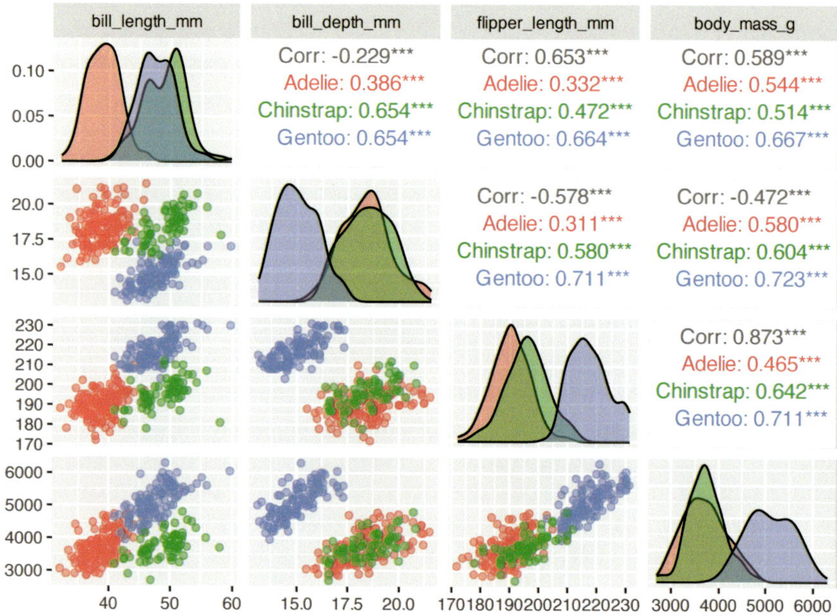

The plots reveal that the physical characteristics of the three species differ significantly. Consequently, these physical characteristics can be used to identify these species of penguins.

To better comprehend the distribution of physical characteristics, we can create box plots. A **box plot** consists of three main components: the box, which represents the **interquartile range** (**IQR**), the distance between the first and third quartiles; the median line, which signifies the median (the 50th percentile); and whiskers extending

ple regression model can be accurately predicted from each other. This condition inflates standard errors, making coefficients unreliable and conclusions misleading. Multicollinearity also makes it difficult to determine the individual effect of each predictor on the dependent variable, complicating model interpretation in machine learning.

from the box to the smallest and largest values that fall within 1.5 times the IQR from the lower and upper quartiles, respectively. Any data points that extend beyond the whiskers are considered potential outliers. An **outlier** is a data point that significantly differs from the rest of the dataset. It can be much higher or lower than the majority, indicating measurement variability, errors, or unusual occurrences. Outliers can skew the results, so their identification and analysis are crucial. Statistical tests and graphical methods, such as box plots and scatter plots, can help identify outliers.

Box plots are particularly useful for comprehending data spread, skewness,[3] and identifying outliers. They provide a concise summary of central tendency, dispersion, and skewness, making them ideal for comparing distributions across different categories. However, it's crucial to remember that box plots don't offer information about the underlying distribution, such as bimodality or multimodality. To gain a more comprehensive understanding, box plots may need to be complemented with other plots like histograms or density plots.

`geom_boxplot()`, a function in the ggplot2 package for R, enables the creation of box plots. To achieve this, we employ the following code, where we modify the variable in y to generate various box plots:

```
data %>%
  ggplot(
    aes(
      x = species,
      y = bill_length_mm
    )
  ) +
  geom_boxplot() +
  xlab("Species") +
  ylab("Bill length (mm)") +
  theme_classic(base_size = 18)
```

[3]Skewness measures how unevenly data spreads around the center, showing whether it extends more to one side than the other.

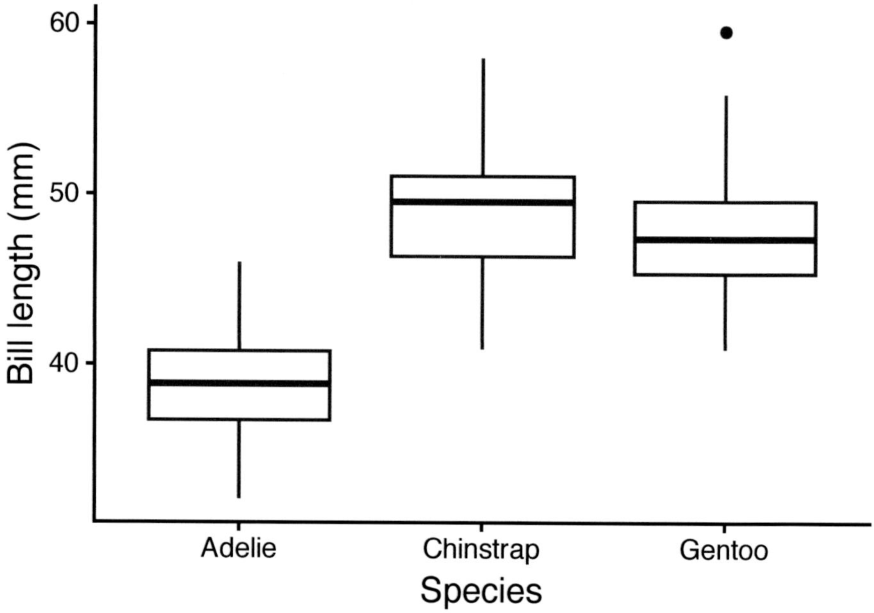

We notice that the values of `bill_length_mm` for *Adélie* are significantly lower compared to those of the *Chinstrap* and *Gentoo* species. Considering the substantial differences in these values, it becomes evident that these measurements can be effectively used to differentiate between the three species.

We also observe the variations in the values of other independent variables among the three species.

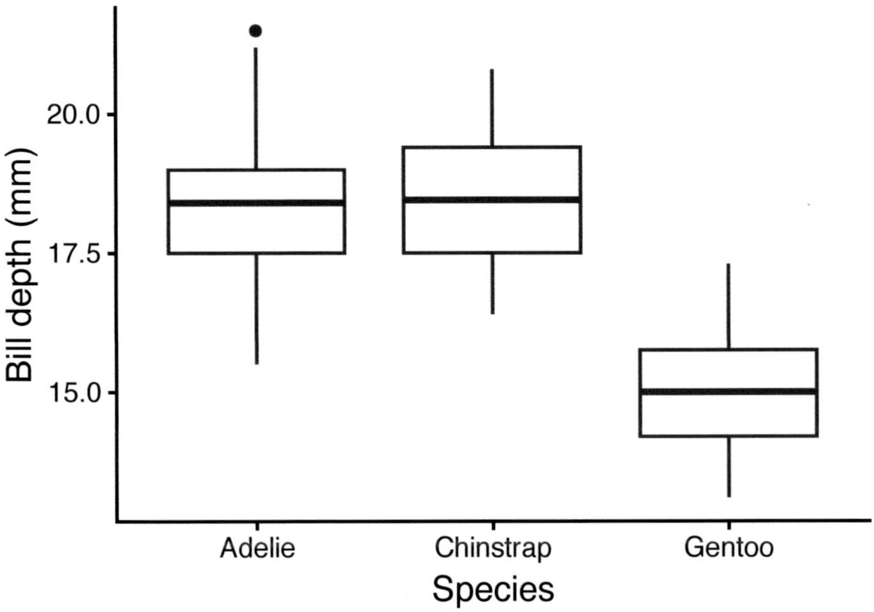
20.0
17.5
15.0
Bill depth (mm)
Adelie
Chinstrap
Gentoo
Species

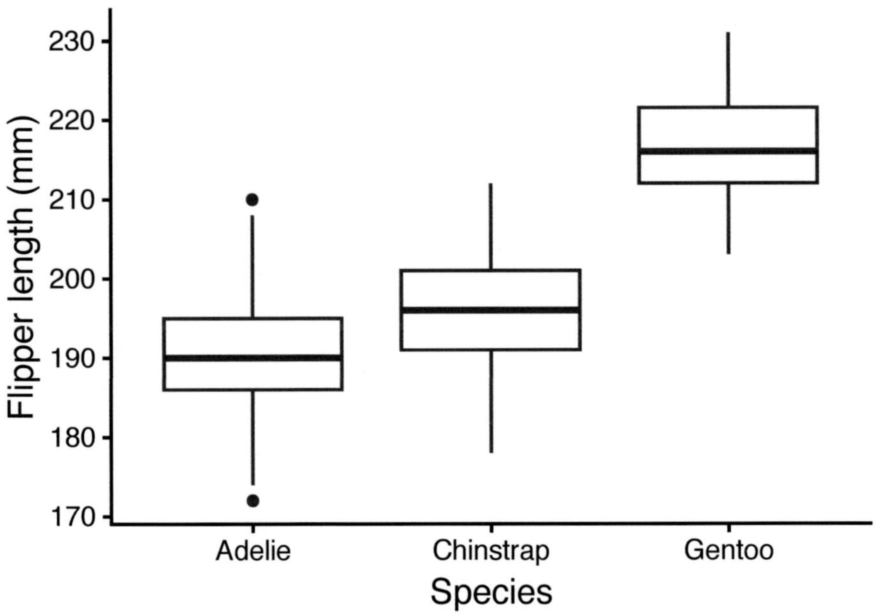
230
220
210
200
190
180
170
Flipper length (mm)
Adelie
Chinstrap
Gentoo
Species

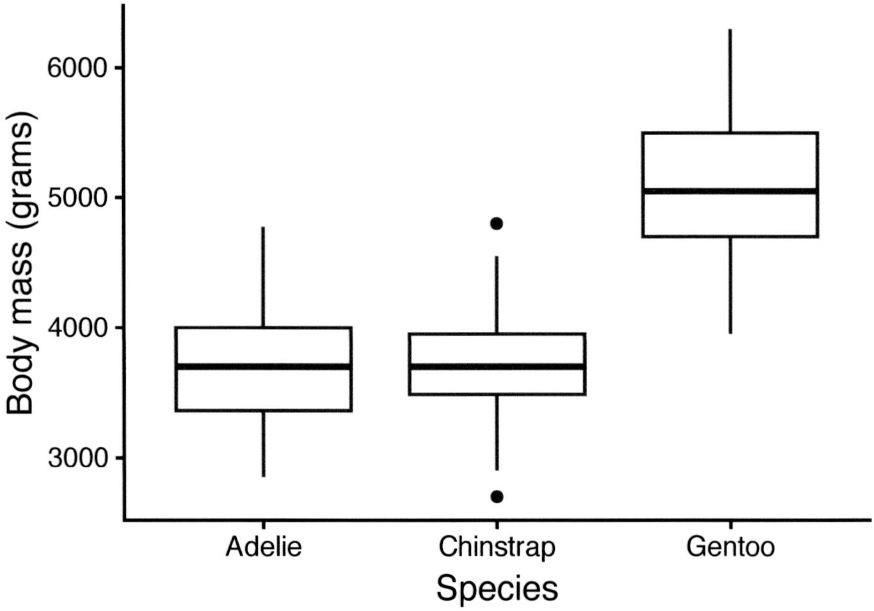

The differences in these values reveal that these measurements can be effectively employed to differentiate between the three species.

Let us now explore the application of SVM using the `tidymodels` package.

8.3 TIDYMODELS EXAMPLE

We will now employ the `tidymodels` package (Kuhn and Wickham, 2020) to perform SVM classification. The primary advantage of utilizing `tidymodels` is that once the coding process has been completed, transitioning from one machine learning algorithm to another becomes an effortless task. Throughout this book, you will observe that we consistently employ comparable codes with various machine learning algorithms.

8.3.1 LOAD DATA

We begin by loading the necessary information into a data file. We also remove any NA (not available; indicating missing data) values from the dataset, shorten column names using `mutate()` function, and select the relevant independent and dependent variables:

We use the pipe operator %>% from the `tidyverse` package to maintain a clean and intuitive code structure. We employ the following code and store the final results in the `data` variable:

```
data <- penguins %>%
  na.omit() %>%
  mutate(
  bill_len = bill_length_mm,
  bill_dep = bill_depth_mm,
  flip_len = flipper_length_mm,
  mass = body_mass_g
) %>%
  select(
  species, bill_len, bill_dep,
  flip_len, mass, sex
)
```

Let us confirm the operations by looking at the `data` variable.

```
data
```

```
## # A tibble: 333 x 6
##    species bill_len bill_dep flip_len  mass sex
##    <fct>      <dbl>    <dbl>    <int> <int> <fct>
##  1 Adelie      39.1     18.7      181  3750 male
##  2 Adelie      39.5     17.4      186  3800 female
##  3 Adelie      40.3     18        195  3250 female
##  4 Adelie      36.7     19.3      193  3450 female
##  5 Adelie      39.3     20.6      190  3650 male
##  6 Adelie      38.9     17.8      181  3625 female
##  7 Adelie      39.2     19.6      195  4675 male
##  8 Adelie      41.1     17.6      182  3200 female
##  9 Adelie      38.6     21.2      191  3800 male
## 10 Adelie      34.6     21.1      198  4400 male
## # i 323 more rows
```

`data` looks good!

8.3.2 SPLIT DATA INTO TRAINING AND TESTING

We partition the data into training and testing sets using the `initial_split()` function from the `rsample` package, a component of `tidymodels`. We retain 70% of the data for training and 30% for testing. These sets are then extracted using `training()` and `testing()` functions:

```
# Split data into training and testing sets
set.seed(2025)
split_7030 <- initial_split(data, prop = 0.7)
data_train <- training(split_7030)
data_test <- testing(split_7030)
```

In this exemplar case, we have set the seed to 2025. `set.seed` is a function used in programming to initialize the random number generator to a specific state, ensuring that the results of random operations are reproducible. By setting the seed, we ensure that the sequence of random numbers generated is consistent each time our code is executed. This consistency facilitates reliable and reproducible results in experiments and simulations, particularly in troubleshooting scenarios. In practice, this line will be removed or commented out upon code deployment, at which point it is not required.

8.3.3 CREATE A RECIPE

We create a recipe using the `recipe()` function. In the recipe, we input the formula `species ~ .` to predict the species using all the available variables in the dataset. We further specify that this recipe should be applied to `data_train`, which is the training data. We also normalize all numeric predictors using the `step_normalize()` function. This step normalizes numeric data to have a standard deviation of one and a mean of zero. The complete recipe code is provided below:

```
recipe_ml <- recipe(
  species ~ .,
  data = data_train
) %>%
  step_normalize(all_numeric())
```

8.3.4 CREATE A MODEL DESIGN

Given our interest in using SVM, we utilize the `svm_linear()` function from `parsnip` package to construct a model design. We further specify the engine as Kernel-based machine learning methods (`kernlab`). Since our objective is to predict the species of individual birds, we set the mode to `classification`. The code that follows demonstrates the creation of the model design:

```
model_svm <- svm_linear() %>%
  set_engine("kernlab") %>%
  set_mode("classification")
```

8.3.5 CREATE A WORKFLOW

We add the recipe and model to the `workflow()` function, and train the model on `data_train` dataset using `fit()` function:

```
workflow_recipe_model <- workflow() %>%
  add_recipe(recipe_ml) %>%
  add_model(model_svm) %>%
  fit(data_train)
```

8.3.6 ASSESS PREDICTION QUALITY WITH TEST DATA

We can now evaluate the performance of our trained and fitted model by using the test data set, `data_test`. We make use of `augment()` function. The `augment()` function, originating from the `broom` package in R, appends predictions and residuals from a statistical model to the original dataset. This facilitates the evaluation of a model's fit by juxtaposing actual and predicted values. The function makes predictions for `data_test` based on the trained model in `workflow_recipe_model`, and adds the predicted classes and class probabilities for species predictions in columns `.pred_class`, `.pred_Adelie`, `.pred_Chinstrap`, and `.pred_Gentoo`, respectively. We use the following code:

```
data_test_with_prediction <- augment(
  workflow_recipe_model, data_test
)
```

To examine the predictions, we select the `species` column and the prediction columns using the `select()` function:

```
data_test_with_prediction %>%
  select(
  species, .pred_class, .pred_Adelie,
```

```
  .pred_Chinstrap, .pred_Gentoo
)
```

```
## # A tibble: 100 x 5
##    species .pred_class .pred_Adelie .pred_Chinstrap
##    <fct>   <fct>              <dbl>           <dbl>
##  1 Adelie  Adelie             0.999       0.000297
##  2 Adelie  Adelie             0.835       0.161
##  3 Adelie  Adelie             0.994       0.000373
##  4 Adelie  Adelie             1.00        0.0000481
##  5 Adelie  Adelie             0.997       0.00136
##  6 Adelie  Adelie             1.00        0.000129
##  7 Adelie  Adelie             1.00        0.0000755
##  8 Adelie  Adelie             0.998       0.000557
##  9 Adelie  Adelie             0.918       0.0802
## 10 Adelie  Adelie             0.989       0.00495
## # i 90 more rows
## # i 1 more variable: .pred_Gentoo <dbl>
```

We observe that the first 10 species were correctly predicted as `Adélie`, indicating the effectiveness of SVM. The probability of a datum being a particular species is given by the `.pred_` columns.

To have a fuller understanding of what is going on, we can create a confusion matrix. A **confusion matrix** is a tabular representation employed to assess the efficacy of a classification model by quantifying the occurrences of true positives, true negatives, false positives, and false negatives. We can create a confusion matrix using the `conf_mat()` function, giving it three arguments—data, truth, and estimate. In this case, data comprises the augmented `data_test_with_prediction`, the truth value is in the column `species`, and the estimated value is in the column `.pred_class`. Thus, we execute the following code:

```
conf_mat(
  data_test_with_prediction,
  truth = species,
  estimate = .pred_class
)
```

```
##            Truth
## Prediction  Adelie Chinstrap Gentoo
##   Adelie        48         0      0
```

```
##    Chinstrap       0          16        0
##    Gentoo          0           0       36
```

We observe from the confusion matrix that all of the predictions are 100% correct!

We can also calculate the accuracy metrics. In machine learning, **accuracy metrics** evaluate a model's performance by comparing its predictions to the actual outcomes. We define the desired metrics in a metric set using the `metric_set()` function in the `yardstick` package, and use the augmented `data_test_with_prediction` dataset to compare the predicted values to the actual results. To achieve this, we employ the following code:

```
metrics_set <- metric_set(
  accuracy,
  sensitivity,
  specificity,
  precision,
  f_meas,
  kap,
  mcc
)
metrics_set(data_test_with_prediction,
               truth = species,
               estimate = .pred_class)
```

```
## # A tibble: 7 x 3
##   .metric     .estimator .estimate
##   <chr>       <chr>          <dbl>
## 1 accuracy    multiclass         1
## 2 sensitivity macro              1
## 3 specificity macro              1
## 4 precision   macro              1
## 5 f_meas      macro              1
## 6 kap         multiclass         1
## 7 mcc         multiclass         1
```

We discover that the SVM's accuracy, sensitivity, specificity, and precision scores are all 100% (or 1), suggesting that it is indeed quite effective in classifying data!

8.4 WHEN TO USE SUPPORT VECTOR MACHINES, AND WHEN NOT TO

SVMs are versatile tools for classification and regression tasks. However, like any algorithm, they have specific scenarios where they excel and others where they may not be the most suitable choice. The following is a guide on when to employ SVMs, and when not to.

8.4.1 WHEN TO USE SVMS

SVMs excel at solving classification problems involving two distinct categories. They are particularly effective when the data points from these classes are clearly separable by a line in two dimensions or a hyperplane in higher dimensions. This implies that SVMs are ideal when the difference between the two groups is evident, and our goal is to identify the optimal line or boundary that divides them. For example, in tasks such as spam email detection, SVMs can effectively distinguish between spam and non-spam emails.

SVMs are renowned for their robustness in handling high-dimensional datasets with a limited number of data points. This capability allows SVMs to effectively manage situations where we have numerous characteristics or properties of data but only a limited number of examples (e.g., thousands of words as features but only a few hundred emails). This scenario is prevalent in various fields, such as text classification (e.g., distinguishing different topics in documents) and medical data analysis (e.g., identifying whether a patient's genetic data points indicate a specific disease). SVMs excel in handling this complexity without overfitting, meaning they don't merely memorize the data but rather strive to discover general patterns applicable to new, unseen data.

SVMs also excel in scenarios where the data cannot be conveniently separated by a straight line. In such cases, SVMs employ a "kernel function" to transform the data into a higher-dimensional space, potentially simplifying the task of separating the classes with a distinct boundary.

SVMs excel at handling outliers, which are those few data points that deviate from the pattern of the majority. Since SVMs prioritize the points closest to the boundary (support vectors), they are less susceptible to the influence of outliers. Even if there are a few "unusual" data points that don't conform to the overall pattern, SVMs can still effectively separate the main groups without being significantly affected by them.

SVMs are particularly advantageous when dealing with limited data. In situations where we have a small to medium-sized dataset, such as a few hundred examples, SVMs excel and can provide a reliable model without the need for an extensive amount of labeled data. This makes them ideal for circumstances where acquiring a substantial amount of labeled data is challenging or costly.

For these reasons, SVMs are widely used in text classification tasks, such as spam detection, sentiment analysis (determining the positive or negative sentiment of a text), and organizing news articles based on their topics. In these scenarios, we often deal with a large number of features, like every word in a text. SVMs excel at handling this high-dimensional data. Their ability to create a separating boundary in this complex space allows them to accurately classify text based on its content.

8.4.2 WHEN NOT TO USE SVMS

While SVMs are powerful for many classification tasks, they may not be the best choice in certain situations. One of their main limitations is computational expense, especially with large datasets containing many samples. Training an SVM on such datasets can be time-consuming and resource-intensive. In such cases, alternative algorithms like logistic regression or decision trees may be more efficient and suitable for handling large volumes of data.

SVMs may struggle with noisy data, particularly when there are overlapping classes or numerous outliers. SVMs aim to find the optimal separating hyperplane, but in the presence of noise, they may end up fitting the noise rather than the underlying patterns in the data. This can lead to overfitting and reduced generalization. In such scenarios, algorithms like Random Forests or simpler models like Naive Bayes may perform better due to their robustness to noisy data.

While SVMs can be adapted for multiclass problems using methods like "one-vs-one" or "one-vs-all," these approaches can make the model more complex and computationally intensive. For multiclass classification tasks, decision trees or neural networks may offer more straightforward solutions. These models are naturally suited to handle multiple classes without the need to split the problem into separate binary classifiers.

Another downside of SVMs is the difficulty in their interpretation, especially when using non-linear kernels. The decision boundaries created by SVMs, particularly in higher-dimensional spaces, can be challenging to comprehend. This lack of interpretability poses a significant drawback in fields where transparency is paramount, such as medical diagnoses or legal decision-making. In such scenarios, simpler models like decision trees or logistic regression, which offer greater clarity, are often preferred.

Choosing the appropriate kernel for an SVM model is a critical yet challenging task. If the kernel is unsuitable for the problem, the model's performance may suffer significantly. This becomes particularly problematic when dealing with complex data, where the wrong kernel could result in inadequate separation between classes. The process of kernel selection and parameter tuning can be time-consuming and computationally intensive. For scenarios where this complexity is not ideal, alternative algorithms such as decision trees or k-nearest neighbors may offer simpler implementations.

SVMs are also not particularly well-suited for real-time or online learning, where the model requires continuous updates as new data arrives. Training an SVM involves solving an optimization problem, which makes it challenging to implement incremental updates.

Lastly, while SVMs can perform well with small datasets, they may become susceptible to overfitting when dealing with extremely limited data points, especially when there are numerous features. In such cases, the model may memorize the training data, and fail to generalize to new examples. Simpler models like Naive Bayes or decision trees may perform better with very small datasets, as they tend to have a lower risk of overfitting in these situations.

8.5 CHAPTER SUMMARY

- SVM is a supervised machine learning algorithm used for classification and regression tasks. It finds the optimal decision boundary that separates data into different categories.
- SVM focuses on maximizing the margin, the distance between the decision boundary and the nearest support vectors. A large margin reduces misclassification.
- For linearly separable data, SVM finds a separating hyperplane. Otherwise, it uses the "kernel trick" to map data into a higher-dimensional space where a hyperplane can more easily separate classes.
- Support vectors determine the separating line's position and orientation, defining the optimal boundary.
- SVM is effective in high-dimensional spaces, and is widely used in image classification, bioinformatics, and text categorization. Its robustness against overfitting makes it suitable for complex tasks with high-dimensional data.
- SVM excels with clearly separable classes, especially in high dimensions. It's also effective with small to medium datasets and quite robust to outliers.
- However, SVM struggles with very large datasets and noisy data with too many outliers. Simpler models like decision trees or logistic regression may be more efficient in such cases. Additionally, SVM can be challenging to interpret, especially with non-linear kernels, limiting its usefulness in fields where transparency is paramount, such as medical diagnoses or legal decision-making.

9 Decision trees and forests

Imagine you're faced with the (daunting) task of choosing what to wear for the day. With a closet brimming with options, the decision hinges on factors like the weather condition, the occasion, and perhaps even your mood. A decision tree, akin to a guiding compass, can be a valuable tool in these situations. It facilitates choices by posing a series of thought-provoking questions and presenting well-reasoned answers. However, its versatility extends far beyond everyday dilemmas, finding applications in diverse fields, from simple decision-making processes to intricate data analysis.

A decision tree starts with a single question at its **root**, which serves as the initial point of decision-making. For example, when deciding what to wear, the initial question could be, "What's the weather like today?" Based on the response, you proceed along different branches of the tree. If the weather is sunny, you follow the path labeled "Sunny." Conversely, if it's rainy, you take the path labeled "Rainy."

Each **branch** of the tree leads to more questions, which help you narrow down your choices further. For example, if you select "Sunny," the next question might be, "Is it hot or cool?" This question helps you decide whether to wear shorts and a t-shirt or a long-sleeve shirt and jeans. The tree continues to branch out based on your answers, guiding you through each step until you reach a final recommendation, known as a **leaf** of the decision tree.

For the "Rainy" path, you might be prompted to ask, "Do you need an umbrella?" The decision tree will then suggest either wearing a raincoat and carrying an umbrella or simply putting on a rain jacket, depending on whether you need one or not. Each path and branch leads to more specific choices based on the answers to previous questions, as illustrated in Figure 9.1.

The beauty of a decision tree lies in its simplicity and clarity. It breaks down complex decisions into smaller, more manageable parts, making them easier to follow and understand. Each question on the tree is carefully crafted to group options that are more similar to each other, guiding you toward a choice that aligns with your circumstances.

Decision trees, beyond everyday choices, have diverse applications in various fields, including finance, healthcare, and machine learning. In finance, they assist in evaluating investment opportunities based on financial metrics. In healthcare, they aid in disease diagnosis by analyzing symptoms and test results. In machine learning, they enable algorithms to make predictions by analyzing data patterns.

Decision trees, with their visual component, are easy to interpret. They are typically depicted as a tree structure, where each node represents a question, and each branch

```
              [What to Wear Today?]
                        |
                        v
                    [Weather]
                   /         \
                 v            v
              [Sunny]      [Rainy]
                 |            |
                 v            v
           [Temperature]    [Need umbrella?]
              /      \           /       \
             v        v         v         v
           [Hot]   [Cool]     [Yes]     [No]
             |        |         |         |
             v        |         |         v
 [Shorts & T-shirt]   |         |     [Rain Jacket]
                      v         v
[Long-sleeve Shirt & Jeans]  [Raincoat (& Carry Umbrella)]
```

FIGURE 9.1: A schematic of a decision tree.

signifies potential answers. The final decisions or classifications are displayed at the end of the branches, which are often referred to as leaves. This visual representation helps users grasp the decision-making process and the factors that influence the final outcome.

However, decision trees also have their fair share of limitations. They can become extremely large and intricate if there are numerous questions and potential answers, which could potentially make them challenging to manage. Additionally, decision trees can be prone to overfitting, especially with small datasets. **Overfitting** occurs when the tree becomes too tailored to the specific data it was trained on, making it less effective on new, unseen data.

Despite their limitations, decision trees remain a valuable tool for decision-making and data analysis. They offer a clear and structured approach to complex decisions, breaking them down into smaller, more manageable parts. By following the tree's branches based on the answers to each question, we can make informed choices and understand the reasoning behind them. Whether for personal decisions or advanced data analysis, decision trees provide a straightforward and effective method for solving problems and making decisions.

9.1 WHAT IS A DECISION TREE?

A **decision tree** is a visual analytical tool that aids decision-making by outlining potential outcomes and their consequences. It allows for the evaluation of various paths and outcomes, ultimately guiding choices. The process involves simplifying a complex decision-making process into a series of binary choices. The resulting model resembles a tree, with each node representing a decision or test on a specific feature. Branches emanate from these nodes, representing the outcomes of those decisions, which may lead to further decisions or final classifications.

At the core of a decision tree lies the concept of **recursive partitioning**. This process begins with the entire dataset and iteratively divides it into subsets based on the values of specific features. Each internal node of the tree represents a test conducted on a particular feature, while each branch signifies the outcome of that test. The leaves of the tree represent the final decision or classification. The primary objective of constructing a decision tree is to develop a model that accurately predicts outcomes by recursively dividing the data in a manner that maximizes the distinction between distinct classes or outcomes.

To construct a decision tree, various algorithms can be employed, including ID3 (Iterative Dichotomiser 3), C4.5, and CART (Classification and Regression Tree). Each algorithm employs distinct criteria to select the feature to split on at each node. For instance, ID3 and its successor C4.5 utilize **information gain** or **gain ratio** to determine the optimal feature, while CART employs measures such as **Gini impurity** or **mean squared error**. These criteria are explained below.

9.1.1 INFORMATION GAIN AND ENTROPY

Imagine we have a large box of toys that are jumbled up, and we want to categorize them into two distinct groups: "Cars" and "Dolls." To make this sorting process easier, we can ask two different questions: "Are the toys large or small?" and "Are the toys red or blue?"

If we ask, "Are the toys large or small?" and find that most of them are either large cars or small dolls, we gain valuable insights. This question categorizes the toys into two distinct groups, providing a comprehensive overview of the contents of the box and simplifying the sorting process.

If we had instead asked, "Are the toys red or blue?" and discovered that both cars and dolls come in various colors, this question might not be as effective in distinguishing them. The toys don't neatly categorize into groups based solely on color, so we don't gain as much useful information.

While "large or small" is a valuable sorting criterion, "red or blue" provides less useful information. The concept of information gain revolves around identifying the most effective question that aids comprehension and effective categorization of information.

This concept is similar to comprehending the clarity gained after asking a question. Initially, we assess the level of confusion in our toy box before asking a question. Subsequently, we categorize the toys based on the question, for instance, sorting them into large and small. Finally, we evaluate the level of confusion within each group after the sorting exercise. The **information gain** quantifies the difference between the initial confusion and the reduced confusion after categorization. If we have successfully reduced the amount of confusion (in this case, between cars and dolls) through the question (in this case, "large or small"), we say that we have gained information. The magnitude of the information gain is proportional to the reduction in confusion. In the current context, for instance, the question "large or small" results in a significant gain of information because one of the outcome buckets is predominantly filled with cars, while the second outcome bucket is predominantly filled with dolls. Conversely, the question "red or blue" results in a small gain of information because both the outcome buckets are a mix of cars and dolls, just as we had initially. In this manner, information gain serves as a measure of the enhanced understanding we acquire about our toys after posing a question.

The concept of **entropy**, which quantifies the level of uncertainty in a dataset, serves as the foundation for calculating information gain. We begin by determining the entropy of the entire dataset, which represents its overall level of uncertainty.[1] Subsequently, we divide the dataset into distinct subsets based on a specific attribute and calculate the entropy for each resulting subset. Next, we compute the weighted average of these entropies, where the weighting factor is the proportion of items in each subset compared to the entire dataset. Finally, we subtract this weighted average entropy from the original entropy. The resulting quantity, known as information gain, quantifies the reduction in uncertainty achieved by utilizing the attribute to partition the data.

In decision tree learning, after calculating the entropy of the entire dataset, we compute the entropy of the subsets formed by splitting the data based on an attribute. To determine the optimal attribute to split on, we calculate the weighted average of the entropies for each branch or subset resulting from the split and compare it to the entropy of the original node.

The entropy of the entire dataset is calculated using the formula:

$$H(S) = -\sum_{i=1}^{n} p_i \log_2(p_i)$$

where p_i is the proportion of data points that belong to class i in the dataset S.

Next, we split the data on an attribute A that has multiple possible values, say $v_1, v_2, \ldots, v_k$. For each value v_i, a subset of the data S_{v_i} is formed, where the at-

[1]Entropy is highest when the dataset contains an equal number of cars and dolls and lowest when the dataset contains only one type. Our objective is to move toward minimum entropy, a state where each *bucket* has only one kind of item.

tribute A takes the value v_i. The entropy for each subset S_{v_i} is computed individually as:

$$H(S_{v_i}) = -\sum_{j=1}^{m} p_{ij} \log_2(p_{ij})$$

where p_{ij} is the proportion of examples in class j in the subset S_{v_i}.

Now, to calculate the weighted average of the entropy for all the subsets, we sum up the entropies of each subset, with the weight being the proportion of instances in that subset. The weighted entropy (or conditional entropy) for a split is:

$$H(S|A) = \sum_{i=1}^{k} \frac{|S_{v_i}|}{|S|} H(S_{v_i})$$

where:

- $|S_{v_i}|$ is the number of examples in the subset S_{v_i},
- $|S|$ is the total number of examples in the original dataset S.

Finally, the information gain from the split on attribute A is calculated by subtracting the conditional entropy from the entropy of the original dataset.

$$IG(S, A) = H(S) - H(S|A)$$

This formula quantifies the reduction in entropy following the split, indicating the extent to which the resulting subsets are more "pure" or "organized" compared to the original dataset.

The attribute with the highest information gain is chosen for splitting at the current node. The objective is to minimize uncertainty in the dataset at each step of the tree-building process.

9.1.2 GINI IMPURITY

Imagine a large box filled with toys, including cars and dolls. **Gini impurity** serves as a measure of how evenly distributed these toys are within the box. If the box predominantly contains cars with only a few dolls, it's relatively unmixed, indicating a low impurity value. Conversely, if the box has almost an equal number of cars and dolls, it's highly mixed, resulting in a high impurity value.

In this manner, the Gini impurity helps us understand the degree of mixing in our toy box, which can guide us in determining the most effective method for sorting them into pure buckets.

To calculate Gini impurity, we first determine the proportion of each class in a dataset. Then, for each class, we square its proportion. Next, we sum up these squared proportions and subtract the result from one. This calculation yields the Gini impurity, which quantifies the degree of mixing or impurity within the dataset. Lower values signify less impurity or greater homogeneity, while higher values indicate more impurity or a mixture of classes.

In decision trees, Gini Impurity serves as a metric that quantifies the degree of homogeneity within a node. This metric helps determine the criteria for splitting the node at each stage. Its value ranges from 0 to 1, with 0 representing a pure node (all samples belong to a single class) and 1 indicating maximum impurity (samples are evenly distributed across all classes). The Gini Impurity for a dataset S is calculated using the following formula:

$$Gini(S) = 1 - \sum_{i=1}^{C} p_i^2$$

where:

- S is the dataset or subset at the node.
- C is the number of possible classes in the target variable.
- p_i is the proportion of samples in class i at the node.

Now, let's consider an example where two splits reduce the Gini Impurity of the parent node, but one split is superior to the other. Imagine we have a dataset with 10 samples: 6 samples belonging to Class A and 4 samples belonging to Class B. The proportions of the classes in the parent node are $p_A = \frac{6}{10} = 0.6$ and $p_B = \frac{4}{10} = 0.4$. The Gini Impurity for the parent node is:

$$Gini(S) = 1 - (0.6^2 + 0.4^2) = 1 - (0.36 + 0.16) = 1 - 0.52 = 0.48$$

Let's now examine two distinct splits of this parent node. In Split 1, the parent node is divided into two child nodes: the left child contains 5 samples of Class A and 1 sample of Class B, while the right child has 1 sample of Class A and 3 samples of Class B. For the left child, the proportions of Class A and Class B are $p_A = \frac{5}{6} \approx 0.833$ and $p_B = \frac{1}{6} \approx 0.167$, respectively. The Gini Impurity for the left child is calculated as follows:

$$Gini(\text{left}) = 1 - (0.833^2 + 0.167^2) = 1 - (0.694 + 0.028) = 1 - 0.722 = 0.278$$

For the right child, the proportions are $p_A = \frac{1}{4} = 0.25$ and $p_B = \frac{3}{4} = 0.75$. The Gini Impurity for the right child is:

$$Gini(\text{right}) = 1 - (0.25^2 + 0.75^2) = 1 - (0.0625 + 0.5625) = 1 - 0.625 = 0.375$$

The weighted Gini Impurity for Split 1 is then calculated as:

$$Gini_{\text{split 1}} = \frac{6}{10} \times 0.278 + \frac{4}{10} \times 0.375 = 0.1668 + 0.15 = 0.3168$$

In Split 2, the parent node is divided into two child nodes. The left child contains four samples of Class A and two samples of Class B, while the right child has two samples of Class A and two samples of Class B. For the left child, the proportions of Class A and Class B are $p_A = \frac{4}{6} \approx 0.667$ and $p_B = \frac{2}{6} \approx 0.333$, respectively. The Gini Impurity for the left child is calculated as follows:

$$Gini(\text{left}) = 1 - (0.667^2 + 0.333^2) = 1 - (0.444 + 0.111) = 1 - 0.555 = 0.445$$

For the right child, the proportions are $p_A = \frac{2}{4} = 0.5$ and $p_B = \frac{2}{4} = 0.5$. The Gini Impurity for the right child is:

$$Gini(\text{right}) = 1 - (0.5^2 + 0.5^2) = 1 - (0.25 + 0.25) = 1 - 0.5 = 0.5$$

The weighted Gini Impurity for Split 2 is then calculated as:

$$Gini_{\text{split 2}} = \frac{6}{10} \times 0.445 + \frac{4}{10} \times 0.5 = 0.267 + 0.2 = 0.467$$

When comparing the two splits, we observe that Split 1 achieves a Gini Impurity of 0.3168, which is lower than the Gini Impurity of the parent node (0.48). In contrast, Split 2 results in a Gini Impurity of 0.467, which is still lower than the Gini Impurity of the parent node, but not as significantly as Split 1. Consequently, Split 1 emerges as the superior split, as it leads to a more substantial reduction in Gini Impurity, resulting in purer child nodes.

This example illustrates how decision trees use Gini Impurity to evaluate potential splits. Even when multiple splits lead to lower impurity than the parent node, the split with the lowest Gini Impurity is chosen, ultimately helping the tree to better classify the data.

9.2 CONSTRUCTING A DECISION TREE

Once the splitting criteria are determined, the recursive decision tree construction process begins. At each node, the chosen criteria are applied to split the data until a **stopping condition** is met. These stopping conditions can include reaching a maximum tree depth, achieving a minimum number of samples in a node, or when further splitting no longer significantly enhances the model's performance.

9.3 PRUNING A DECISION TREE

Pruning is a crucial step in decision tree construction that involves removing insignificant parts of the tree to prevent overfitting. **Overfitting** happens when the tree model becomes overly complex and fits noise in the training data instead of capturing underlying patterns. There are two main types of pruning: pre-pruning and post-pruning. **Pre-pruning** involves halting the tree's growth early, while **post-pruning** involves growing the entire tree and then removing nodes that have minimal impact on the model's accuracy.

9.4 THE BREAST CANCER DATASET

The Breast Cancer dataset in the `mlbench` package (Blake and Merz, 1998) is a widely used dataset for machine learning tasks related to classification. The dataset is used for predicting whether a breast tumor is malignant[2] or benign[3] based on various features extracted from cell images. It provides a valuable resource for testing and evaluating classification algorithms due to its well-structured data and clear classification objective.

9.4.1 DATASET DESCRIPTION

The dataset is originally from the UCI Machine Learning Repository and has been adapted for use in the `mlbench` package in R. It contains a variety of features derived from images of breast tissue samples. Each feature represents some aspect of the tissue image, such as the size or shape of the cell nuclei. There are two classes in the dataset:

1. Malignant, indicating that the tumor is cancerous, and
2. Benign, indicating that the tumor is non-cancerous.

In the `mlbench` package, the dataset is provided as a data frame with the following structure:

---

[2]A malignant tumor is a type of cancerous growth that can invade nearby tissues and spread to other parts of the body. It is characterized by uncontrolled cell division and has the potential to metastasize, forming secondary tumors in distant organs or tissues. Malignant tumors are generally more dangerous than benign tumors due to their aggressive nature and ability to spread.

[3]A benign tumor is a non-cancerous growth that does not invade nearby tissues or spread to other parts of the body. While it can grow and may cause problems depending on its size and location, it typically remains localized and does not have the potential to metastasize. Benign tumors are usually less dangerous than malignant tumors and can often be removed surgically without significant risk of recurrence.

1. Inputs (Features) are a number of attributes related to the characteristics of the tumors, typically including measurements like radius, texture, smoothness, and compactness.
2. Target (Class) is the classification label indicating whether the tumor is malignant or benign.

9.4.2 USING THE DATASET

We begin by (installing and) loading all the required packages and libraries:

```
if (!require("tidymodels")){
  install.packages("tidymodels")
}
library(tidymodels)

if (!require("tidyverse")){
  install.packages("tidyverse")
}
library(tidyverse)

if (!require("janitor")){
  install.packages("janitor")
}
library(janitor)

if (!require("rpart.plot")){
  install.packages("rpart.plot")
}
library(rpart.plot)

if (!require("kableExtra")){
  install.packages("kableExtra")
}
library(kableExtra)

if (!require("parallel")){
  install.packages("parallel")
}
library(parallel)

if (!require("xgboost")){
  install.packages("xgboost")
```

```
}
library(xgboost)

if (!require("mlbench")){
  install.packages("mlbench")
}
library(mlbench)

if (!require("dslabs")){
  install.packages("dslabs")
}
library(dslabs)
```

The Breast Cancer dataset can be loaded from the `mlbench` package using the `data()` function as follows:

```
data("BreastCancer", package = "mlbench")
```

Let us explore this dataset. We begin by looking at the top few values using the `glimpse()` function:

```
BreastCancer %>%
  glimpse(width = 50)
```

```
## Rows: 699
## Columns: 11
## $ Id              <chr> "1000025", "1002945", "1~
## $ Cl.thickness    <ord> 5, 5, 3, 6, 4, 8, 1, 2, ~
## $ Cell.size       <ord> 1, 4, 1, 8, 1, 10, 1, 1,~
## $ Cell.shape      <ord> 1, 4, 1, 8, 1, 10, 1, 2,~
## $ Marg.adhesion   <ord> 1, 5, 1, 1, 3, 8, 1, 1, ~
## $ Epith.c.size    <ord> 2, 7, 2, 3, 2, 7, 2, 2, ~
## $ Bare.nuclei     <fct> 1, 10, 2, 4, 1, 10, 10, ~
## $ Bl.cromatin     <fct> 3, 3, 3, 3, 3, 9, 3, 3, ~
## $ Normal.nucleoli <fct> 1, 2, 1, 7, 1, 7, 1, 1, ~
## $ Mitoses         <fct> 1, 1, 1, 1, 1, 1, 1, 1, ~
## $ Class           <fct> benign, benign, benign, ~
```

We find that the dataset has 699 rows of data, and that the data is arranged in 11 columns, of which the first is the sample code number, followed by 9 sample characteristics (input features for our purpose), and the last column the target class. In particular, following are the column names and their meanings:

1. `Id`: Sample code number.
2. `Cl.thickness`: Clump Thickness.
3. `Cell.size`: Uniformity of Cell Size.
4. `Cell.shape`: Uniformity of Cell Shape.
5. `Marg.adhesion`: Marginal Adhesion.
6. `Epith.c.size`: Single Epithelial Cell Size.
7. `Bare.nuclei`: Bare Nuclei.
8. `Bl.cromatin`: Bland Chromatin.
9. `Normal.nucleoli`: Normal Nucleoli.
10. `Mitoses`: Mitoses.
11. `Class`: Class.

In this chapter, we wish to predict the class (column 11) based on the properties of the clump (columns 2–10). We will make use of the `tidymodels` package to this end.

9.5 TIDYMODELS EXAMPLE

9.5.1 LOAD DATA

We start by loading the necessary information into a data file. This is also an opportune moment to convert the raw data into a tibble format for improved visualization and to exclude any missing values (NA) that may exist in the dataset. While decision trees are quite resilient to missing data, if the missing data pertain to a crucial characteristic of tumors and we are certain that we will have access to that characteristic when we apply machine learning to classify samples, we should remove the NA values to train the algorithm with all the anticipated data it will utilize later on.

We also convert variable names to lowercase and rename them to more intuitive ones. This optional step is highly recommended for better understanding in subsequent steps and for improved visualization in tables and charts.

As before, we employ the pipe operator %>% from `tidyverse` to keep our code intuitive and clean. We make use of the following code, and save the final values into `data`:

```
data <- BreastCancer %>%
  # Convert to tibble
  as_tibble() %>%
```

```
# Remove NA values
na.omit() %>%
# Select pertinent columns and
# rename them
dplyr::select(
  "thickness" = "Cl.thickness",
  "uni_size" = "Cell.size",
  "uni_shape" = "Cell.shape",
  "adhesion" = "Marg.adhesion",
  "size" = "Epith.c.size",
  "bare_nuclei" = "Bare.nuclei",
  "chromatin" = "Bl.cromatin",
  "nucleoli" = "Normal.nucleoli",
  "mitoses" = "Mitoses",
  "class" = as.factor("Class")
)
```

Here # is used to insert comments—lines used to make the code more human-readable and understandable, but which are not processed by R.

Let us confirm the operations by exploring the data variable:

```
data
```

```
## # A tibble: 683 x 10
##    thickness uni_size uni_shape adhesion size  bare_nuclei
##    <ord>     <ord>    <ord>     <ord>    <ord> <fct>
##  1 5         1        1         1        2     1
##  2 5         4        4         5        7     10
##  3 3         1        1         1        2     2
##  4 6         8        8         1        3     4
##  5 4         1        1         3        2     1
##  6 8         10       10        8        7     10
##  7 1         1        1         1        2     10
##  8 2         1        2         1        2     1
##  9 2         1        1         1        2     1
## 10 4         2        1         1        2     1
## # i 673 more rows
## # i 4 more variables: chromatin <fct>, nucleoli <fct>,
## #   mitoses <fct>, class <fct>
```

We find that the number of rows has reduced from 699 to 683, meaning that there are 16 ($699 - 683$) rows with NA values. Before we move forward, let us look at those rows. We can do that by defining a function that finds if any of the values in a row is NA, and applying that function to the BreastCancer dataset using the following code:

```
# Define function to return true
# if any value is NA in a row
na_row <- function(x){
  return(any(is.na(x)))
}

# Apply function to dataset to
# extract rows with NA values
rows <- apply(
  BreastCancer, # the dataset
  MARGIN = 1, # apply over rows of dataset
  na_row
)
```

We can find the number of rows with NA values using the `sum()` function:

```
# Find number of rows with NA
sum(rows)
```

```
## [1] 16
```

We discover that there are 16 rows with missing values, as we had previously observed. To identify which columns contain these missing values, we can utilize the `summary()` function[4] as follows:

```
# Summarize NA values
summary(is.na(BreastCancer))
```

[4]The summary() function in base R provides quick overviews of data based on its class. For vectors, data frames, or fitted models, it offers an instant snapshot of the data. For numeric data, it returns key statistics. For factors, it returns level frequencies. For linear models, it yields coefficients, residuals, significance levels, and goodness-of-fit. Essential for exploratory data analysis, it reveals data distribution and characteristics.

```
##       Id             Cl.thickness     Cell.size
##  Mode :logical   Mode :logical   Mode :logical
##  FALSE:699       FALSE:699       FALSE:699
##
##  Cell.shape      Marg.adhesion   Epith.c.size
##  Mode :logical   Mode :logical   Mode :logical
##  FALSE:699       FALSE:699       FALSE:699
##
##  Bare.nuclei     Bl.cromatin     Normal.nucleoli
##  Mode :logical   Mode :logical   Mode :logical
##  FALSE:683       FALSE:699       FALSE:699
##  TRUE :16
##   Mitoses          Class
##  Mode :logical   Mode :logical
##  FALSE:699       FALSE:699
##
```

We find that all the NA values are located in the `Bare.nuclei` column. **Bare nuclei** are cell nuclei devoid of cytoplasm. They are frequently observed in malignant tumors, making them crucial factors in determining whether a sample is malignant or benign.

Let us also check the distribution of bare nuclei in the complete `data` set by making a bar chart.

```
data %>%
  # Do grouping by bare nuclei and class
  group_by(bare_nuclei, class) %>%
  # Find number of samples
  summarise(n = n()) %>%
  # Send to ggplot
  ggplot(aes(x = bare_nuclei,
             y = n,
             fill = class)) +
  # Make bar chart
  geom_bar(stat = 'identity') +
  # Set labels
  labs(x = "Bare nuclei",
       y = "Number of samples") +
  # Use grey color palette
  scale_fill_grey() +
  # Use classic theme for uncluttered plot
  theme_classic(base_size = 18)
```

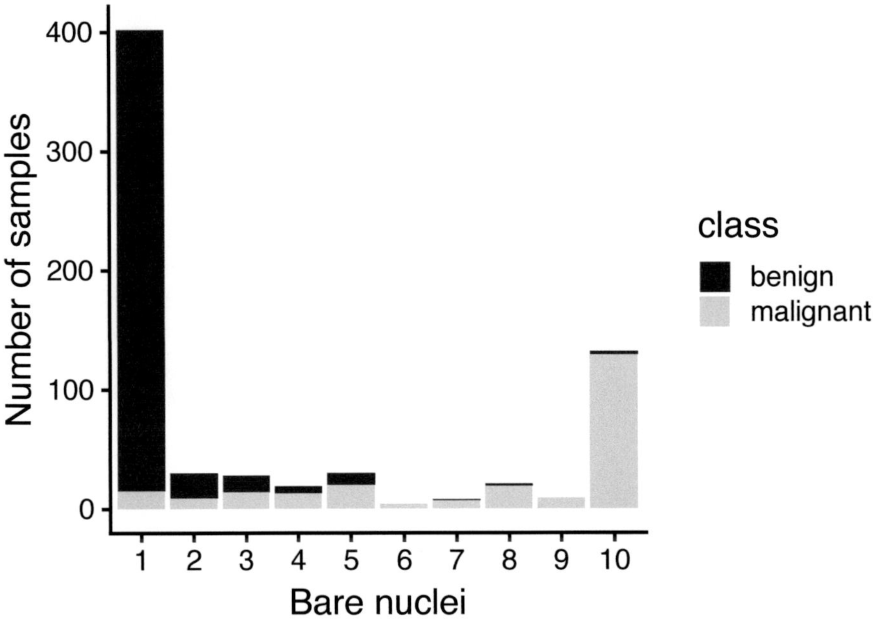

We discover that bare nuclei are a crucial indicator of malignancy. Low values indicate benign tumors, while high values suggest malignant tumors. Therefore, it's beneficial that we removed NA values during the pre-processing stage. As a result, our `data` variable is now free of NA values.

9.5.2 SPLIT DATA INTO TRAINING AND TESTING

We divide our data into training and testing sets using the `initial_split()` function from the `rsample` package, which is part of the `tidymodels` library. We allocate 70% of the data for training and 30% for testing. Subsequently, we extract these two sets using the `training()` and `testing()` functions.

```
# Split data into training and testing sets
set.seed(2025)
split_7030 <- initial_split(data, prop = 0.7)
data_train <- training(split_7030)
data_test <- testing(split_7030)
```

In this exemplary case, we have set the seed to 2025. The `set.seed` function in programming initializes the random number generator to a specific state, ensuring reproducibility of random operations. By setting the seed, we guarantee that the sequence of random numbers generated is consistent each time we run our code. This consistency is crucial for achieving repeatable results in experiments and simulations, particularly for troubleshooting. However, when we deploy the code, this line will be removed or commented out, since at that point it will not be required.

9.5.3 CREATE A RECIPE

In the recipe, we input the formula `class ~ .` to predict the class using all the available variables in the dataset. Remember that since we don't expect the `Id` variable to be related to the final classification, we've already removed it using the `select()` function from the `dplyr` library while extracting the dataset into the variable `data`. We further specify that this recipe should be applied to `data_train`, which is the training data. The complete recipe code is provided below:

```
recipe_ml <- recipe(
  formula = class ~ .,
  data = data_train
)
```

9.5.4 CREATE A MODEL DESIGN

We define a `decision_tree` model using the `rpart` engine. `rpart` constructs decision trees employing the **CART** algorithm. This algorithm is versatile and can be used for both classification and regression tasks. In this particular scenario, we intend to classify data into two categories: `malignant` and `benign`. Therefore, we set the `mode` parameter to `classification`.

In the model design, we also define `tree_depth` to be 3. **Tree depth**, which refers to the length of the longest path from the root node to a leaf node in a decision tree, indicates the number of splits or levels that a tree has. In this case, by setting `tree_depth = 3`, we are specifying that we do not want very tall trees. The following is the code for the model design:

```
model_dt <- decision_tree(
  tree_depth = 3
) %>%
  set_engine("rpart") %>%
  set_mode("classification")
```

9.5.5 CREATE A WORKFLOW AND TRAIN IT

We add the recipe and model to the `workflow()` function, and train the model on `data_train` dataset:

```
workflow_recipe_model <- workflow() %>%
  add_recipe(recipe_ml) %>%
  add_model(model_dt) %>%
  fit(data_train)
```

9.5.6 PLOT THE DECISION TREE

We can visualize decision trees using the `rpart.plot()` function from the `rpart.plot` package (Milborrow, 2024). This function offers a graphical representation of decision trees generated using the `rpart` package. By visualizing the structure of the decision tree, including the splits, node labels, and predicted outcomes, we can gain a clearer understanding and can more effectively analyze the model. We use the following code:

```
rpart.plot(
  extract_fit_engine(workflow_recipe_model),
  yes.text="YES",
  no.text="NO",
  roundint=FALSE
)
```

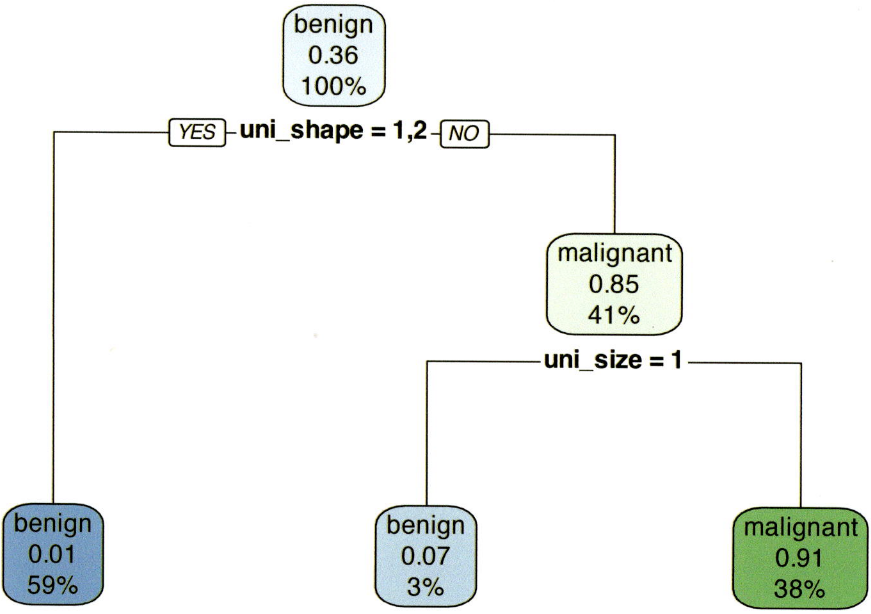

The root node, which is light blue in color, contains three entries:

1. benign indicates that the majority class in the root node is benign.
2. 0.36 represents the proportion of malignant samples at the root node, which is 36%.
3. 100% signifies that all the samples in the dataset data_train have been accounted for.

Let us verify the accuracy of these values. We gather all the data from data_train, categorize it based on class, and calculate the fraction for each class using the following code:

```
data_train %>%
  # Group by class into benign and malignant
  group_by(class) %>%
  # Find number of entries in each class
  summarize(n = n()) %>%
  # Convert number of entries into fraction
  mutate(fraction = n / sum(n))
```

```
## # A tibble: 2 x 3
##   class         n fraction
##   <fct>     <int>    <dbl>
```

```
## 1 benign       305     0.638
## 2 malignant    173     0.362
```

The value 0.36 is indeed correct.

The initial decision point in the tree is whether uni_shape=1,2. Samples that satisfy this condition are placed on the left branch, while those that do not are placed on the right branch.

Let's focus on the left branch, which contains samples with either uni_shape of 1 or 2. This leads us to a leaf (terminal) node that appears blue with a darker shade. The blue color indicates that this node has a majority of samples belonging to the benign class. This is reflected in the first entry of the node, which is benign. The shade represents the purity of the data points bucket at that specific node. This is also shown in the second entry of the node, which is 0.01. Therefore, only 1% of the samples in this node are malignant, compared to 36% at the root node. This makes for a much purer bucket compared to the root node, and this fact is represented by the darker shade of the leaf node. The last entry, 59%, indicates that this node comprises 59% of all data points in the data_train dataset.

Let us check if these values are correct. We use the filter() function of tidyverse to filter data where the value of uni_shape lies in the vector comprised of two entries: 1 and 2. We next group the filtered results by class into benign and malignant, and calculate the fraction for each class using the following code:

```
data_train %>%
  # Filter entries with uni_shape of 1 or 2
  filter(uni_shape %in% c(1, 2)) %>%
  # Group by class into benign and malignant
  group_by(class) %>%
  # Find number of entries in each class
  summarize(n = n()) %>%
  # Convert number of entries into fraction
  mutate(fraction = n / sum(n))
```

```
## # A tibble: 2 x 3
##   class         n fraction
##   <fct>     <int>    <dbl>
## 1 benign      276   0.986
## 2 malignant     4   0.0143
```

We've confirmed that the value of 0.01 is indeed accurate. Now, let's calculate the percentage of entries in this node. We'll do this by dividing the number of entries in the node by the total number of entries in the training dataset, and then multiplying the result by 100. Here's the code we'll use for this calculation:

```
data_train %>%
  # Filter entries with uni_shape of 1 or 2
  filter(uni_shape %in% c(1, 2)) %>%
  # Find percentage of total entries in the node
  summarize(percentage = n() / nrow(data_train) * 100)
```

```
## # A tibble: 1 x 1
##   percentage
##        <dbl>
## 1       58.6
```

The value of 59% is indeed correct.

Next let us focus on the right branch, which contains samples for which `uni_shape` is neither 1 nor 2. This leads us to a non-leaf (non-terminal) node that is light green. The green color indicates that this node has a majority of samples belonging to the other class, the `malignant` class. This is also reflected in the first entry of the node, which is `malignant`. The shade represents the `purity` of the data points bucket at that specific node. This is shown in the second entry of the node, which is `0.85`. Therefore, 85% of the samples in this node are malignant, compared to 36% at the root node. The last entry, 41%, indicates that this node comprises 41% of all data points in the `data_train` dataset.

Let us check if these values are correct. In this case, we use the `filter()` function of `tidyverse` to filter data where the value of `uni_shape` does not lie in the vector comprised of two entries: 1 and 2, by employing the NOT operator `!`. We next group the filtered results by `class` into benign and malignant, and calculate the fraction for each class using the following code:

```
data_train %>%
  # Filter entries with uni_shape neither 1 nor 2
  filter(!uni_shape %in% c(1, 2)) %>%
  # Group by class into benign and malignant
  group_by(class) %>%
  # Find number of entries in each class
  summarize(n = n()) %>%
  # Convert number of entries into fraction
  mutate(fraction = n / sum(n))
```

```
## # A tibble: 2 x 3
##   class         n fraction
##   <fct>     <int>    <dbl>
## 1 benign       29    0.146
## 2 malignant   169    0.854
```

We've confirmed that the value of 0.85 is indeed accurate. Now, let's calculate the percentage of entries in this node. We'll do this by dividing the number of entries in the node by the total number of entries in the training dataset, and then multiplying the result by 100. Here's the code we'll use for this calculation:

```
data_train %>%
  # Filter entries with uni_shape neither 1 nor 2
  filter(!uni_shape %in% c(1, 2)) %>%
  # Find percentage of total entries in the node
  summarize(percentage = n() / nrow(data_train) * 100)
```

```
## # A tibble: 1 x 1
##   percentage
##        <dbl>
## 1       41.4
```

The value of 41% is indeed correct.

The second decision question in the tree is whether uni_size=1. Samples where this question is true are placed on the left branch, while samples where it is false are placed on the right branch.

Let's focus on the left branch, which contains samples with a uni_size of 1. This leads us to a leaf (terminal) node that appears blue with a light shade. The blue color indicates that this node has a majority of samples belonging to the benign class. This is also reflected in the first entry of the node, which is benign. The shade represents the purity of the data points bucket at this specific node. This is shown in the second entry of the node, which is 0.07. Therefore, only 7% of the samples in this node are malignant, compared to 85% at the previous node. The last entry, 3%, indicates that this node comprises 3% of all data points in the data_train dataset.

Let us check if these values are correct. As before, we use the filter() function of tidyverse to filter data where the value of uni_shape does not lie in the vector comprised of two entries: 1 and 2, but where the value of uni_size equals 1. We next group the filtered results by class into benign and malignant, and calculate the fraction for each class using the following code:

```
data_train %>%
  # Filter entries with uni_shape neither 1 nor 2
  filter(!uni_shape %in% c(1, 2)) %>%
  # Filter entries where uni_size equals 1
  filter(uni_size == 1) %>%
  # Group by class into benign and malignant
  group_by(class) %>%
  # Find number of entries in each class
  summarize(n = n()) %>%
  # Convert number of entries into fraction
  mutate(fraction = n / sum(n))
```

```
## # A tibble: 2 x 3
##   class         n fraction
##   <fct>     <int>    <dbl>
## 1 benign       13   0.929
## 2 malignant     1   0.0714
```

We've confirmed that the value of `0.07` is indeed accurate. Now, let's calculate the percentage of entries in this node. We'll do this by dividing the number of entries in the node by the total number of entries in the training dataset, and then multiplying the result by 100. Here's the code we'll use for this calculation:

```
data_train %>%
  # Filter entries with uni_shape neither 1 nor 2
  filter(!uni_shape %in% c(1, 2)) %>%
  # Filter entries where uni_size equals 1
  filter(uni_size == 1) %>%
  # Find percentage of total entries in the node
  summarize(percentage = n() / nrow(data_train) * 100)
```

```
## # A tibble: 1 x 1
##   percentage
##        <dbl>
## 1       2.93
```

The value of 3% is indeed correct.

Let us now concentrate on the right branch, having samples for which `uni_size` is not 1. This brings us to a leaf (terminal) node that is deep green. Green color represents the fact that this node has a majority of samples belonging to the `malignant` class. This is also represented in the first entry in the node, `malignant`. The shade represents the `purity` of the data points bucket at the particular node. This is also shown in the second entry of the node, `0.91`. Thus, 91% of the samples in this node are malignant, as against 85% at the previous node. The last entry, 38% shows that this node comprises 38% of all data points in the data_train dataset.

Let us check if these values are correct. As before, we use the `filter()` function of `tidyverse` to filter data where the value of `uni_shape` does not lie `in` the vector comprised of two entries: 1 and 2, and where the value of `uni_size` does not equal 1. We next group the filtered results by `class` into benign and malignant, and calculate the fraction for each class using the following code:

```
data_train %>%
  # Filter entries with uni_shape neither 1 nor 2
  filter(!uni_shape %in% c(1, 2)) %>%
  # Filter entries where uni_size is not 1
  filter(uni_size != 1) %>%
  # Group by class into benign and malignant
  group_by(class) %>%
  # Find number of entries in each class
  summarize(n = n()) %>%
  # Convert number of entries into fraction
  mutate(fraction = n / sum(n))
```

```
## # A tibble: 2 x 3
##   class         n fraction
##   <fct>     <int>    <dbl>
## 1 benign       16   0.0870
## 2 malignant   168   0.913
```

We've confirmed that the value of `0.91` is indeed accurate. Now, let's calculate the percentage of entries in this node. We'll do this by dividing the number of entries in the node by the total number of entries in the training dataset, and then multiplying the result by 100. Here's the code we'll use for this calculation:

```
data_train %>%
  # Filter entries with uni_shape neither 1 nor 2
  filter(!uni_shape %in% c(1, 2)) %>%
  # Filter entries where uni_size is not 1
  filter(uni_size != 1) %>%
  # Find percentage of total entries in the node
  summarize(percentage = n() / nrow(data_train) * 100)
```

```
## # A tibble: 1 x 1
##   percentage
##        <dbl>
## 1       38.5
```

The value of 38% is indeed correct. We've discovered that each decision step brings about an increase in information gain and decreases impurity, resulting in more homogeneous buckets in terms of our target `class`.

9.5.7 ASSESS PREDICTION QUALITY WITH TEST DATA

We can now evaluate the performance of our trained and fitted model by using the test data set. To achieve this, we employ the `augment()` function from the `broom` package in R. The `augment()` function extends the original dataset by including predictions and residuals generated by a statistical model. This visualization approach provides a direct comparison between the actual values and the predicted values, enabling us to assess the model's accuracy. We use the following code:

```
data_test_with_prediction <- augment(
  workflow_recipe_model, data_test
)
```

Let's take a look at the augmented file, `data_test_with_prediction`.

```
data_test_with_prediction %>%
  glimpse(width = 50)
```

```
## Rows: 205
## Columns: 13
## $ .pred_class     <fct> benign, benign, benign, ~
## $ .pred_benign    <dbl> 0.98571429, 0.98571429, ~
## $ .pred_malignant <dbl> 0.01428571, 0.01428571, ~
## $ thickness       <ord> 3, 1, 2, 1, 7, 10, 1, 1,~
## $ uni_size        <ord> 1, 1, 1, 1, 3, 5, 1, 1, ~
## $ uni_shape       <ord> 1, 1, 1, 1, 2, 5, 1, 3, ~
## $ adhesion        <ord> 1, 1, 1, 1, 10, 3, 1, 1,~
## $ size            <ord> 2, 2, 2, 2, 5, 6, 2, 2, ~
## $ bare_nuclei     <fct> 2, 10, 1, 3, 10, 7, 1, 1~
## $ chromatin       <fct> 3, 3, 1, 3, 5, 7, 3, 1, ~
```

```
## $ nucleoli        <fct> 1, 1, 1, 1, 4, 10, 1, 1,~
## $ mitoses         <fct> 1, 1, 5, 1, 4, 1, 1, 1, ~
## $ class           <fct> benign, benign, benign, ~
```

We notice that three new columns have been added to the dataset:

1. The `.pred_class` column contains the predictions made by the model. For the first three samples, it correctly classifies them as `benign`, which aligns with the values in the `class` column.
2. The `.pred_benign` column provides the probability of a sample being benign, calculated by the decision tree.
3. The `.pred_malignant` column offers the probability of a sample being malignant, also computed by the decision tree.

The two probabilities sum up to 1. Next, we construct a **confusion matrix** to understand how many samples were correctly and incorrectly classified. To achieve this, we utilize the `conf_mat()` function. This function, part of the `yardstick` package in R, generates a confusion matrix from model predictions. This matrix aids in evaluating the performance of a classification model by presenting the counts of true positive, true negative, false positive, and false negative predictions. This enables us to assess metrics such as accuracy, precision, recall, and F1 score. We employ the following code:

```
conf_mat(
  data_test_with_prediction,
  truth = class,
  estimate = .pred_class
)
```

```
##               Truth
## Prediction  benign malignant
##   benign       129         5
##   malignant     10        61
```

We discover that out of the 134 samples predicted by the decision tree to be benign (129 + 5), 129 were indeed benign, while 5 were malignant. Similarly, out of the 71 samples predicted by the decision tree to be malignant (10 + 61), 61 were malignant, and 10 were benign.

We can also calculate the accuracy metrics. In machine learning, **accuracy metrics** assess a model's performance by comparing its predictions to the actual outcomes. We define the desired metrics in a metrics set and use the augmented data_test_with_prediction dataset to compare the predicted values to the actual results. We use the following code to this end:

```
metrics_set <- metric_set(
  accuracy,
  sensitivity,
  specificity,
  precision,
  f_meas,
  kap,
  mcc
)
metrics_set(data_test_with_prediction,
            truth = class,
            estimate = .pred_class)
```

```
## # A tibble: 7 x 3
##   .metric     .estimator .estimate
##   <chr>       <chr>          <dbl>
## 1 accuracy    binary         0.927
## 2 sensitivity binary         0.928
## 3 specificity binary         0.924
## 4 precision   binary         0.963
## 5 f_meas      binary         0.945
## 6 kap         binary         0.836
## 7 mcc         binary         0.837
```

We discover that the decision tree's accuracy, sensitivity, specificity, and precision scores exceed 90%, suggesting that it is indeed quite effective in classifying data!

9.6 INSTABILITY OF DECISION TREES

Despite their widespread use and interpretability, decision trees are often criticized for their **instability**, which refers to their sensitivity to variations in the training data. Even a slight change in the training dataset can lead to different tree structures and predictions, significantly impacting their performance and reliability.

Instability in decision trees stems from the way they split the data at each node. The process of constructing a decision tree involves selecting the most suitable feature to split on at each step, based on criteria like information gain or Gini impurity. However, minor alterations in the data can result in different splits, leading to a different tree structure.

A key factor contributing to instability is the recursive nature of the tree-building process. Each decision made at a node influences subsequent splits and the overall tree structure. Consequently, minor variations in the data can lead to substantial changes in the tree's branches and leaves. For instance, if a training dataset contains noise or outliers, the decision tree may create branches that overfit to these anomalies, resulting in a model that performs poorly on new, unseen data.

9.6.1 CONSEQUENCES OF INSTABILITY

The instability of decision trees leads to several significant consequences. One major problem is **overfitting**, where the decision tree becomes excessively trained to the training data, capturing noise and minor details that don't generalize well to new data. Overfitting results in a model with high variance, performing well on the training set but poorly on test sets. This issue occurs because the tree may construct overly intricate structures to account for every minute detail in the training data, instead of focusing on the underlying patterns.

Decision trees can also suffer from **poor generalization**. Their sensitivity to data variations means that even a slight change in the dataset can result in an entirely different tree. This lack of robustness makes decision trees less reliable in situations where the data is subject to random fluctuations or when the training set is not representative of the broader population.

9.6.2 MITIGATING INSTABILITY

Several strategies can be employed to address the instability of decision trees and enhance their performance. One prevalent approach is **tree pruning**, which entails removing insignificant portions of the tree that contribute little to the model's predictive accuracy. Pruning simplifies the tree, thereby reducing overfitting and improving generalization.

There are two primary types of pruning: pre-pruning and post-pruning. **Pre-pruning** involves halting the tree's growth before it becomes overly intricate, while **post-pruning** involves cultivating a full tree and then trimming it to simplify the model.

Another effective technique is to employ ensemble methods. An **ensemble method** in machine learning combines multiple models to enhance overall performance and robustness. Instead of relying solely on a single model, ensemble methods capitalize on the strengths of various models to make more accurate predictions. Common ensemble techniques include:

1. **Bagging (Bootstrap Aggregating):** This technique involves training multiple models on different subsets of the training data and then averaging

their predictions. For instance, Random Forest is an example of a bagging algorithm.
2. **Boosting**: Boosting is a sequential training approach where each new model is designed to correct the errors made by the previous models. **Gradient Boosting** and **AdaBoost** are two popular boosting algorithms.
3. **Stacking**: Stacking combines predictions from multiple models using a meta-model to make a final prediction.

Ensemble methods often enhance performance by reducing variance and/or bias, leading to more reliable and accurate predictions.

9.7 EXEMPLIFYING INSTABILITY IN A DECISION TREE

We can easily illustrate instability in decision trees by altering the seed used to split the dataset into training and testing datasets. If the `prop` parameter remains unchanged, we observe a bifurcation with the same number of rows in both datasets, albeit with a different random selection. Consequently, it's possible that a particular row of `data` was previously a part of `data_train`, but due to a change in the seed, it now belongs to `data_test`. Let us now recreate the steps of section 9.5, with the only difference being that the seed is set to 1987 in place of 2025.

9.7.1 SPLIT DATA INTO TRAINING AND TESTING

We divide our data into training and testing sets using the `initial_split()` function from the `rsample` package, which is part of the `tidymodels` library. We allocate 70% of the data for training and 30% for testing. Subsequently, we extract these two sets using the `training()` and `testing()` functions.

```
# Split data into training and testing sets
set.seed(1987)
split_7030 <- initial_split(data, prop = 0.7)
data_train <- training(split_7030)
data_test <- testing(split_7030)
```

9.7.2 CREATE A RECIPE

In the recipe, we input the formula `class ~ .` to predict the class using all the available variables in the dataset. Remember that since we don't expect the `Id` variable to be related to the final classification, we've already removed it using the `select()` function from the `dplyr` library while extracting the dataset into the variable `data`.

We further specify that this recipe should be applied to data_train, which is the training data. The complete recipe code is provided below:

```
recipe_ml <- recipe(
  class ~ .,
  data = data_train
)
```

9.7.3 CREATE A MODEL DESIGN

We define a decision_tree model using the rpart engine. rpart constructs decision trees employing the **CART** algorithm. This algorithm is versatile and can be used for both classification and regression tasks. In this particular scenario, we intend to classify data into two categories: malignant and benign. Therefore, we set the mode parameter to classification.

In the model design, we also define tree_depth to be 3. **Tree depth**, which refers to the length of the longest path from the root node to a leaf node in a decision tree, indicates the number of splits or levels that a tree has. In this case, by setting tree_depth = 3, we are specifying that we do not want very tall trees. The following is the code for the model design:

```
model_dt <- decision_tree(
  tree_depth = 3
) %>%
  set_engine("rpart") %>%
  set_mode("classification")
```

9.7.4 CREATE A WORKFLOW

We add the recipe and model to the workflow() function, and train the model on data_train dataset:

```
workflow_recipe_model <- workflow() %>%
  add_recipe(recipe_ml) %>%
  add_model(model_dt) %>%
  fit(data_train)
```

9.7.5 PLOT A DECISION TREE

We can visualize decision trees using the rpart.plot() function from the rpart.plot package (Milborrow, 2024). This function offers a graphical representation of decision trees generated using the rpart package. By visualizing the structure of the decision tree, including the splits, node labels, and predicted outcomes, we can gain a clearer understanding and can more effectively analyze the model. We use the following code:

```
rpart.plot(
  extract_fit_engine(workflow_recipe_model),
  yes.text="YES",
  no.text="NO",
  roundint=FALSE
)
```

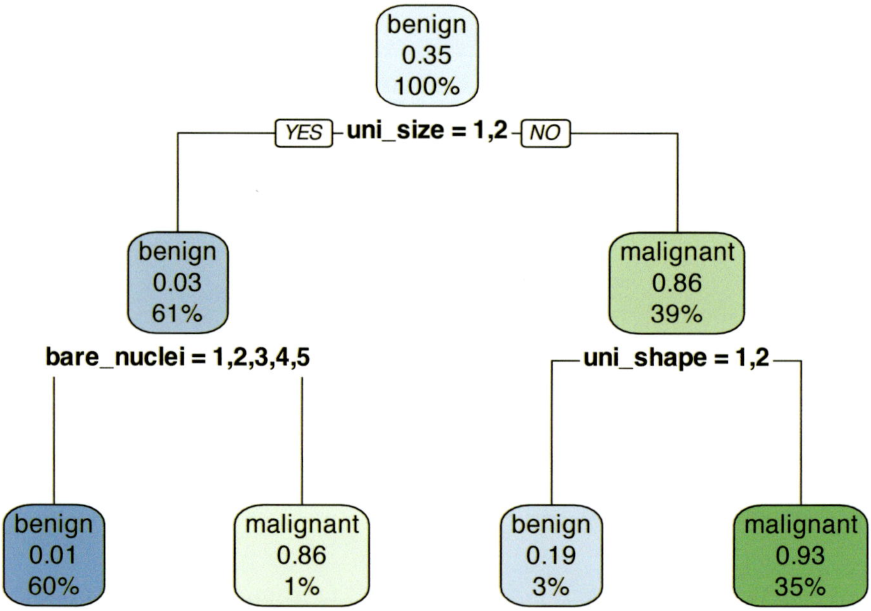

We find that the shape of the tree is now very different from the previous tree. Earlier, we had three leaf nodes; now we have four leaf nodes. Earlier only two parameters were considered for classification, uni_shape and uni_size. Now, in addition to these two, we also have bare_nuclei (thus also showing that it was a good idea after all to remove the rows having NA values for bare_nuclei!). The distribution of samples in the nodes is also quite different from the previous tree.

Let us now consider the prediction quality of this tree.

9.7.6 ASSESS PREDICTION QUALITY WITH TEST DATA

We evaluate the performance of our trained and fitted model by using the test data set. To achieve this, we employ the augment() function from the broom package in R. The augment() function extends the original dataset by including predictions and residuals generated by a statistical model. This visualization approach provides a direct comparison between the actual values and the predicted values, enabling us to assess the model's accuracy. We use the following code:

```
data_test_with_prediction <- augment(
  workflow_recipe_model, data_test
)
```

Next, we construct a **confusion matrix** to understand how many samples were correctly and incorrectly classified. To achieve this, we utilize the conf_mat() function. This function, part of the yardstick package in R, generates a confusion matrix from model predictions. This matrix aids in evaluating the performance of a classification model by presenting the counts of true positive, true negative, false positive, and false negative predictions. This enables us to assess metrics such as accuracy, precision, recall, and F1 score. We employ the following code:

```
conf_mat(
  data_test_with_prediction,
  truth = class,
  estimate = .pred_class
)
```

```
##             Truth
## Prediction  benign malignant
##   benign       127         4
##   malignant      8        66
```

We discover that out of the 131 samples predicted by the decision tree to be benign (127 + 4), 127 were indeed benign, while 4 were malignant. Similarly, out of the 74 samples predicted by the decision tree to be malignant (8 + 66), 66 were malignant, and 8 were benign.

Let us also compute the accuracy metrics. In machine learning, **accuracy metrics** assess a model's performance by comparing its predictions to the actual outcomes. We define the desired metrics in a metrics set and use the augmented data_test_with_prediction dataset to compare the predicted values to the actual results. We use the following code to this end:

```
metrics_set <- metric_set(
  accuracy,
  sensitivity,
  specificity,
  precision,
  f_meas,
  kap,
  mcc
)
metrics_set(data_test_with_prediction,
            truth = class,
            estimate = .pred_class)
```

```
## # A tibble: 7 x 3
##   .metric     .estimator .estimate
##   <chr>       <chr>          <dbl>
## 1 accuracy    binary         0.941
## 2 sensitivity binary         0.941
## 3 specificity binary         0.943
## 4 precision   binary         0.969
## 5 f_meas      binary         0.955
## 6 kap         binary         0.872
## 7 mcc         binary         0.872
```

We discover that the accuracy, sensitivity, specificity, and precision scores exceed 90%, suggesting that this decision tree is also highly effective in classifying data!

This exercise also demonstrates the fact that even with their instability, decision trees are extremely useful in correctly predicting the right classes. In fact, we may even use the instability of decision trees to our advantage by creating a `forest` of trees through bagging ensemble.

9.8 FROM TREES TO (RANDOM) FORESTS

Decision Forests, also known as **Random Forests**, are an ensemble method that combines the predictions of multiple decision trees to enhance the model's accuracy, stability, and overall performance.

Decision forests, a concept rooted in the idea of constructing a collection of decision trees, each trained on slightly different data and features, offer a solution to the limitations of single decision trees. Single decision trees can be susceptible to variations in training data and prone to overfitting. Decision forests mitigate these issues through aggregation and averaging. By combining the strengths of multiple decision trees, decision forests achieve better generalization and robustness, making them a powerful tool for predictive modeling.

To build a decision forest, the process starts with creating multiple training datasets using a technique called bootstrapping [Fig. 9.2]. In **bootstrapping**, random samples are taken from the original dataset with replacement, producing several distinct subsets. Each decision tree in the forest is then trained on one of these subsets. This approach ensures that each tree learns from different variations of the data, enhancing the diversity within the forest. This process utilizes the fact that slight variations in the data produce very different decision trees due to the inherent instability of decision trees. Thus, each decision tree will incorporate the unique features present in one random subset of the data, increasing the overall robustness of the random forest.

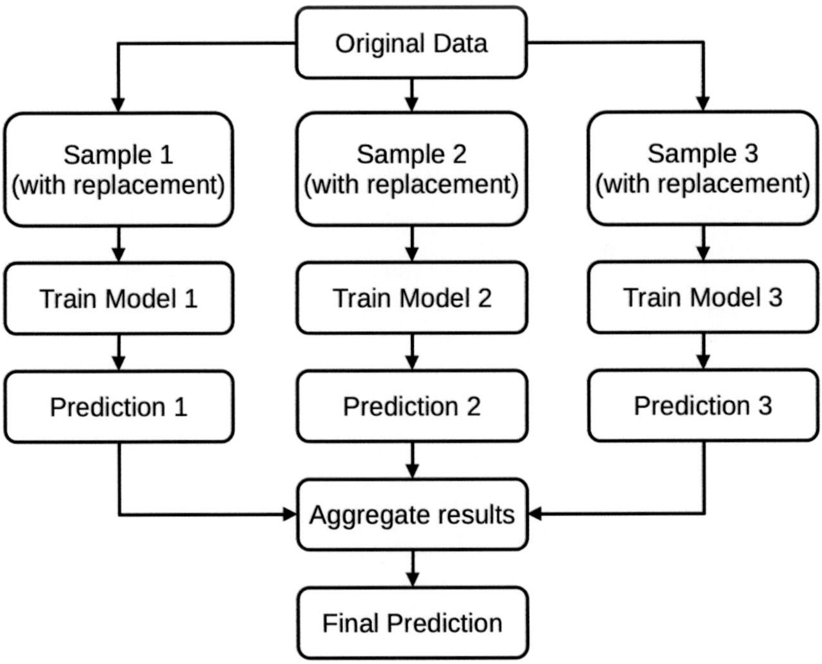

FIGURE 9.2: A schematic of bootstrapping.

In addition to bootstrapping, **feature randomization** is another crucial component of constructing decision forests. During the training of each decision tree, only a random subset of features is considered for each split at every node. This randomness prevents any single tree from dominating the model and reduces the correlation between the trees in the forest. By limiting the features considered at each decision point, the algorithm fosters a variety of decision trees, which collectively contribute to the robustness of the forest.

Each decision tree in the forest is typically grown to its full size without pruning. This means that the trees are allowed to fully develop their branches based on the training data. The rationale behind this approach is that the ensemble method of decision forests inherently reduces overfitting through the process of averaging predictions. Consequently, individual trees might be complex, but the combination of many trees leads to a more stable and generalized model.

Once all decision trees are trained, they aggregate their predictions to arrive at a final decision. For classification tasks, this is done through **majority voting**, where each tree "votes" for a class, and the class receiving the most votes is selected as the final prediction. In regression tasks, the forest's prediction is the average of the predictions made by each tree. This aggregation process smooths out individual trees' biases and errors, resulting in a more accurate overall prediction.

One of the primary benefits of decision forests is their increased accuracy compared to single decision trees. By averaging the predictions of multiple trees, decision forests reduce the variance and bias associated with individual trees. This leads to a model that generalizes better to new, unseen data. Furthermore, decision forests are robust to overfitting. The ensemble approach ensures that the model is less influenced by noise and anomalies in the training data, improving its performance on test sets.

Decision forests also provide valuable insights into feature importance. By analyzing how much each feature contributes to reducing impurity across all trees, one can identify which features are most significant in the prediction process. This **feature importance analysis** can be useful in understanding the underlying patterns in the data and in feature selection for model refinement.

Despite their advantages, decision forests are not without limitations. They can be computationally intensive, especially when dealing with large datasets or a high number of trees. However, random forests are especially suited to parallel processing, and with the availability of multiple processor cores, random forests can often surpass other machine learning methods. The models can also become complex and less interpretable compared to individual decision trees. While the collective strength of the trees improves prediction accuracy, the resulting forest can be challenging to interpret as a whole.

In summary, decision forests represent a powerful enhancement of decision tree models. By leveraging the collective predictions of multiple trees, decision forests achieve greater accuracy and stability while addressing prominent limitations associated with single decision trees. They are versatile and effective for a wide range of predictive tasks, though considerations around computational efficiency and model interpretability should be kept in mind.

That said, let us now create a random forest! The steps are similar to those for creating a decision tree.

9.8.1 SPLIT DATA INTO TRAINING AND TESTING

We divide our data into training and testing sets using the `initial_split()` function from the `rsample` package, which is part of the `tidymodels` library. We allocate 70% of the data for training and 30% for testing. Subsequently, we extract these two sets using the `training()` and `testing()` functions.

```
# Split data into training and testing sets
set.seed(2025)
split_7030 <- initial_split(data, prop = 0.7)
data_train <- training(split_7030)
data_test <- testing(split_7030)
```

In this exemplary case, we have set the seed to 2025. The `set.seed` function in programming initializes the random number generator to a specific state, ensuring reproducibility of random operations. By setting the seed, we guarantee that the sequence of random numbers generated is consistent each time we run our code. This consistency is crucial for achieving repeatable results in experiments and simulations, particularly for troubleshooting. However, when we deploy the code, this line will be removed or commented out, since at that point it will not be required.

9.8.2 CREATE A RECIPE

In the recipe, we input the formula `class ~ .` to predict the class using all the available variables in the dataset. Remember that since we don't expect the `Id` variable to be related to the final classification, we've already removed it using the `select()` function from the `dplyr` library while extracting the dataset into the variable `data`. We further specify that this recipe should be applied to `data_train`, which is the training data. The complete recipe code is provided below:

```
recipe_ml <- recipe(
  class ~ .,
  data = data_train
)
```

9.8.3 CREATE A MODEL DESIGN

To construct the model, we leverage the rand_forest() function, which is part of the parsnip package within the tidymodels library. This function encompasses several crucial arguments. The mode parameter is a string that determines the prediction outcome mode. It can take on four possible values: "unknown," "regression," "classification," or "censored regression." Depending on the nature of the problem we aim to solve, we can select the appropriate value for mode. In our specific scenario, since we intend to predict the class of the dependent variable, we set mode to "classification."

The engine argument, a string, specifies the computational engine to be used for model fitting. This enables us to select the underlying method or package for computations. In our case, we set it to "ranger." Additionally, we set num.threads=detectCores() to enable parallel computing, utilizing all the cores of the computer's processor.

The trees argument specifies the number of trees to be included in the ensemble. An integer value determines the number of individual decision trees that will be built and combined to form the random forest. Let us keep this value as 2000.

The mtry argument, an integer, specifies the number of predictors randomly sampled at each split during the creation of tree models. This parameter controls the randomness and diversity of the trees within the forest. Let us keep this value as 2.

Finally, the min_n argument is an integer that specifies the minimum number of data points needed for a node to be split further. This parameter plays a crucial role in controlling the size of the nodes and, consequently, the complexity of the trees. Let us keep this value as 5.

Following is the complete code that emerges:

```
model_rf <- rand_forest(
  min_n = 5,
  mtry = 2,
  trees = 2000
) %>%
  set_engine("ranger", num.threads=detectCores()) %>%
  set_mode("classification")
```

9.8.4 CREATE A WORKFLOW AND TRAIN IT

We add the recipe and model to the workflow() function, and train the model on data_train dataset:

```
workflow_recipe_model <- workflow() %>%
  add_recipe(recipe_ml) %>%
  add_model(model_rf) %>%
  fit(data_train)
```

9.8.5 ASSESS PREDICTION QUALITY WITH TEST DATA

We can now evaluate the performance of our trained and fitted model by using the test data set. To achieve this, we employ the augment() function from the broom package in R. The augment() function extends the original dataset by including predictions and residuals generated by a statistical model. This visualization approach provides a direct comparison between the actual values and the predicted values, enabling us to assess the model's accuracy. We use the following code:

```
data_test_with_prediction <- augment(
  workflow_recipe_model, data_test
)
```

Next, we construct a **confusion matrix** to understand how many samples were correctly and incorrectly classified. To achieve this, we utilize the conf_mat() function. This function, part of the yardstick package in R, generates a confusion matrix from model predictions. This matrix aids in evaluating the performance of a classification model by presenting the counts of true positive, true negative, false positive, and false negative predictions. This enables us to assess metrics such as accuracy, precision, recall, and F1 score. We employ the following code:

```
conf_mat(
  data_test_with_prediction,
  truth = class,
  estimate = .pred_class
)
```

```
##             Truth
## Prediction  benign malignant
##   benign       136         2
##   malignant      3        64
```

We discover that out of the 138 samples predicted by the decision tree to be benign (136 + 2), 136 were indeed benign, while 2 were malignant. Similarly, out of the 67 samples predicted by the decision tree to be malignant (3 + 64), 64 were malignant, and 3 were benign.

We can also calculate the accuracy metrics. In machine learning, **accuracy metrics** assess a model's performance by comparing its predictions to the actual outcomes. We define the desired metrics in a metrics set and use the augmented `data_test_with_prediction` dataset to compare the predicted values to the actual results. We use the following code to this end:

```
metrics_set <- metric_set(
  accuracy,
  sensitivity,
  specificity,
  precision,
  f_meas,
  kap,
  mcc
)
metrics_set(data_test_with_prediction,
            truth = class,
            estimate = .pred_class)
```

```
## # A tibble: 7 x 3
##   .metric     .estimator .estimate
##   <chr>       <chr>          <dbl>
## 1 accuracy    binary         0.976
## 2 sensitivity binary         0.978
## 3 specificity binary         0.970
## 4 precision   binary         0.986
## 5 f_meas      binary         0.982
## 6 kap         binary         0.944
## 7 mcc         binary         0.944
```

We can observe that the metrics have improved substantially! This clearly demonstrates the importance of utilizing an ensemble of decision trees in the form of a random forest.

9.9 BOOSTED TREES

While decision trees are intuitive and easy to interpret, they sometimes struggle with complex data patterns, resulting in less accurate predictions. Random forests of decision trees overcome this limitation by combining multiple decision trees to enhance accuracy. But as we have discussed, in random forests, each tree is randomly formed. Consequently, the efficiency of random forests may not be the most optimal.

Boosted trees aim to enhance efficiency by progressively improving each decision tree compared to the previous one. They construct sequential models to rectify errors. The process commences with a fundamental decision tree, similar to the one employed in random forests, which generates initial predictions on the data. However, this initial tree may not be highly accurate and could potentially make errors in predicting the outcomes. The fundamental concept behind boosting lies in learning from these errors.

After the initial decision tree makes its predictions, we calculate the errors it made. Boosting then creates a new decision tree specifically designed to correct these errors. This new tree focuses on the data points that were misclassified or poorly predicted by the initial tree. By doing so, the second tree improves the overall model's accuracy.

The process is repeated multiple times, with each new tree working to correct the mistakes of the combined model from the previous trees. Each new tree contributes its predictions to the accumulated predictions, progressively enhancing the model's accuracy. This iterative process builds upon the strengths of the preceding trees, resulting in a robust and reliable final model.

Boosted trees are valued for their ability to create highly accurate models by combining the strengths of multiple decision trees. They are widely used in various applications, such as predicting customer behavior, financial forecasting, and medical diagnosis, because they are effective in capturing complex patterns in data with improved prediction accuracy.

9.9.1 POPULAR BOOSTING ALGORITHMS

There are several popular tree boosting algorithms widely used for various machine learning tasks. Among them are the following well-known ones.

1. **Adaptive Boosting (AdaBoost)** is a machine learning algorithm that merges multiple weak learners (simple models that perform slightly better than random guessing) to create a robust learner. It operates by sequentially training a series of weak learners on modified versions of the data. In each iteration, it allocates higher weights to incorrectly predicted instances, compelling subsequent weak learners to concentrate more on these challenging cases. The final prediction is achieved by combining the predictions of all weak learners, weighted by their individual performance during training. AdaBoost proves effective for tasks such as classification, where it enhances accuracy by prioritizing instances that are difficult to classify.

2. **Gradient Boosting Machine (GBM)** is a fundamental boosting algorithm that constructs decision trees sequentially. Each new tree aims to rectify the errors made by the preceding ones. The model employs **gradient descent**, a technique in machine learning that gradually adjusts the model's parameters based on the direction that minimizes prediction errors the most. Through this iterative process, the model progressively improves accuracy.
3. **XGBoost (Extreme Gradient Boosting)** is an advanced and highly optimized version of gradient boosting. It incorporates several enhancements, including regularization to prevent overfitting, parallel processing for faster training, and more efficient handling of missing values.
4. **LightGBM (Light Gradient Boosting Machine)** is a highly efficient and scalable machine learning algorithm, particularly well-suited for large datasets. It employs techniques such as **Gradient-Based One-Side Sampling (GOSS)** and **Exclusive Feature Bundling (EFB)** to enhance speed and minimize memory usage. LightGBM stands out for its remarkable speed and lower resource consumption, making it an ideal choice for tackling large-scale problems.
5. **CatBoost (Categorical Boosting)**, a specialized boosting algorithm, excels in handling categorical features more effectively compared to other algorithms. It employs an innovative technique called **Ordered Boosting** to manage categorical data, thereby mitigating the risk of overfitting. CatBoost is user-friendly and can accommodate diverse data types without requiring extensive preprocessing.

Each of these algorithms possesses unique strengths and may be preferred in various scenarios, considering factors such as dataset size, data characteristics, and computational resources. They are widely employed in machine learning competitions, industry applications, and academic research due to their capability of generating high-quality predictive models.

9.9.2 ADVANTAGES OF BOOSTED TREES

Tree boosting, a popular tool in machine learning, offer a significant advantage in improving prediction accuracy. By combining multiple decision trees into a powerful model, each tree is designed to address the errors made by the preceding ones. This **incremental error correction** process enables boosted trees to capture complex patterns and relationships in the data that might be overlooked by a single decision tree.

Another key advantage of boosted trees is their robustness to overfitting. **Overfitting** happens when a model learns the noise and details in the training data excessively, resulting in poor performance on new, unseen data. Boosting algorithms are specifically designed to mitigate this risk. For instance, they employ techniques such as **regularization** to control the model's complexity, preventing it from excessively fitting the noise in the data. This approach enhances the models' generalizability and improves their ability to handle novel data effectively.

Boosted trees are highly adaptable and versatile. They can accommodate various data types, including numerical and categorical features, and are applicable to a broad spectrum of problems. Whether we're dealing with classification tasks, where the objective is to categorize data, regression tasks, where the goal is to predict continuous values, or ranking tasks, where the objective is to arrange items based on specific criteria, boosted trees can be tailored to meet the specific requirements of each problem.

These algorithms excel at modeling intricate relationships between features and outcomes. Each tree in the ensemble concentrates on different aspects of the data, enabling the overall model to capture complex interactions and patterns. This makes boosted trees highly effective in comprehending and predicting intricate scenarios.

Another advantage of boosted trees is that they provide valuable insights into the significance of various features. By analyzing how each feature contributes to the model's predictions, we can gain a deeper understanding of the most influential variables. This knowledge is particularly useful for **feature selection**, where we decide which features to include in our model, and for interpreting a model's behavior.

Boosting algorithms also adapt well to different sizes of data. For instance, algorithms like XGBoost and LightGBM are specifically designed to handle large datasets efficiently. They use techniques such as parallel processing and optimized data handling to manage big data, making them suitable for tasks involving large volumes of information.

Some boosting algorithms incorporate built-in mechanisms to handle missing values, simplifying the data preprocessing step. For instance, XGBoost and LightGBM can manage missing values directly during training, thereby reducing the necessity for extensive data **imputation** (the process of filling in missing values in a dataset with estimates based on other available data) and cleaning.

Finally, boosted trees often require less preprocessing of data compared to other methods. For instance, CatBoost can effectively manage categorical features without the need for complex encoding, saving time and effort in data preparation.

Overall, boosted trees offer numerous advantages, including enhanced accuracy, reduced risk of overfitting, flexibility, the ability to handle intricate relationships, and effective management of diverse data types. These qualities make them a potent and extensively utilized tool in machine learning. They can generate high-quality models that perform exceptionally well across a wide range of predictive tasks, making them a valuable asset in data science.

9.9.3 LIMITATIONS OF BOOSTED TREES

While boosted trees are powerful tools in the machine learning toolkit, they come with several notable limitations that users should be aware of. These include susceptibility to overfitting, especially when used with suboptimal datasets; high computational costs; challenges in interpretability; sensitivity to hyperparameters; reliance on data quality; and difficulties with imbalanced datasets. Understanding these limitations is crucial for practitioners who want to deploy boosted trees effectively and responsibly in real-world applications.

One significant limitation of boosted trees is their susceptibility to overfitting. Overfitting happens when a model not only learns the underlying patterns in the training data but also the noise and fluctuations. Boosting, by its design, aims to minimize the errors of previous models by giving more weight to instances that are difficult to predict. While this can result in high accuracy on the training data, it can also lead to an overly complex model, resulting in poor performance on unseen data. This issue becomes particularly pronounced when the training dataset is small or noisy. To address overfitting, practitioners often resort to strategies such as early stopping or careful tuning of hyperparameters, which can complicate the modeling process.

Another limitation of boosting trees is the computational cost involved in training them. Training a boosted model can be resource-intensive, especially when dealing with large datasets or complex trees. Boosting involves sequentially training multiple models, where each new model aims to correct the errors of its predecessors. This iterative process can be time-consuming and requires substantial computational power. Therefore, in situations where quick model training and prediction are crucial, boosted trees may not be the most suitable choice. Furthermore, their complexity can slow down **inference**,[5] which is a critical factor in real-time applications.

Interpretability is another challenge when using boosted trees. While individual decision trees are relatively easy to understand, a boosting ensemble combines multiple trees, making it significantly harder to comprehend the overall decision-making process. This lack of interpretability can be a major drawback in fields where understanding the model's reasoning is crucial, such as healthcare or finance. Stakeholders may find it difficult to trust the predictions of a complex model if they cannot grasp how the model arrived at its conclusions. This can hinder the adoption of boosted trees in certain applications where transparency is essential.

Furthermore, boosted trees can be sensitive to the choice of hyperparameters. Parameters like learning rate, number of estimators, and tree depth significantly impact model performance. Finding the optimal hyperparameter combination involves extensive experimentation and can be challenging for practitioners. A poor

[5]Inference in machine learning refers to the process of using a trained model to make predictions or draw conclusions about new, unseen data based on what the model has learned from the training data.

choice of hyperparameters, on the other hand, can result in suboptimal performance.

In addition to the aforementioned challenges, boosted trees can also be influenced by issues related to data quality. The performance of boosted models heavily relies on the quality of the input data. If the data contains biases, missing values, or irrelevant features, the model may produce unreliable predictions. Since boosting algorithms primarily focus on correcting errors in the training data, any inherent issues in the data can be amplified, resulting in biased outcomes. Consequently, meticulous preprocessing and data cleansing become crucial to ensure reliable results, necessitating an additional step in the modeling process.

Lastly, boosting methods can face challenges when dealing with imbalanced datasets. When one class in the data is significantly underrepresented, boosting algorithms tend to focus on learning from the majority class, which can lead to poor performance for the minority class. While techniques such as re-weighting or employing specific evaluation metrics can help mitigate this issue, they often require additional effort and expertise. In situations where class balance is paramount, users may need to consider alternative approaches or meticulously adjust their boosting models to ensure equitable treatment of all classes.

9.10 COMPARISON: IDENTIFYING HANDWRITING

Let us use random forests to identify handwriting. We had used **MNIST dataset** previously to identify handwriting using K-Nearest Neighbors [Chapter 7]. In this section, we shall examine if random forests can produce better results.

To recap, MNIST, which stands for **Modified National Institute of Standards and Technology**, is a famous dataset used in machine learning and computer vision. It consists of 70,000 images of handwritten digits from 0 to 9, with each image being 28 by 28 pixels. The dataset is divided into 60,000 images for training the model and 10,000 images for testing its accuracy. MNIST is widely used by researchers and developers to train and evaluate algorithms that recognize and classify handwritten numbers, making it a great starting point for learning about image processing and machine learning techniques.

We can access MNIST dataset using the `read_mnist()` function from the `dslabs` package (Irizarry and Gill, 2024) as follows:

```
mnist <- read_mnist()
```

We can explore the structure of the dataset using the `str()` function:

```
str(mnist)
```

```
## List of 2
##  $ train:List of 2
##   ..$ images: int [1:60000, 1:784] 0 0 0 0 0 0 0 0 0 0 ...
##   ..$ labels: int [1:60000] 5 0 4 1 9 2 1 3 1 4 ...
##  $ test :List of 2
##   ..$ images: int [1:10000, 1:784] 0 0 0 0 0 0 0 0 0 0 ...
##   ..$ labels: int [1:10000] 7 2 1 0 4 1 4 9 5 9 ...
```

The dataset is divided into two sub-datasets. The training set comprises 60,000 images and their corresponding 60,000 labels. The test set contains 10,000 images and their corresponding 10,000 labels.

The images are stored as **digital values** for each pixel. These can be converted to 28×28 matrix for interpretation [Figure 9.3].

0	0	0	0	0	0	0	0	0	0	0	0	0	0	0	0	0	0	0	0	0	0	0	0	0	0	0	0
0	0	0	0	0	0	0	0	0	0	0	0	0	0	0	0	0	0	0	0	0	0	0	0	0	0	0	0
0	0	0	0	0	0	0	0	0	0	0	0	0	0	0	0	0	0	0	0	0	0	0	0	0	0	0	0
0	0	0	0	0	0	0	0	0	0	0	0	0	0	0	0	0	0	0	0	0	0	0	0	0	0	0	0
0	0	0	0	0	0	0	0	0	0	0	0	0	0	0	0	0	0	0	0	0	0	0	0	0	0	0	0
0	0	0	0	0	0	0	0	0	0	0	0	3	18	18	18	126	136	175	26	166	255	247	127	0	0	0	0
0	0	0	0	0	0	0	0	30	36	94	154	170	253	253	253	253	253	225	172	253	242	195	64	0	0	0	0
0	0	0	0	0	0	0	49	238	253	253	253	253	253	253	253	253	251	93	82	82	56	39	0	0	0	0	0
0	0	0	0	0	0	0	18	219	253	253	253	253	253	198	182	247	241	0	0	0	0	0	0	0	0	0	0
0	0	0	0	0	0	0	0	80	156	107	253	253	205	11	0	43	154	0	0	0	0	0	0	0	0	0	0
0	0	0	0	0	0	0	0	0	14	1	154	253	90	0	0	0	0	0	0	0	0	0	0	0	0	0	0
0	0	0	0	0	0	0	0	0	0	0	139	253	190	2	0	0	0	0	0	0	0	0	0	0	0	0	0
0	0	0	0	0	0	0	0	0	0	0	11	190	253	70	0	0	0	0	0	0	0	0	0	0	0	0	0
0	0	0	0	0	0	0	0	0	0	0	0	35	241	225	160	108	1	0	0	0	0	0	0	0	0	0	0
0	0	0	0	0	0	0	0	0	0	0	0	0	81	240	253	253	119	25	0	0	0	0	0	0	0	0	0
0	0	0	0	0	0	0	0	0	0	0	0	0	0	45	186	253	253	150	27	0	0	0	0	0	0	0	0
0	0	0	0	0	0	0	0	0	0	0	0	0	0	0	16	93	252	253	187	0	0	0	0	0	0	0	0
0	0	0	0	0	0	0	0	0	0	0	0	0	0	0	0	0	249	253	249	64	0	0	0	0	0	0	0
0	0	0	0	0	0	0	0	0	0	0	0	0	0	46	130	183	253	253	207	2	0	0	0	0	0	0	0
0	0	0	0	0	0	0	0	0	0	0	0	39	148	229	253	253	253	250	182	0	0	0	0	0	0	0	0
0	0	0	0	0	0	0	0	0	0	24	114	221	253	253	253	253	201	78	0	0	0	0	0	0	0	0	0
0	0	0	0	0	0	0	0	23	66	213	253	253	253	253	198	81	2	0	0	0	0	0	0	0	0	0	0
0	0	0	0	0	0	18	171	219	253	253	253	253	195	80	9	0	0	0	0	0	0	0	0	0	0	0	0
0	0	0	0	55	172	226	253	253	253	253	244	133	11	0	0	0	0	0	0	0	0	0	0	0	0	0	0
0	0	0	0	136	253	253	253	212	135	132	16	0	0	0	0	0	0	0	0	0	0	0	0	0	0	0	0
0	0	0	0	0	0	0	0	0	0	0	0	0	0	0	0	0	0	0	0	0	0	0	0	0	0	0	0
0	0	0	0	0	0	0	0	0	0	0	0	0	0	0	0	0	0	0	0	0	0	0	0	0	0	0	0
0	0	0	0	0	0	0	0	0	0	0	0	0	0	0	0	0	0	0	0	0	0	0	0	0	0	0	0

FIGURE 9.3: Pixel values for the digit 5 in MNIST dataset.

The values can also be converted to an image using the `image()` function using the following code (for the first image in the training dataset):

```
image(matrix(mnist$train$images[1,], 28, 28, byrow = TRUE))
```

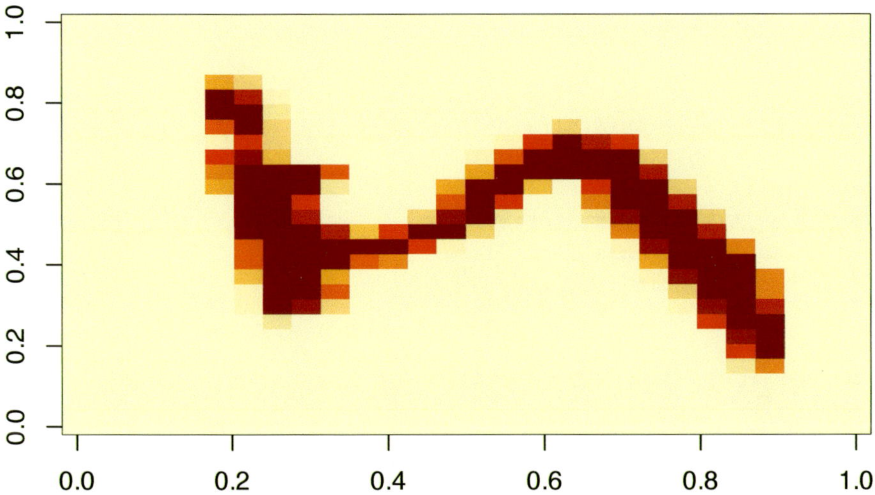

Since the image appears rotated counter-clockwise, we can rotate the same using a `rotate()` function as follows:

```
rotate <- function(x) t(apply(x, 2, rev))
image(
  rotate(matrix(mnist$train$images[1,], 28, 28, byrow = TRUE))
)
```

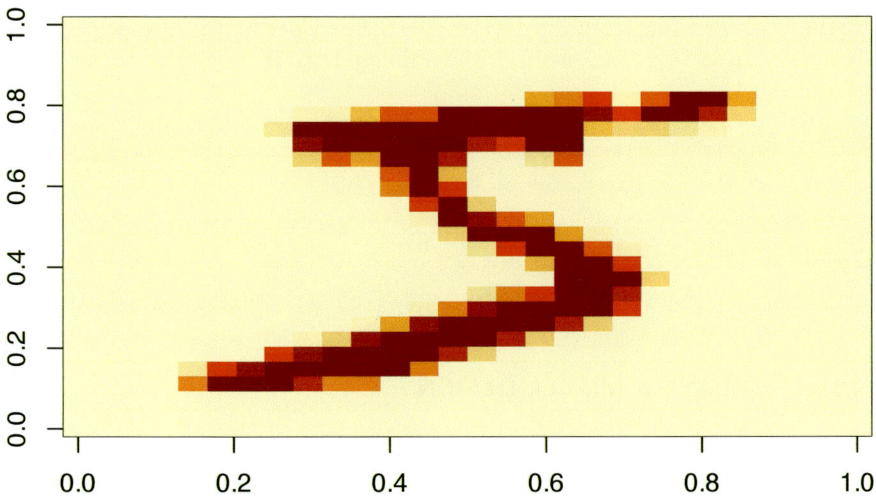

Unlike K-nearest neighbors (KNN), we leverage the entire MNIST dataset here. Random forests employing parallel processing are significantly faster than KNN. We combine images with their corresponding labels in both training and testing datasets by converting their columns into tibbles and binding them together using the `cbind()` function from base R to concatenate columns:

```
im <- as_tibble(mnist$train$images)
lb <- as_tibble(mnist$train$labels)
data_train <- cbind(im, lb)
im <- as_tibble(mnist$test$images)
lb <- as_tibble(mnist$test$labels)
data_test <- cbind(im, lb)
```

We convert the digit values to factors for classification purposes using the `as_factor()` function:

```
data_train$value <- as_factor(data_train$value)
data_test$value <- as_factor(data_test$value)
```

Rest of the steps are similar to what we did before in this chapter with the breast cancer dataset.

9.10.1 CREATE A RECIPE

In the recipe, we input the formula `value ~ .` to predict the value (digit) using all the available variables in the dataset. We further specify that this recipe should be applied to `data_train`, which is the training data. The complete recipe code is provided below:

```
recipe_ml <- recipe(
  value ~ .,
  data = data_train
)
```

9.10.2 CREATE A MODEL DESIGN

To construct the model, we leverage the `rand_forest()` function, which is part of the `parsnip` package within the `tidymodels` library. This function encompasses several crucial arguments. The mode parameter is a character string that determines the prediction outcome mode. It can take on four possible values: "unknown," "regression," "classification," or "censored regression." Depending on the nature of the problem we aim to solve, we can select the appropriate value for mode. In our specific scenario, since we intend to predict the class of the dependent variable, we set mode to "classification."

The engine argument, a character string, specifies the computational engine to be used for model fitting. This enables us to select the underlying method or package for the computations. In our case, we set it to "ranger". Additionally, we set num.threads=detectCores() to enable parallel computing, utilizing all the cores of the computer's processor.

The trees argument specifies the number of trees to be included in the ensemble. An integer value determines the number of individual decision trees that will be built and combined to form the random forest. We will maintain this value at 2000.

The mtry argument, an integer, specifies the number of predictors randomly sampled at each split during the creation of tree models. This parameter controls the randomness and diversity of the trees within the forest. We will maintain its value at 2.

Finally, the min_n argument is an integer that specifies the minimum number of data points needed for a node to be split further. This parameter plays a crucial role in controlling the size of the nodes and, consequently, the complexity of the trees. We will maintain this value at 5.

Following is the complete code that emerges:

```
model_rf <- rand_forest(
  min_n = 5,
  mtry = 2,
  trees = 2000
) %>%
  set_engine("ranger", num.threads=detectCores()) %>%
  set_mode("classification")
```

9.10.3 CREATE A WORKFLOW AND TRAIN IT

We add the recipe and model to the workflow() function, and train the model on data_train dataset:

```
workflow_recipe_model <- workflow() %>%
  add_recipe(recipe_ml) %>%
  add_model(model_rf) %>%
  fit(data_train)
```

9.10.4 ASSESS PREDICTION QUALITY WITH TEST DATA

We can now evaluate the performance of our trained and fitted model by using the test data set. To achieve this, we employ the `augment()` function from the `broom` package in R. The `augment()` function extends the original dataset by including predictions and residuals generated by a statistical model. This visualization approach provides a direct comparison between the actual values and the predicted values, enabling us to assess the model's accuracy. We use the following code:

```
data_test_with_prediction <- augment(
  workflow_recipe_model, data_test
)
```

Next, we construct a **confusion matrix** to understand how many digits were correctly and incorrectly classified. To achieve this, we utilize the `conf_mat()` function. This function, part of the `yardstick` package in R, generates a confusion matrix from model predictions. This matrix aids in evaluating the performance of a classification model by presenting the counts of true positive, true negative, false positive, and false negative predictions. This enables us to assess metrics such as accuracy, precision, recall, and F1 score. We employ the following code:

```
conf_mat(
  data_test_with_prediction,
  dnn = c("Pred", "Truth"),
  truth = value,
  estimate = .pred_class
)
```

```
##       Truth
## Pred     0    1    2    3    4    5    6    7    8    9
##    0   959    0   15    1    1   12   10    2   15   13
##    1     0 1125    6    6    9    7    5   22   12   10
##    2     0    2  956   15    2    4    1   14    7    2
```

```
##      3    2    2   12  946    0   69    0    3   31   13
##      4    0    0    6    0  904    5    3    5   10   17
##      5    2    0    0    2    0  756    6    0    8    0
##      6    8    5    5    2   16   21  932    0   10    1
##      7    2    0   13   16    3    4    0  948   10   25
##      8    7    1   19   15    2    7    1    4  858    5
##      9    0    0    0    7   45    7    0   30   13  923
```

We discover that the majority of the digits are correctly predicted. Next, we calculate the **accuracy metrics**. We define the desired metrics in a metrics set and use the augmented `data_test_with_prediction` dataset to compare the predicted values to the actual results. To achieve this, we employ the following code:

```
metrics_set <- metric_set(
  accuracy,
  sensitivity,
  specificity,
  precision,
  f_meas,
  kap,
  mcc
)
metrics_set(data_test_with_prediction,
               truth = value,
               estimate = .pred_class)
```

```
## # A tibble: 7 x 3
##   .metric     .estimator .estimate
##   <chr>       <chr>          <dbl>
## 1 accuracy    multiclass     0.931
## 2 sensitivity macro          0.929
## 3 specificity macro          0.992
## 4 precision   macro          0.932
## 5 f_meas      macro          0.930
## 6 kap         multiclass     0.923
## 7 mcc         multiclass     0.923
```

We discover that the metrics are significantly better than those obtained using KNN, and more importantly, these superior results were achieved in a fraction of the time required for KNN! This remarkable achievement underscores the advantages of random forests. Consequently, random forests have become widely adopted and utilized in machine learning.

9.11 WHEN TO USE TREE-BASED MODELS, AND WHEN NOT TO

Decision trees are advantageous because they're interpretable and easy to visualize. Their structure allows users to easily follow the decision-making process, making it straightforward to understand how predictions are made. However, they also have limitations. Decision trees can become very large and complex, making them less interpretable if not properly pruned. Furthermore, they're sensitive to changes in the data; small variations can lead to different tree structures, which can affect the stability and reliability of the model.

Despite the challenges, decision trees remain a fundamental technique in machine learning. They form the foundation for more sophisticated ensemble methods like Random Forests and GBM, which combine multiple decision trees to enhance predictive accuracy and robustness. A simple decision tree to understand whether or not to use decision trees is shown in Figure 9.4.

```
[Should I Use a Decision Tree?]
            |
<Need Easy Explanation?>
            |
   +--------+-------------+
   |                      |
   | [Yes]                | [No]
   |                      |
   v                      v
[Use a Decision Tree] <Need to handle complex patterns?>
                          |
             +----------+----------+
             |                     |
             | [Yes]               | [No]
             |                     |
             v                     |
  <Need to handle missing data?>   |
             |                     v
             |        [Do Not Use a Decision Tree]
       +-------+-----------+
       |                   |
       | [Yes]             | [No]
       |                   |
       v                   v
 [Use a Decision Tree]  [Do Not Use a Decision Tree]
```

FIGURE 9.4: A simple decision tree to decide whether or not to use a decision tree.

In the following sub-sections, a list of situations where tree-based models excel and those where they are outshone is presented.

9.11.1 WHEN TO USE TREE-BASED MODELS

1. Interpretability requirements:

- Why: Decision trees are inherently interpretable. They mimic human decision-making and can be easily visualized and understood.
- Example: When we need to explain to stakeholders or regulatory bodies how certain decisions are made (e.g., loan approvals).

2. Handling non-linearity and interactions:

- Why: Trees can capture non-linear relationships and interactions between variables without requiring data transformations.
- Example: Predicting customer churn where complex interactions between different customer attributes affect the outcome.

3. Handling missing data:

- Why: Trees can handle missing data well by making splits based on available data, reducing the need for imputation.[6]
- Example: Predictive modeling with datasets containing many missing values or where data imputation might be challenging.

4. Scalability and speed:

- Why: Random forests and GBM can efficiently handle large datasets with a large number of features in a short time.
- Example: Analyzing large-scale customer data in real-time for personalized recommendations.

5. Feature importance:

- Why: Tree-based models naturally provide a feature importance ranking, which can aid in feature selection and understanding which variables drive predictions.
- Example: Identifying key drivers of sales performance based on customer demographics and purchasing history.

[6] Imputation is a method used to fill in missing values in a dataset with estimated or calculated values based on existing data. It helps to maintain the integrity of the dataset and improve the quality of statistical analyses and machine learning models by replacing gaps in the data with plausible values.

9.11.2 WHEN NOT TO USE TREE-BASED MODELS

1. Linear relationships:

- Why not: If relationships in our data are predominantly linear, simpler models like linear regression might be more appropriate and easier to interpret.
- Example: Predicting temperature based on historical weather data where a linear model might capture the trend adequately.

2. Highly sparse data:

- Why not: Tree-based models might struggle with very sparse datasets where the signal is weak and noise dominates.
- Example: Text classification tasks with extremely sparse feature vectors.

3. Risk of overfitting:

- Why not: Decision trees, especially deep ones, can easily overfit noisy datasets, leading to poor generalization to new data.
- Example: Modeling with a small dataset where a complex tree might perfectly fit the training data but fail to generalize to unseen data.

4. Computationally expensive inference:

- Why not: While training is efficient, inference (making predictions on new data) can be slower for complex tree ensembles compared to simpler models.
- Example: Real-time applications where predictions need to be made rapidly, such as high-frequency trading.

5. Non-interpretable predictions:

- Why not: While individual decision trees are interpretable, ensembles like random forests or GBMs can be harder to interpret due to their complexity.
- Example: Legal or medical applications where predictions need to be explainable and transparent.

9.12 CHAPTER SUMMARY

- Decision trees, visual tools for decision-making, map potential outcomes and consequences.

- They simplify complex decisions into binary choices, creating a tree-like structure with nodes for decisions and branches for outcomes.
- Information gain measures uncertainty reduction after splitting data based on an attribute, guiding decision tree construction.
- Gini impurity quantifies class homogeneity within a node, helping select the best splitting criteria.
- Decision trees are built recursively, and pruned to prevent overfitting and improve generalization.
- Decision trees are unstable, meaning slight changes in training data can lead to different tree structures and predictions. Techniques like tree pruning and ensemble methods like bagging, boosting, and stacking can help mitigate instability.
- Random forests combine multiple decision trees trained on different data and features, averaging predictions for improved accuracy and robustness. This reduces overfitting and enhances generalization, but can be computationally intensive and less interpretable.
- Boosted trees improve efficiency by sequentially improving each decision tree to correct errors.

10 Neural networks and deep learning

Imagine you have a friend named Lily, an avid reader and animal enthusiast. Lily aspires to become an exceptional animal identifier solely based on visual cues.

To aid her in this pursuit, you embark on a creative journey and compile a special book for her. This book is a treasure trove of animal photographs, featuring creatures like majestic lions, awesome elephants, graceful giraffes, and many more. As you present each picture to Lily, you accompany it with its corresponding name.

Initially, the task of memorizing a myriad of animals and their names may seem daunting. However, with consistent reading and exposure to these captivating images, Lily begins to discern subtle characteristics of each animal. She discovers that lions boast magnificent manes, elephants possess enormous trunks, and that giraffes have distinctive spots.

Lily's animal recognition skills improve significantly after reading the book repeatedly and seeing its diverse photographs. Even when presented with animals not as photographs but as sketches, cartoons, or paintings, she effortlessly identifies them as lions, elephants, giraffes, and so on, based on the knowledge she's gained from the book's photographs.

In essence, Lily embodies the concept of a neural network. The book, with its visual representations and animal names, serves as the training data for this neural network. Each time Lily encounters an animal and its corresponding image, she gains valuable insights into recognizing and classifying various species. Similarly, a neural network learns from a vast collection of examples (such as animal photographs) to enhance its ability to recognize and categorize objects.

10.1 NEURONS: BUILDING BLOCKS OF THE NERVOUS SYSTEM

Computer neural networks aim to mimic biological neural networks. In Biology, neurons are the fundamental building blocks of our brain and nervous system. They play a pivotal role in processing and transmitting information. They form an intricate network of communication, enabling us to interact, interpret, and adapt to our surroundings. Neurons facilitate the transmission of information through both electrical impulses and chemical signals, underpinning a range of functions, from basic reflexive actions to complex cognitive tasks like decision-making, memory formation, and problem-solving. This functionality is not only essential for survival but also contributes to the sophistication of human experience.

10.1.1 KINDS OF NEURONS

Neurons are classified into three primary types based on their functions: sensory neurons, motor neurons, and interneurons. **Sensory neurons** are responsible for transmitting information from sensory receptors to the central nervous system, enabling us to perceive external stimuli like light, sound, and touch. They play a pivotal role in informing our brain about the external world, allowing us to respond appropriately. In computer neural networks, we have **input layer** with similar functions.

Motor neurons, on the other hand, transmit signals from the central nervous system to effectors such as muscles and glands, facilitating movement and physical responses. They are instrumental in executing voluntary movements and reflexive actions, like pulling your hand away from a hot surface. They manifest the output of neural processing. In computer neural networks, we have **output layer** with similar functions.

Interneurons are perhaps the most numerous type of neurons. They serve as connectors that facilitate communication between sensory and motor neurons. They are primarily found within the brain and spinal cord, playing a crucial role in interpreting sensory signals and coordinating appropriate responses. This intricate network enables the processing of complex information and the execution of coordinated responses. In computer neural networks, we have **intermediate layers** with similar functions.

10.1.2 STRUCTURE OF NEURONS

The unique structure of a neuron [Fig. 10.1] is perfectly suited to its function of efficiently transmitting information. The **cell body**, also known as **soma**, houses the nucleus and other essential organelles involved in the cell's metabolic activities. **Dendrites**, which extend from the cell body, resemble antennas, receiving signals from other neurons and conveying them to the soma.

The **axon**, a long and slender projection, carries the nerve impulse away from the cell body. Its length can vary significantly, with some axons extending over several centimeters in humans, like the sciatic nerve. The axon is often covered in a **myelin sheath**, a fatty substance that insulates the axon and significantly enhances the speed and fidelity of signal transmission.

At the end of the axon, the **axon terminals**, or **synaptic boutons**, are responsible for transmitting the signal to the next cell. These terminals release **neurotransmitters** into the **synapse**—the junction between neurons—converting the electrical signal into a chemical one and passing it to the adjacent neuron.

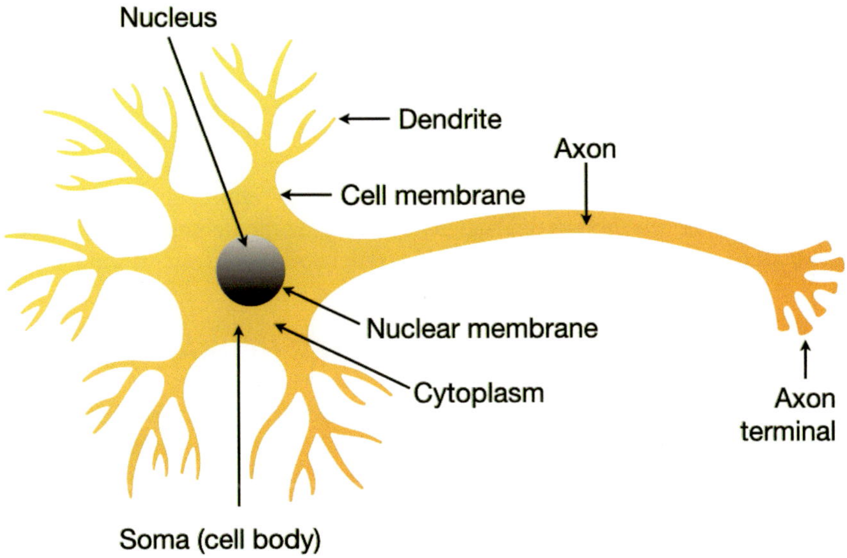

FIGURE 10.1: A schematic of the structure of a neuron.

10.1.3 WORKING MECHANISM OF NEURONS

The intricate process of neuron communication is both highly efficient and complex. It commences when a neuron receives a signal through its dendrites, typically from another neuron or a sensory receptor. These signals manifest as excitatory or inhibitory postsynaptic potentials.

The cell body integrates these signals, and if the cumulative electrical charge surpasses a threshold, an action potential is generated. **Action potential** is a vital electrical event in neuronal communication, and represents a rapid change in the electrical charge across the neuron's membrane. This event facilitates the swift propagation of signals across the nervous system. The rapid and abrupt voltage change across the neuron's membrane traverses the axon to the axon terminals. This electrical impulse serves as the fundamental mechanism for transmitting signals over extended distances within the body.

Upon reaching the axon terminal, the action potential triggers the release of neurotransmitters into the synapse. **Neurotransmitters**, diverse chemical substances, are crucial for communication between neurons across the synapse. Well-known examples include **dopamine**, **serotonin**, and **acetylcholine**, each of which also plays a distinct role in regulating mood, arousal, and cognitive functions. These chemical messengers bind to receptors on the neighboring neuron's dendrites, initiating the subsequent phase of communication. This meticulously coordinated system enables the nervous system to execute its essential functions with remarkable efficiency.

Neuroplasticity is the brain's ability to reorganize and adapt. It is a fundamental characteristic that enables learning and memory. This adaptability is facilitated through the formation of new synaptic connections and the strengthening of existing ones in response to experiences and learning. Training a computer neural network mimics this process of plasticity by changing the weights and biases of individual computer neurons, letting the neural network learn from its inputs.

10.2 FROM NEURONS TO COMPUTER NEURAL NETWORKS

The concept of neural networks in artificial intelligence (AI) was inspired by the biological neural networks found in the human brain. The development of **artificial neural networks** (**ANN**) commenced with the intention of replicating the way human brains process information.

The initial model of a neural network was the "**perceptron**," introduced by Frank Rosenblatt in the late 1950s. Its primary objective was to simulate the thought processes of a human brain by forming connections and learning from them. However, it faced limitations, primarily due to its inability to solve non-linear problems.

A breakthrough came with the introduction of the **backpropagation** algorithm in the 1980s. This algorithm enabled neural networks to adjust the weights of each connection based on the error in their output, leading to the development of more accurate models.

Today's neural networks are far more sophisticated and capable of tackling intricate tasks such as image recognition, language processing, and autonomous driving. There are several types of computer neural networks, each designed for specific tasks:

1. **Feedforward Neural Networks**: This simplest type of ANN is characterized by connections between nodes that don't form cycles. They are employed in applications like facial recognition and basic classification tasks.
2. **Convolutional Neural Networks** (**CNN**): Primarily used for image processing, CNNs employ a mathematical operation called **convolution**. They have revolutionized computer vision, enabling breakthroughs in tasks like object detection and facial recognition.
3. **Recurrent Neural Networks** (**RNN**): Designed to recognize patterns in sequences of data, such as text, genomes, handwriting, or spoken words, these are particularly strong in natural language processing (NLP) tasks.
4. **Deep Neural Networks** (**DNN**): These networks feature multiple layers between the input and output layers, allowing for more complex processing.

Deep learning, a subset of machine learning, relies on DNNs and has been pivotal in advancing AI technologies.

5. **Generative Adversarial Networks** (**GAN**): Comprising two networks, a **generator** and a **discriminator**, GANs are capable of generating new data instances that closely resemble the training data. They are utilized in fields like art, fashion, and gaming for creating novel and unique content.

10.2.1 APPLICATIONS OF NEURAL NETWORKS

Neural networks have found applications in various industries, revolutionizing them by enabling previously unthinkable tasks. In healthcare, they aid in disease diagnosis from medical images, predicting patient outcomes, and personalizing treatment plans. For instance, CNNs can analyze X-rays, MRIs, and other medical imaging formats to detect anomalies like tumors with accuracy comparable to human experts.

In finance, neural networks are employed for fraud detection and algorithmic trading, identifying unusual transaction patterns indicative of fraudulent activity. They also model complex financial data to predict stock movements, enhancing risk management. Neural networks have been crucial in developing autonomous vehicles in the automotive industry. By processing data from cameras, RADAR, LiDAR, and other sensors, they enable vehicles to comprehend their surroundings, anticipate the behavior of other road users, and make real-time driving decisions.

In entertainment, GANs are utilized in gaming and movie production to create realistic textures, environments, characters, and voices. They have contributed to the development of immersive virtual and augmented reality experiences. Neural networks enhance speech recognition, translation, and sentiment analysis in NLP. They are also responsible for the operation of virtual assistants like Siri and Alexa, allowing them to comprehend and respond effectively to human queries.

10.2.2 CHALLENGES AND THE FUTURE

Despite their successes, neural networks encounter several challenges. Firstly, they require substantial amounts of data to function effectively, posing a limitation in domains where data is scarce or costly to acquire. Secondly, training intricate neural networks demands substantial computational resources, often necessitating specialized hardware like Graphics Processing Units (GPUs). Moreover, neural networks are often perceived as "**black boxes**," hindering comprehension of their decision-making processes. This lack of transparency poses challenges in critical applications such as healthcare and finance.

Overcoming these challenges is pivotal for further progress. Researchers are actively investigating novel architectures, including **spiking neural networks** that closely resemble biological neural networks, and **quantum neural networks** that harness the power of quantum computing.

10.3 DEEP LEARNING NETWORKS

Deep learning, a revolutionary approach in AI, mimics the neural activity of the human brain to interpret complex data patterns. As a subset of machine learning, it has significantly advanced capabilities in image and speech recognition.

10.3.1 HISTORICAL BACKGROUND

The journey of deep learning commenced in the 1940s when pioneering researchers like Warren McCulloch and Walter Pitts introduced models inspired by biological neural networks. However, it wasn't until the late 1980s that deep learning experienced significant progress with the advent of **backpropagation** algorithms. The 2000s witnessed a resurgence of interest, fueled by advancements in computational power, data storage capabilities, and the accessibility of substantial datasets.

10.3.2 STRUCTURE OF DEEP LEARNING NETWORKS

Deep learning networks, inspired by the human brain, are composed of multiple layers of artificial neurons. These layers are usually categorized into three primary types [Fig. 10.2].

The **input layer** receives raw data from the external environment. The number of neurons in this layer aligns with the number of features present in the dataset. This is similar to a sensory layer.

Hidden layers are the heart of deep learning, where data processing takes place in multiple dimensions. The depth of a model, determined by the number of hidden layers, allows it to learn increasingly abstract features at higher layers. This is similar to a layer of intermediate neurons.

The **output layer** generates the final output of the network, which can either be a single value or a class, depending on whether the network is solving a regression or classification problem. This is akin to a layer of motor neurons that output final results.

10.3.3 CHALLENGES IN DEEP LEARNING

Deep learning models are known for their ability to learn from vast amounts of data, but this also means they require substantial datasets for effective training. This requirement can be a significant barrier for applications where data is scarce, expensive to collect, or sensitive due to privacy concerns. In many industries, acquiring

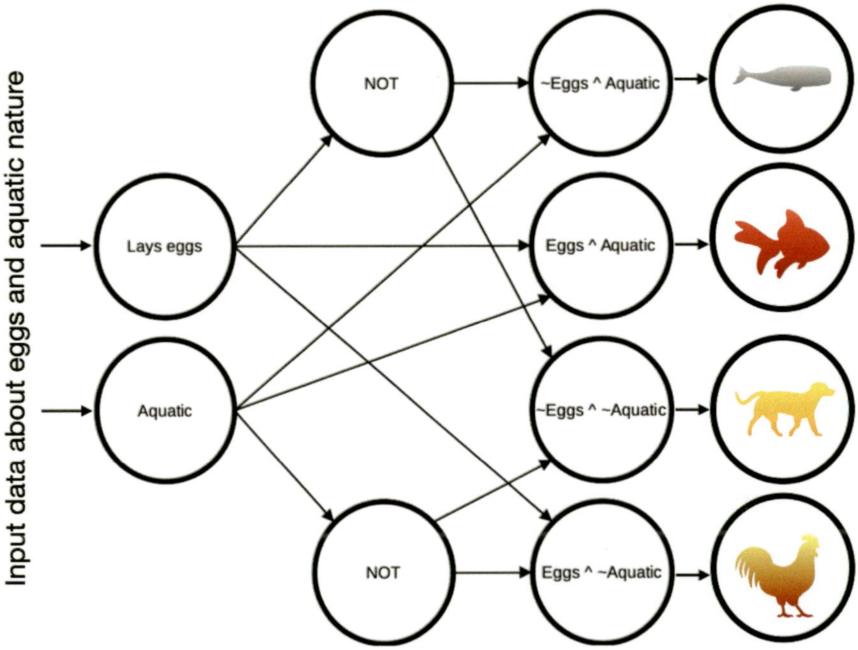

FIGURE 10.2: A schematic of a simple deep learning network with one input, two hidden, and one output layers.

quality and diverse datasets can be a bottleneck, hindering the deployment of deep learning techniques.

Training deep networks is computationally intensive, often requiring significant computational power and specialized hardware such as GPUs or Tensor Processing Units (TPUs). This requirement can lead to high costs in terms of time and financial resources, making it challenging for smaller companies or research teams to fully exploit deep learning. Additionally, the energy consumption of running these computations can be substantial, raising concerns about sustainability and cost-effectiveness.

One of the critical issues with deep learning models is their lack of interpretability. They are often regarded as "**black boxes**," since their decision-making processes are difficult to explain. This opacity can be a significant concern in fields such as healthcare, finance, and law, where **transparency** and understanding are crucial. The inability to interpret model decisions can create trust issues and hinder the adoption of deep learning in these areas.

Despite these challenges, the field of deep learning continues to evolve rapidly. Researchers are actively working on various solutions, such as **transfer learning**, which leverages pre-trained models to be adapted for new tasks with limited data.

This can significantly reduce the data and computational requirements for training new models. **Federated learning** enables models to be trained across decentralized devices while keeping data local, enhancing privacy and reducing the need for centralized data collection. **Neural Architecture Search** (**NAS**) is an automated process that designs neural network architectures that can outperform manually designed ones, potentially leading to more efficient models.

As deep learning becomes more integrated into various aspects of life, ethical considerations are becoming increasingly important. Ensuring that models do not perpetuate existing **biases** in training data is essential to maintain **fairness** and **equity**. Protecting the **privacy** of individuals whose data is used in training is crucial, especially with the growing concerns over **data security**. **Defining responsibility for decisions made by AI systems** is vital, particularly in critical sectors like healthcare and law enforcement.

10.4 TIMELINE OF SIGNIFICANT EVENTS

The history of neural networks is a captivating tale of the evolution of AI, mirroring the intricate workings of the human brain. This journey commenced in the 1940s when neurophysiologist Warren McCulloch and mathematician Walter Pitts laid the groundwork by constructing a model of artificial neurons. Their pioneering work demonstrated how networks of artificial neurons could be used to perform computation in principle, establishing the theoretical foundations that would later support advances in neural network research.

In 1958, Frank Rosenblatt introduced the **Perceptron**, one of the first practical neural network models. Inspired by biological neurons, the Perceptron could learn simple classification tasks. However, it faced limitations—most notably, its inability to solve non-linearly separable problems—which significantly restricted its practical utility.

The 1969 publication of **Perceptrons** by Marvin Minsky and Seymour Papert highlighted these limitations, especially the Perceptron's failure to solve the **XOR problem**. Although the analysis applied only to single-layer networks, it was widely misinterpreted as a critique of neural networks as a whole. This misunderstanding contributed to a decline in interest and funding for neural network research, often referred to as the **AI winter**.

Nevertheless, in 1986, the development and popularization of the **backpropagation** algorithm by David Rumelhart, Geoffrey Hinton, and Ronald Williams reignited interest in neural networks. Backpropagation enabled multi-layer neural networks to adjust weights and biases through gradient-based learning, marking a significant resurgence in neural network research and expanding their practical applications.

The 1990s witnessed remarkable advancements in AI, fueled by enhanced computational power and more sophisticated algorithms. Neural networks gained traction in domains such as speech recognition and image processing. In 1997, Sepp Hochreiter and Jürgen Schmidhuber introduced **Long Short-Term Memory** (**LSTM**) networks, providing an effective solution to the **vanishing gradient problem**.[1] LSTMs significantly improved the performance of RNN[2], enabling breakthroughs in tasks involving sequence learning, such as language modeling and time-series forecasting.

In 2006, Geoffrey Hinton and his team introduced a new approach for training DNN, utilizing **layer-wise pre-training** to initialize deep architectures more effectively. By 2009, advancements in hardware—particularly the widespread adoption of **GPUs**—made it feasible to train DNN efficiently, accelerating progress across a wide range of AI applications.

A pivotal moment arrived in 2012 with the development of **AlexNet**, a deep CNN created by Alex Krizhevsky, Ilya Sutskever, and Geoffrey Hinton. AlexNet's decisive victory in the **ImageNet** competition demonstrated the extraordinary potential of deep learning for complex image recognition tasks and propelled deep learning to the forefront of machine learning research.

Building on this momentum, Ian Goodfellow introduced **GANs** in 2014. GANs opened new possibilities for creative and generative AI by producing highly realistic images, audio, and text, laying the groundwork for applications in art, design, entertainment, and simulation.

In 2017, the introduction of the **Transformer** model revolutionized NLP. With its innovative **self-attention** mechanism,[3] the Transformer enabled more effective modeling of long-range dependencies while improving efficiency. This breakthrough led to state-of-the-art models such as **BERT** and **GPT-3**, which set new benchmarks in translation, summarization, and conversational AI.

As we move further into the **2020s**, neural networks continue to evolve, driving innovations across diverse industries. From **autonomous vehicles** that rely on neural networks for perception and control, to personalized recommendations in e-commerce and streaming, neural networks are profoundly reshaping technology and society. In healthcare, they support early diagnosis, treatment planning, and drug discovery, demonstrating their growing impact on improving lives.

In essence, neural networks have transformed from a theoretical concept into a cornerstone of modern AI, paving the way for ongoing research and developments that promise to further expand their capabilities and applications.

---

[1]The vanishing gradient problem arises when gradients diminish as they propagate backward through layers, impairing the learning of earlier layers.

[2]RNNs are designed to process sequential data by retaining information from previous time steps through feedback loops.

[3]Self-attention allows each element in a sequence to weigh the importance of others, enhancing context understanding in tasks like translation and text generation.

10.5 TRAINING A NEURAL NETWORK

In a neural network, **weights** and **biases** are key numbers learned during training to connect input features to output. Weights show how strong and in what direction one neuron's output affects another. They are changed using methods like back-propagation to reduce prediction mistakes. Biases help the model work better by allowing a neuron to activate even if all inputs are zero. Weights and biases together help the neural network learn complex patterns and make accurate predictions.

Training neural networks aims at adjusting the weights and biases to minimize the difference between predicted and actual outputs. This process is essential for the network to learn from data and enhance its performance. **Backpropagation**, a pivotal technique, facilitates this learning process by enabling the network to adjust its parameters, including weights and biases, based on errors in its predictions. The objective is to minimize the **loss function** by iteratively updating the network's parameters, enabling it to learn and make more accurate predictions. This process involves two main phases: the forward pass and the backward pass.

10.5.1 THE FORWARD PASS

The **forward pass**, the initial phase of the network's operation, commences with the input of data, such as an image, a sequence of text, or a set of numerical values. This data is then transmitted through the network's layers, each comprising neurons that execute a series of calculations. These calculations involve multiplying the input by a weight, adding a bias, and subsequently applying an activation function to generate the output.

Each layer extracts distinct features from the input data, enabling the network to progressively develop more abstract representations of the data. In the final layer, the network generates its output prediction. For instance, in a classification task, the output could manifest as a probability distribution encompassing various classes (e.g., the probability of an image being classified as a cat or a dog).

Following the generation of the output, it is subsequently compared to the actual label using a loss function. The **loss function** quantifies the disparity between the predicted value and the true value. For instance, in binary classification scenarios, a commonly employed loss function is **binary cross-entropy**. In regression tasks, **mean squared error** (**MSE**) may be utilized to calculate the difference between the predicted and actual values.

10.5.2 THE BACKWARD PASS (BACKPROPAGATION)

After calculating the error (or loss) at the output layer, the network initiates the **backward pass** to modify its parameters. This is where **backpropagation** enters the scene. Backpropagation is a procedure for efficiently computing the gradient of the loss function with respect to every parameter (weight and bias) in the network. These gradients are then used by optimization algorithms such as gradient descent.

Backpropagation begins by computing the error at the output layer, based on the difference between the predicted output and the true label as measured by the loss function. For example, if a network predicts a high probability for the "dog" class but the actual class is "cat," the loss function quantifies the mismatch.

Once the loss is computed at the output layer, backpropagation propagates the gradients backward through the network. Using the **chain rule**,[4] backpropagation determines how changes in each intermediate value affect the final loss. This allows the algorithm to compute how much each weight and bias contributed to the overall error.

For each layer, the gradient of the loss with respect to the weights and biases is calculated. This involves:

1. **The derivative of the activation function**, which indicates how sensitive a neuron's output is to changes in its input.
2. **The gradient passed from the next layer**, which indicates how sensitive the overall loss is to that neuron's output.

Together, these components yield the gradient for each neuron and its parameters.

Once gradients are computed, the weights are updated using an optimization algorithm. The simplest such algorithm is **gradient descent**. In gradient descent, weights are updated by moving them in the opposite direction of the gradient, thereby reducing the loss. The update rule for each weight W is:

$$W_{\text{new}} = W_{\text{old}} - \eta \cdot \frac{\partial L}{\partial W}$$

where:

- W_{new} is the updated weight,
- W_{old} is the current weight,

[4]The chain rule in calculus is a formula used to determine the derivative of a composite function. It states that the derivative of a composite function (f(g(x))) is equal to the derivative of the outer function (f'(g(x))) multiplied by the derivative of the inner function (g'(x)). For instance, let's consider the function $f(x) = (2x + 3)^2$. In this case, the outer function is squaring, and the inner function is $2x + 3$. To find the derivative of f(x), we first differentiate the outer function to get $2 \times (2x + 3)$. Then, we multiply this result by the derivative of the inner function, which is 2. Therefore, $f'(x) = 4(2x + 3)$.

- η is the learning rate (a hyperparameter controlling update size), and
- $\frac{\partial L}{\partial W}$ is the gradient of the loss with respect to the weight.

The **learning rate** is a crucial **hyperparameter**. If it is too high, updates may overshoot and destabilize training. If it is too low, training may become slow, and the optimizer may struggle to move through flat or poorly conditioned regions of the loss landscape.

10.5.3 ACTIVATION FUNCTIONS

Activation functions are crucial components of neural networks, introducing non-linearity that enables them to learn complex patterns. Without activation functions, neural networks would essentially perform only linear transformations, limiting their ability to model intricate relationships.

Some commonly used activation functions include:

1. **Sigmoid**: The sigmoid function outputs values between 0 and 1, making it suitable for binary classification tasks where the output represents a probability. Its formula is:

$$\sigma(z) = \frac{1}{1 + e^{-z}}$$

It is important to remember that the sigmoid function faces the vanishing gradient problem when the inputs are very large or very small, which can slow down training.

2. **ReLU (Rectified Linear Unit)**: ReLU is widely used due to its computational efficiency and mitigates the vanishing gradient problem for positive inputs. It outputs the input directly if it's positive and zero if it's negative:

$$\text{ReLU}(z) = \max(0, z)$$

While ReLU is effective, it can cause some neurons to become "inactive" and stop learning, a problem known as the "**dying ReLU**" problem.

3. **Leaky ReLU**: Leaky ReLU is a variation of ReLU designed to address the "dying ReLU" problem. It allows a small negative slope for inputs less than zero, preventing neurons from becoming inactive:

$$\text{Leaky ReLU}(z) = \begin{cases} z & \text{if } z > 0 \\ \alpha z & \text{if } z \leq 0 \end{cases}$$

where α is a small constant (usually 0.01).

4. **Softmax**: The softmax function is commonly used in the output layer of classification networks, particularly for multi-class classification. It transforms raw outputs into a probability distribution, ensuring that all output values sum to 1:

$$\sigma(z)_i = \frac{e^{z_i}}{\sum_{j=1}^{K} e^{z_j}}$$

where K is the number of classes, and z_i is the output for class i.

10.5.4 OPTIMIZERS

Optimizers are algorithms used to update the weights of a neural network during training. They determine how the parameters should change based on the gradients computed during backpropagation. While basic gradient descent is the simplest approach, several more advanced optimizers have been developed to improve training speed, stability, and performance.

1. **Stochastic Gradient Descent (SGD)**:
   In SGD, weights are updated after each training example or a small batch of examples, rather than using the entire dataset. This leads to faster updates and injects beneficial noise into the training process. The noise can help the optimizer escape *saddle points* and *sharp minima*, which are common in high-dimensional loss landscapes, although SGD does introduce more variability in the updates.
2. **Momentum**:
   Momentum accelerates learning by accumulating a "velocity" vector that reflects past gradients. This helps push the parameters in consistent directions and reduces oscillations in regions where gradients change rapidly. The classical momentum update is:
   $$v_{\text{new}} = \beta v_{\text{old}} + \frac{\partial L}{\partial W}$$
   $$W_{\text{new}} = W_{\text{old}} - \eta\, v_{\text{new}}$$
   Here, v_{old} is the previous velocity, and β (commonly 0.9) controls how strongly past gradients influence the current update.
3. **Adam (Adaptive Moment Estimation)**:
   Adam is an advanced optimizer. It maintains exponentially weighted moving averages of the gradients and the squared gradients, and it applies bias correction to both estimates. By adapting the learning rate for each parameter individually, Adam performs well on problems with noisy gradients or

sparse features. Due to its efficiency and reliability, it is one of the most widely used optimizers in deep learning.

10.5.5 SUMMARY OF THE TRAINING PROCESS

Backpropagation is a fundamental algorithm that allows neural networks to compute the gradients needed for learning. During training, the network performs a **forward pass** to generate predictions and compute the loss, followed by a **backward pass** where errors are propagated through the layers to determine how each weight and bias contributed to the loss. An optimizer then uses these gradients to update the parameters in a direction that minimizes the loss.

The effectiveness of this process depends heavily on key design choices such as the activation function, loss function, and optimizer. Repeating forward and backward passes over many iterations (epochs) gradually improves the network's performance, enabling it to learn complex patterns and relationships from data.

Enhancements such as momentum, adaptive learning-rate methods, and advanced activation functions further improve the efficiency and stability of training. Together with backpropagation and optimization, these techniques have enabled neural networks to achieve state-of-the-art results in tasks such as image and speech recognition, NLP, and game-playing AI. Understanding how backpropagation works is therefore essential for building and optimizing modern neural networks.

10.6 THE BOSTON DATASET

We looked at the Boston housing dataset while discussing polynomial regression. To recapitulate, the Boston Housing Dataset is a seminal dataset in machine learning and statistics. It offers valuable historical insights into various characteristics of homes in Boston. It is widely utilized for predicting the historical median value of owner-occupied homes across diverse neighborhoods.

The dataset includes various features that describe aspects of the housing market and neighborhood characteristics. These features include the per capita crime rate in each town, the percentage of residential land zoned for large lots, the proportion of non-retail business acres per town, and a dummy variable indicating proximity to the Charles River. Other features include nitric oxides concentration, the average number of rooms per dwelling, the proportion of owner-occupied units built before 1940, weighted distances to five major Boston employment centers, an index of accessibility to radial highways, the full-value property tax rate per $10,000, the pupil-teacher ratio by town, and the percentage of the population with a low socioeconomic status.

The dataset's target variable is the median value of owner-occupied homes, expressed in thousands of dollars. The primary objective of utilizing this dataset is to construct predictive models capable of estimating home values based on the offered features. It serves as an excellent illustration for practicing linear regression, polynomial regression, and other machine learning techniques, providing a balanced combination of simplicity and practical relevance.

Originally published in 1978 by David Harrison Jr. and Daniel L. Rubinfeld, the Boston Housing dataset has become a widely recognized benchmark in the field of machine learning. Its inclusion in data science libraries and educational materials makes it a frequently used tool for teaching and learning.

However, it's crucial to be aware of the dataset's historical context. Since it reflects conditions from the 1970s, certain features may be biased by historical events. When using this dataset, especially in contemporary contexts, it's essential to consider these ethical considerations and comprehend how they might influence the analysis and conclusions.

10.6.1 USING THE DATASET

We begin by (installing and) loading all the required packages and libraries:

```
if (!require("tidymodels")){
  install.packages("tidymodels")
}
library(tidymodels)

if (!require("tidyverse")){
  install.packages("tidyverse")
}
library(tidyverse)

if (!require("janitor")){
  install.packages("janitor")
}
library(janitor)

if (!require("nnet")){
  install.packages("nnet")
}
library(nnet)

if (!require("brulee")){
```

```
  install.packages("brulee")
}
library(brulee)

library(MASS)
```

The MASS (Modern Applied Statistics with S) package (Venables and Ripley, 2002) offers access to the Boston dataset, which contains housing values in the suburbs of Boston. Let's delve into this dataset. We commence by examining the top few values using the `glimpse()` function:

```
Boston %>%
  glimpse(width = 50)
```

```
## Rows: 506
## Columns: 14
## $ crim    <dbl> 0.00632, 0.02731, 0.02729, 0.032~
## $ zn      <dbl> 18.0, 0.0, 0.0, 0.0, 0.0, 0.0, 1~
## $ indus   <dbl> 2.31, 7.07, 7.07, 2.18, 2.18, 2.~
## $ chas    <int> 0, 0, 0, 0, 0, 0, 0, 0, 0, 0, 0,~
## $ nox     <dbl> 0.538, 0.469, 0.469, 0.458, 0.45~
## $ rm      <dbl> 6.575, 6.421, 7.185, 6.998, 7.14~
## $ age     <dbl> 65.2, 78.9, 61.1, 45.8, 54.2, 58~
## $ dis     <dbl> 4.0900, 4.9671, 4.9671, 6.0622, ~
## $ rad     <int> 1, 2, 2, 3, 3, 3, 5, 5, 5, 5, 5,~
## $ tax     <dbl> 296, 242, 242, 222, 222, 222, 31~
## $ ptratio <dbl> 15.3, 17.8, 17.8, 18.7, 18.7, 18~
## $ black   <dbl> 396.90, 396.90, 392.83, 394.63, ~
## $ lstat   <dbl> 4.98, 9.14, 4.03, 2.94, 5.33, 5.~
## $ medv    <dbl> 24.0, 21.6, 34.7, 33.4, 36.2, 28~
```

We find that this dataset contains 506 rows and 14 columns. The details of the columns as read from the help file using `?Boston` are as follows:

1. `crim`: per capita crime rate by town.
2. `zn`: proportion of residential land zoned for lots over 25,000 sq.ft.
3. `indus`: proportion of non-retail business acres per town.
4. `chas`: Charles River dummy variable (= 1 if tract bounds river; 0 otherwise).
5. `nox`: nitrogen oxides concentration (parts per 10 million).
6. `rm`: average number of rooms per dwelling.

7. age: proportion of owner-occupied units built prior to 1940.
8. dis: weighted mean of distances to five Boston employment centers.
9. rad: index of accessibility to radial highways.
10. tax: full-value property-tax rate per $10,000.
11. ptratio: pupil-teacher ratio by town.
12. black: $1000(Bk - 0.63)^2$ where Bk is the proportion of blacks by town.
13. lstat: lower status of the population (percent).
14. medv: median value of owner-occupied homes in $1000s.

In this chapter, we shall explore if we can predict median value of owner-occupied homes ($medv$) using various parameters.

10.7 TIDYMODELS EXAMPLE

We will use tidymodels (Kuhn and Wickham, 2020) to use neural networks on the Boston dataset. The biggest benefit of using tidymodels is that once the steps have been coded, moving from one machine learning algorithm to another becomes a breeze. Throughout this book, you will find that we use very similar codes with various machine learning algorithms.

10.7.1 LOAD DATA

We start by loading the necessary information into a data file. This is also an opportune moment to convert the raw data into a tibble format using the as_tibble() function, which enhances visualization capabilities. Additionally, we remove any missing values (NA) from the dataset.

We employ the pipe operator %>% from tidyverse to keep our code intuitive and clean:

```
data <- Boston %>%
  na.omit() %>%
  as_tibble()
```

Let us confirm the operations by examining the data variable. Since we've converted our data into a tibble, there is no longer a need to use the glimpse() function:

```
data
```

```
## # A tibble: 506 x 14
##        crim    zn indus  chas   nox    rm   age   dis   rad
##       <dbl> <dbl> <dbl> <int> <dbl> <dbl> <dbl> <dbl> <int>
##  1 0.00632  18     2.31     0 0.538  6.58  65.2  4.09     1
##  2 0.0273    0     7.07     0 0.469  6.42  78.9  4.97     2
##  3 0.0273    0     7.07     0 0.469  7.18  61.1  4.97     2
##  4 0.0324    0     2.18     0 0.458  7.00  45.8  6.06     3
##  5 0.0690    0     2.18     0 0.458  7.15  54.2  6.06     3
##  6 0.0298    0     2.18     0 0.458  6.43  58.7  6.06     3
##  7 0.0883   12.5   7.87     0 0.524  6.01  66.6  5.56     5
##  8 0.145    12.5   7.87     0 0.524  6.17  96.1  5.95     5
##  9 0.211    12.5   7.87     0 0.524  5.63 100    6.08     5
## 10 0.170    12.5   7.87     0 0.524  6.00  85.9  6.59     5
## # i 496 more rows
## # i 5 more variables: tax <dbl>, ptratio <dbl>,
## #   black <dbl>, lstat <dbl>, medv <dbl>
```

We observe that the number of rows remains unchanged, indicating the absence of any rows containing missing values.

10.7.2 SPLIT DATA INTO TRAINING AND TESTING

We partition the data into training and testing sets using the `initial_split()` function from the `rsample` package, a component of `tidymodels`. We retain 70% of the data for training and 30% for testing. These sets are then extracted using `training()` and `testing()` functions:

```
# Split data into training and testing sets
set.seed(2025)
split_7030 <- initial_split(data, prop = 0.7)
data_train <- training(split_7030)
data_test <- testing(split_7030)
```

In this exemplar case, we have set the seed to 2025. `set.seed` is a function used in programming to initialize the random number generator to a specific state, ensuring that the results of random operations are reproducible. By setting the seed, we ensure that the sequence of random numbers generated is consistent each time our code is executed. This consistency facilitates reliable and reproducible results in experiments and simulations, particularly in troubleshooting scenarios. In practice, this line will be removed or commented out upon code deployment, at which point it is not required.

10.7.3 CREATE A RECIPE

In the recipe, we input the formula as `medv ~ .` since we wish to predict the median value of the house using all the available variables in the dataset. We further specify that this recipe should be used on `data_train`, i.e. the training data. We also normalize all the predictor variables. The complete recipe code is provided below:

```
recipe_ml <- recipe(
  medv ~ .,
  data = data_train
) %>%
  step_normalize(
    all_predictors()
  )
```

10.7.4 CREATE A MODEL DESIGN

In tidymodels, the `nnet` (neural net) engine (Venables and Ripley, 2002) is used to create neural networks. We define a `mlp` (multi-layer perceptron) model using the `nnet` engine. To begin with, we specify that we need 20 hidden units, and 10,000 epochs. An **epoch** in neural networks refers to a complete pass through the entire training dataset during the learning process. We also set `penalty` to 0 to specify that we do not wish to apply any regularization at this stage.

Since we wish to estimate the values of median value using independent variables, we set the mode to `regression`. The code that follows demonstrates the creation of the model design:

```
model_nnet <- mlp(
  hidden_units = 20,
  epochs = 10000,
  penalty = 0
) %>%
  set_engine("nnet") %>%
  set_mode("regression")
```

10.7.5 CREATE A WORKFLOW

We add the recipe and model to the `workflow()` function, and train the model on `data_train` dataset using `fit()` function:

```
workflow_recipe_model <- workflow() %>%
  add_recipe(recipe_ml) %>%
  add_model(model_nnet) %>%
  fit(data_train)
```

10.7.6 ASSESS PREDICTION QUALITY WITH TEST DATA

We can now utilize our trained and fitted model to predict median value of homes using the test data set to assess model efficacy. To accomplish this, we employ the augment() function. The augment() function, originating from the broom package in R, appends predictions and residuals from a statistical model to the original dataset. This facilitates the evaluation of a model's fit by juxtaposing actual and predicted values. We use the following code:

```
data_test_with_prediction <- augment(
  workflow_recipe_model, data_test
)
```

Let us now have a glimpse of the augmented file, data_test_with_prediction. We select the columns that we are interested in, and consider the first few rows using the head() function:

```
data_test_with_prediction %>%
  dplyr::select(
    medv,
    .pred
  ) %>%
  head()
```

```
# A tibble: 6 x 2
   medv .pred
  <dbl> <dbl>
1  24    23.5
2  34.7  28.8
3  16.5  16.5
4  23.1  21.2
5  18.2  18.9
6  18.4  13.7
```

We find that the predicted values (.pred) are in the general vicinity of the actual measured values (medv), albeit with some difference.

Additionally, we can compute accuracy metrics. **Accuracy metrics** in machine learning assess a model's performance by comparing its predictions to the actual outcomes. We utilize the augmented data_test_with_prediction dataset to evaluate the estimated values against the actual truth using the metrics() function as follows:

```
metrics(
  data_test_with_prediction,
  truth = medv,
  estimate = .pred
)
```

```
## # A tibble: 3 x 3
##   .metric .estimator .estimate
##   <chr>   <chr>          <dbl>
## 1 rmse    standard       7.96
## 2 rsq     standard       0.482
## 3 mae     standard       4.92
```

We find that the root mean squared error (RMSE) is around 7.96 thousand dollars (it was 5.36 thousand dollars in the case of polynomial regression), the mean absolute error is around 4.92 thousand dollars (it was 3.7 thousand dollars in the case of polynomial regression), and we have been able to explain around 48.2% of the results (R squared = 0.482), whereas we were able to explain around 66.5% of the results using polynomial regression. This suggests that our neural network needs more work!

10.8 HYPERPARAMETER TUNING

We selected 20 hidden units for our neural network. But the number of hidden units could as well have been 30, 40, or some higher value. As we increase the number of hidden units, the neural network can get a better fit to the data points. But this benefit also has a limit, after which a very high number makes for a very complicated neural network that can accommodate the noise in the dataset, reducing the predictive powers of the model. But what should be the best value for these hyperparameters? One way to find that is through hyperparameter tuning.

Hyperparameter tuning is the process of adjusting the settings of a machine learning model (called **hyperparameters**) to improve its performance. In this case, we can create a grid with different values of hidden_units, use them to fit models, and test those models. The value of `hidden_units` that has the lowest error, or that is best able to explain the majority of the data is then taken as the best value.

We can also employ hyperparameter tuning to several hyperparameters at once. In this section, we'll employ a grid of hyperparameters and tune the workflow to compute the best values of `hidden_units` and `dropout`. **Dropout** is a **regularization** technique in neural networks that randomly sets a portion of the activations (usually in hidden layers) to zero during training to reduce overfitting. Most of the steps are similar to what we did before, with the addition of `tune()` being used to hold the value of `hidden_units` and `dropout` in the model design, and `tune_grid()` function used to tune the workflow. We'll begin with the splitting step, and continue till the computation of accuracy metrics.

10.8.1 SPLIT DATA INTO TRAINING AND TESTING

As before, we split data into training and testing sets using `initial_split()` function from `rsample` package, a constituent of `tidymodels`. We keep 70% of data for training, and 30% for testing purposes, and extract these two sets using `training()` and `testing()` functions.

```
# Split data into training and testing sets
set.seed(2025)
split_7030 <- initial_split(data, prop = 0.7)
data_train <- training(split_7030)
data_test <- testing(split_7030)
```

As before, we have set the seed to 2025 to enable comparison with the previous results. The `set.seed` function in programming initializes the random number generator to a specific state, ensuring reproducibility of random operations. By setting the seed, we guarantee that the sequence of random numbers generated is consistent each time we run our code. This consistency is crucial for achieving repeatable results in experiments and simulations, particularly for troubleshooting. However, when we deploy the code, this line will be removed or commented out, since at that point it will not be required.

10.8.2 CREATE A RECIPE

As before, we input the formula as `medv ~ .` since we wish to predict the median value of the house using all the available variables in the dataset. We further specify that this recipe should be used on `data_train`, i.e. the training data. We also normalize all the predictor variables. The complete recipe code is provided below:

```
recipe_ml <- recipe(
  medv ~ .,
  data = data_train
) %>%
  step_normalize(
    all_predictors()
  )
```

10.8.3 CREATE A MODEL DESIGN

We define a `mlp` (multi-layer perceptron) model, this time using the `brulee` engine. We specify `hidden_units` to be `tune()`, which is a placeholder that will be replaced by the value(s) of `hidden_units` that we get after tuning the model. We also specify `dropout` to be `tune()`, and set 10,000 epochs. An epoch in neural networks refers to a complete pass through the entire training dataset during the learning process. We also set `penalty` to 0 to specify that we do not wish to apply any other regularization at this stage.

In tidymodels, the `brulee` package (Kuhn and Falbel, 2025) contains several basic modeling functions that use the `torch` package (Falbel and Luraschi, 2025) infrastructure, permitting fine tuning of several machine learning algorithms. Since we wish to make a numerical estimation of the median value using independent variables, we set the mode to `regression`. The code that follows demonstrates the creation of model design:

```
model_brulee <- mlp(
  hidden_units = tune(),
  dropout = tune(),
  epochs = 10000,
  penalty = 0
) %>%
  set_engine("brulee") %>%
  set_mode("regression")
```

Key tuning parameters provided by the `brulee` package are as follows:

1. **Hidden units** determine the number of neurons in each hidden layer of a multi-layer perceptron (MLP). Adjusting this parameter influences the model's ability to learn intricate features from the data. In the `brulee` framework, we can specify `hidden_units` to define the architecture of the

MLP, ensuring a balance between model complexity and the risk of overfitting.

2. **Dropout** is a regularization technique that involves randomly selecting and ignoring neurons during training. This helps prevent overfitting by encouraging the network to rely on alternative pathways. In `brulee`, we can specify `dropout` to set the dropout rate, which typically ranges from 0 to 1. A value of 0 indicates no dropout, while a value of 1 means dropping all neurons.
3. **Epochs** refer to the number of times the entire dataset is traversed forward and backward through the neural network during training. Increasing the number of epochs allows the model to learn more from the data, but it can also lead to overfitting if not monitored closely. In `brulee`, we can specify `epochs` to control the number of training iterations.
4. **Penalty** is a regularization technique that involves penalizing large weights in the model. This helps prevent overfitting by discouraging the model from becoming excessively complex. In `brulee`, we can specify `penalty` to apply regularization during training. Additionally, we can adjust the regularization strength (`lambda1`, `lambda2`) to strike a balance between model complexity and generalization. We discuss these parameters in detail in the next chapter.

10.8.4 CREATE A WORKFLOW

We add the recipe and model to the `workflow()` function, but leave training for later. Following is the code at this stage:

```
workflow_recipe_model <- workflow() %>%
  add_recipe(recipe_ml) %>%
  add_model(model_brulee)
```

10.8.5 CREATE A HYPER-PARAMETER GRID

We create a hyper-parameter grid of values to be used in `hidden_units` and `dropout` using the `expand.grid()` function to create a grid of values. Here's the code we use:

```
hyper_parameter_grid <- expand.grid(
  hidden_units=c(10, 20, 50, 100),
  dropout=c(0, 0.25)
)
```

We check the values in the hyper_parameter_grid variable.

```
hyper_parameter_grid
```

```
##   hidden_units dropout
## 1           10    0.00
## 2           20    0.00
## 3           50    0.00
## 4          100    0.00
## 5           10    0.25
## 6           20    0.25
## 7           50    0.25
## 8          100    0.25
```

10.8.6 CREATE RESAMPLES FOR CROSS-VALIDATION

V-fold cross-validation, also known as **k-fold cross-validation**, is a method used to assess the performance of a model by dividing the data into V groups of roughly equal size, which are called "**folds**." In each iteration of the analysis, V-1 folds are used to train the model, while the remaining fold is used to test its performance. In its basic form, without any repeats, the number of resamples is equal to V, which provides a thorough evaluation of the model's capabilities.

The vfold_cv() function performs V-fold cross-validation on a data frame. It takes two arguments: v, which determines the number of partitions, and repeats, which specifies how many times the partitioning should be repeated. By default, v is set to 10 and repeats is set to 1, which corresponds to a standard cross-validation process. The function also supports stratified sampling, where a variable in the data frame is used to ensure that each resample maintains a similar distribution across different categories. This is achieved by using the strata argument.

The vfold_cv() function offers several customization options. The data parameter specifies the input data frame, while the v parameter determines the number of partitions for the data set. The strata parameter enables the use of a variable for stratified sampling, while the breaks parameter specifies the number of bins for numeric stratification variables. The pool parameter assesses the proportion of data used to evaluate if a group is too small and should be combined with another group. It is strongly advised to maintain the default value of 0.1 for this argument to avoid the risks of stratifying groups that are too small. The dots argument, ..., is reserved for future extensions and should remain empty when using the function.

In the vfold_cv() function, we specify that the strata should be generated using the data_train dataset. We want to create 10 such strata, and medv is the variable in the dataset that we'll use for stratified sampling. To accomplish this, we execute the following code:

```
set.seed(2025)
folds_for_tuning <- vfold_cv(
  data_train,
  v = 10,
  strata = medv
)
```

We can use the following code to observe the number of training and testing data in each of the cross-validation folds:

```
folds_for_tuning
```

```
## #  10-fold cross-validation using stratification
## # A tibble: 10 x 2
##    splits           id
##    <list>           <chr>
##  1 <split [317/37]> Fold01
##  2 <split [318/36]> Fold02
##  3 <split [318/36]> Fold03
##  4 <split [318/36]> Fold04
##  5 <split [318/36]> Fold05
##  6 <split [318/36]> Fold06
##  7 <split [319/35]> Fold07
##  8 <split [319/35]> Fold08
##  9 <split [320/34]> Fold09
## 10 <split [321/33]> Fold10
```

10.8.7 TUNE THE WORKFLOW

To find out the best values of hyperparameters, we tune the workflow with all the values in the hyper-parameter grid. To do this, we utilize the `tune_grid()` function which tests different sets of model settings to see how well each one performs using various data samples, measuring metrics like accuracy or error.

We pass our workflow `workflow_recipe_model` to the `tune_grid()` function, along with `folds_for_tuning` as `resamples`, `hyper_parameter_grid` as `grid`, and specify the metrics on which to optimize using the following code:

```
tune_results <- tune_grid(
  workflow_recipe_model,
  resamples = folds_for_tuning,
  grid = hyper_parameter_grid,
  metrics = metric_set(rmse, rsq, mae)
)
```

We save the results of tuning to `tune_results`.

10.8.8 EXTRACT THE BEST VALUE(S) OF HYPER-PARAMETER(S)

We can use the `show_best()` function from the `tune` package to extract the optimal hyperparameter values. For instance, to find the best values for RMSE, we can run the following code:

```
show_best(
  tune_results,
  metric = "rmse"
)
```

```
# A tibble: 5 x 8
  hidden_units dropout .metric .estimator  mean     n
         <dbl>   <dbl> <chr>   <chr>      <dbl> <int>
1           50    0    rmse    standard    3.72    10
2           20    0    rmse    standard    4.08    10
3           10    0    rmse    standard    4.08    10
4          100    0    rmse    standard    4.25    10
5          100    0.25 rmse    standard    5.74    10
# i 2 more variables: std_err <dbl>, .config <chr>
```

We discover that the best value for rmse is achieved when `hidden_units` is set to 50, and `dropout` is set to 0.

To identify the most suitable values for rsq, we can utilize the following code:

```
show_best(
  tune_results,
  metric = "rsq"
)
```

```
# A tibble: 5 x 8
  hidden_units dropout .metric .estimator  mean     n
         <dbl>   <dbl> <chr>   <chr>      <dbl> <int>
1           50    0    rsq     standard   0.840    10
2           20    0    rsq     standard   0.809    10
3           10    0    rsq     standard   0.805    10
4          100    0    rsq     standard   0.789    10
5          100    0.25 rsq     standard   0.639    10
# i 2 more variables: std_err <dbl>, .config <chr>
```

We discover that the best value for rsq is achieved when `hidden_units` is set to 100, and `dropout` is set to 0.

To identify the most suitable values for mae, we can utilize the following code:

```
show_best(
  tune_results,
  metric = "mae"
)
```

```
# A tibble: 5 x 8
  hidden_units dropout .metric .estimator  mean     n
         <dbl>   <dbl> <chr>   <chr>      <dbl> <int>
1           50    0    mae     standard    2.60    10
2           20    0    mae     standard    2.78    10
3          100    0    mae     standard    2.83    10
4           10    0    mae     standard    2.85    10
5          100    0.25 mae     standard    4.05    10
# i 2 more variables: std_err <dbl>, .config <chr>
```

We discover that the best value for mae is achieved when `hidden_units` is set to 100, and `dropout` is set to 0.

We can use the `select_best()` function, a component of the tune package, to select the optimal hyperparameter value. For example, to find the best value of hyperparameters that yield the best rmse, we can execute the following code:

```
best_val <- select_best(
  tune_results,
  metric = "rmse"
)
best_val
```

```
## # A tibble: 1 x 3
##   hidden_units dropout .config
##          <dbl>   <dbl> <chr>
## 1           50       0 pre0_mod5_post0
```

10.8.9 FINALIZE AND TRAIN THE BEST MODEL

We can now finalize the workflow by using the `finalize_workflow()` function from the `tune` package. We pass the best value `best_val` as an argument to the function. Additionally, we train the model by fitting it to the `data_train` dataset using the following code:

```
best_model <- workflow_recipe_model %>%
  finalize_workflow(best_val) %>%
  fit(data_train)
```

10.8.10 ASSESS PREDICTION QUALITY WITH TEST DATA

We can now utilize our trained and fitted model to predict `medv` using the test data set to assess model efficacy. To accomplish this, we employ the `augment()` function. The `augment()` function, originating from the `broom` package in R, appends predictions and residuals from a statistical model to the original dataset. This facilitates the evaluation of the model's fit by juxtaposing actual and predicted values directly with the data. We use the following code:

```
data_test_with_prediction <- augment(
  best_model, data_test
)
```

Let us now have a glimpse of the augmented file, `data_test_with_prediction`. We select the columns that we are interested in, and consider the first few rows using the `head()` function:

```
data_test_with_prediction %>%
  dplyr::select(
    medv,
    .pred
  ) %>%
  head()
```

```
# A tibble: 6 x 2
   medv .pred
  <dbl> <dbl>
1  24    26.0
2  34.7  33.0
3  16.5  16.3
4  23.1  20.0
5  18.2  19.4
6  18.4  14.4
```

We find that the predicted values (`.pred`) are in the general vicinity of the actual measured values (`medv`), albeit with some difference.

We also compute the accuracy metrics. **Accuracy metrics** in machine learning assess a model's performance by comparing its predictions to the actual outcomes. We utilize the augmented `data_test_with_prediction` dataset to evaluate the estimated values against the actual truth using the `metrics()` function as follows:

```
metrics(
  data_test_with_prediction,
  truth = medv,
  estimate = .pred
)
```

```
## # A tibble: 3 x 3
##   .metric .estimator .estimate
##   <chr>   <chr>          <dbl>
## 1 rmse    standard       4.22
## 2 rsq     standard       0.795
## 3 mae     standard       2.85
```

We find that the RMSE is around 4.22 thousand dollars (it was 7.96 thousand dollars before tuning the hyperparameters, and 5.36 thousand dollars in the case of polynomial regression), the mean absolute error is around 2.85 thousand dollars (it was 4.92 thousand dollars before tuning the hyperparameters, and 3.7 thousand dollars in the case of polynomial regression), and we have been able to explain around 79.5% of the results (R squared = 0.795), whereas we were only able to explain around 48.2% of the results (R squared = 0.482) before tuning the hyperparameters, and 66.5% of the results using polynomial regression. The rest of the variations are probably due to variables that are absent from the `Boston` dataset, or because the neural network as defined by us is still insufficient to cover those. But in any case, now the neural network is much better than polynomial regression in predicting the median value of homes, something that is extremely difficult to model and predict otherwise.

10.9 WHEN TO USE NEURAL NETWORKS, AND WHEN NOT TO

Neural networks, particularly multilayer perceptrons (MLPs), are formidable tools in machine learning, excelling at various intricate tasks by learning complex patterns from data. However, there are instances where neural networks may not be the most appropriate choice. Let's delve into when to consider using neural networks, and when alternative methods should be explored.

10.9.1 WHEN TO USE NEURAL NETWORKS

Neural networks excel when the relationship between inputs and outputs is highly non-linear and complex. They can automatically learn hierarchical representations of data, making them suitable for tasks like image recognition, NLP, and speech recognition.

However, to do this, neural networks often require large datasets to generalize effectively. Thus, they shine when ample labeled data is available, for they can leverage this information to learn and generalize patterns robustly, producing reliable models.

Neural networks can significantly reduce the need for extensive manual feature engineering. They can learn relevant features from raw data, eliminating the effort required to preprocess and engineer input features manually. Thus, they are the method of choice when manual feature engineering is difficult or costly.

In domains where achieving state-of-the-art performance is paramount, such as computer vision or language modeling, neural networks, particularly deep learning architectures like CNNs or transformers, are often the preferred choice due to their ability to capture intricate patterns and dependencies.

10.9.2 WHEN NOT TO USE NEURAL NETWORKS

Neural networks, especially deep architectures, often require substantial amounts of data to generalize effectively and avoid overfitting. However, with limited datasets, they may struggle to discern meaningful patterns and can easily overfit, resulting in poor performance on novel, unseen data. Thus, we should refrain from using neural networks with very small training datasets.

Neural networks, particularly deep architectures, are often considered black-box models due to their lack of transparency in decision-making processes. This lack of interpretability can be a disadvantage in domains where understanding model decisions is crucial, such as healthcare, law and finance. Thus, these sectors may have less use for neural networks.

Training DNN can be computationally intensive, necessitating powerful hardware and substantial computational resources. For applications with strict latency or resource constraints, simpler models or model compression techniques may be more suitable.

Similarly, in cases where the relationship between inputs and outputs is linear or can be effectively captured by simpler models like linear regression, decision trees, or support vector machines (SVMs), using neural networks may be unnecessary and could lead to unnecessarily complex models.

In certain domains, regulatory or domain-specific constraints may require simpler, interpretable models. For instance, in healthcare and legal fields, where transparency and regulatory compliance are paramount, simpler models may be preferred over neural networks.

10.10 CHAPTER SUMMARY

- Neural networks, inspired by the human brain, are computational models made of units similar to biological neurons. They consist of interconnected nodes, each representing a neuron, and learn from training data to perform tasks like image recognition and language processing.
- Different types of neural networks, such as feedforward, convolutional, recurrent, and deep neural networks, are tailored for specific applications.
- Deep learning, a type of neural network, has revolutionized various industries by mimicking human brain activity to interpret complex data. It enables tasks like disease diagnosis, fraud detection, autonomous driving, and realistic content creation. However, challenges like data scarcity, computational intensity, and interpretability need to be addressed.
- Training neural networks involves adjusting weights and biases to minimize prediction errors, facilitated by backpropagation and optimization algorithms. Activation functions introduce non-linearity, while optimizers control weight updates for efficient learning.
- The Boston Housing dataset, used for predicting historical home values, includes features like crime rate, land zoning, and median value. The dataset, originally from the 1970s, requires ethical considerations since it incorporates historical biases.
- The tidymodels package simplifies neural network implementation, while allowing for easy transition between algorithms.
- Neural networks are powerful for complex tasks, but require large datasets to do so and may overfit when used with small datasets. They excel at automatically learning features, even those with non-linear relationships, making them suitable for image and speech recognition. However, simpler models are preferred when data is limited or the relationship is simple or there are regulatory requirements.

Part III

Epilogue

11 What to do next?

Congratulations on completing the operational part of the book! By now, I'm sure you have a good understanding of various classic algorithms used in machine learning, and perhaps even more importantly, that machine learning is relatively easy to grasp and execute on your own machine. Most of the codes are concise and straightforward to comprehend. Additionally, with packages like tidymodels, which provide recipes and routines, implementing machine learning on any dataset is a simple and hassle-free process.

That said, this book, with its objective of being an easy, hands-on introduction to machine learning, has explored and employed readily available datasets. And it is up to you, the reader, to hone your skills through practice and personal expeditions deep into the realms of machine learning and artificial intelligence. In this section, I've included some helpful tips for your upcoming journeys.

11.1 EXPLORING STACKS

Stacking, also known as **stacked generalization** or **meta-ensembling**, is a technique in machine learning that involves combining multiple and diverse models to enhance predictive performance. It capitalizes on the unique strengths of each model to enhance accuracy compared to any single model's capabilities. Here's a concise explanation of stacking in the context of machine learning.

11.1.1 HOW STACKING WORKS

Stacking involves training multiple base models on the same dataset [Fig. 11.1]. These base models come in various forms, such as linear models, tree-based models, and neural networks. They can also be variations of the same type, each with its own set of hyperparameters. By utilizing a diverse range of models, stacking aims to capture distinct patterns in the data, ultimately improving the overall predictive accuracy.

After training the base models, their predictions are used as input features, known as **meta-features**, for a final **meta learner** or **meta model**. This meta learner, typically a simple model like logistic regression or linear regression, combines the predictions of the base models. The primary objective of this meta learner is to leverage the strengths of the individual base models to enhance the overall accuracy of the final model. This two-stage training process enables the meta learner to understand how to effectively combine the predictions from the various base models.

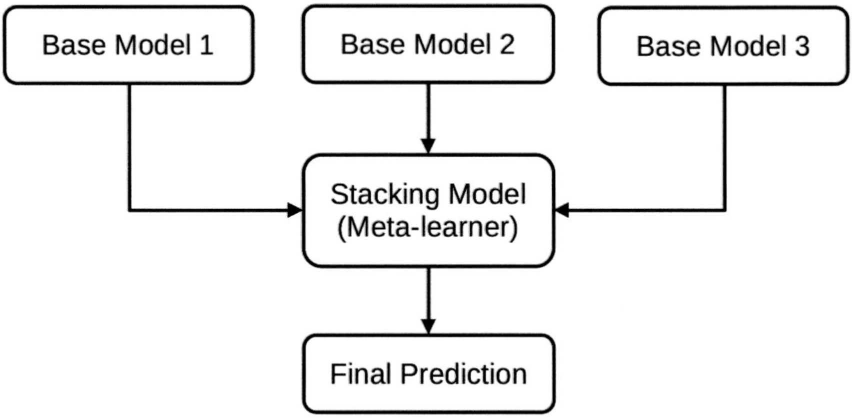

FIGURE 11.1: A schematic of the process of stacking.

During prediction, new data is passed through each base model to generate predictions. These predictions are then fed into the trained meta learner, which produces the final ensemble prediction, usually through weighted averages. This approach combines the strengths of multiple models, leading to more accurate and reliable predictions.

11.1.2 ADVANTAGES OF STACKING

Stacking offers several key advantages. Firstly, it enhances performance by combining the strengths of multiple models, resulting in higher predictive accuracy compared to using individual models alone. By leveraging the unique capabilities of each model, the ensemble makes better overall predictions.

Secondly, stacking can help reduce **overfitting**. Overfitting occurs when a model learns excessively from the training data, including noise. However, errors made by one model often get compensated for by other models in the ensemble, ensuring better **generalization** to unseen data.

Thirdly, stacking provides flexibility in model selection. Practitioners can choose various algorithms and configurations tailored to specific data aspects. This adaptability allows different models to be selected based on their strengths, leading to more effective and customized solutions for complex problems.

Lastly, stacking can improve the generalization ability of a model. By training multiple models and combining their predictions, stacking reduces the likelihood of a model being biased toward specific patterns in the training data. This is particularly beneficial in complex problems where the data may have diverse characteristics or patterns that a single model might overlook.

11.1.3 CONSIDERATIONS AND BEST PRACTICES

When implementing stacking in machine learning, several key considerations must be taken into account to ensure its effectiveness. First and foremost, selecting the appropriate base models is crucial. A diverse set of models, such as combining linear models with tree-based models or utilizing various types of neural networks, is essential for stacking's success. This diversity enables the capturing of diverse data patterns, potentially leading to improved overall performance. However, if all base models are overly similar, they may inadvertently replicate the same errors, diminishing the potential benefits of stacking.

Secondly, the choice of the meta learner is equally important. The meta learner, responsible for combining the predictions from the base models, is typically simpler compared to the base models. Its primary function is to learn the optimal method of combining their outputs. Common meta learner options include logistic regression, linear regression, or even another decision tree. The meta learner should effectively aggregate the predictions while avoiding overfitting to the base model predictions. Experimenting with various meta learners is crucial to determine the most suitable one for a specific problem.

Thirdly, proper management of the training and validation data is also vital in stacking. A common approach involves using k-fold cross-validation. This process helps ensure that the predictions made by the base models are not biased by the specific data they were trained on. During the meta learner's training phase, it can utilize the predictions generated during this cross-validation process to effectively learn how to combine the outputs from the base models.

It is also crucial to prevent information leakage during the stacking process. **Information leakage** occurs when the training data contains information about the validation data, resulting in overly optimistic performance metrics. To avoid this, the predictions of the base models used as inputs to the meta learner must be generated as **out-of-fold predictions** so that each prediction comes from a model that has not seen that specific data point during training. Maintaining a strict separation between the training, validation, and test sets is essential for obtaining a realistic estimate of how the stacked model will perform on unseen data.

Hyperparameter tuning is another crucial aspect of stacking. Each base model and the meta learner have their own set of hyperparameters that need to be optimized for optimal performance. Systematically tuning these hyperparameters is essential to find the optimal settings for each model. Techniques like grid search or random search can be used to explore various combinations of hyperparameters. Additionally, using **cross-validation** during this tuning process is advisable to ensure that the chosen hyperparameters lead to models that generalize well.

Evaluating the performance of the stacked model is equally important. When assessing the accuracy of the stacked ensemble, selecting the appropriate metrics is crucial. Depending on the problem, this may involve metrics such as accuracy, precision, recall, or the F1 score for classification tasks, or mean squared error and R-squared for regression tasks. Comparing the performance of the stacked model against individual base models and other ensemble techniques offers valuable insights into its effectiveness.

Lastly, stacking can significantly increase computational complexity. Training multiple models and a meta learner can consume substantial computational resources and time, especially when dealing with large datasets. Therefore, it's crucial to consider the computational costs and select models that strike a balance between performance and efficiency. **Parallel processing** techniques can help expedite the training process for the base models.

In essence, successfully implementing stacking involves careful model selection, choosing the right meta learner, managing data splits to prevent information leakage, optimizing hyperparameters, and effectively evaluating performance. Additionally, being mindful of computational complexity is crucial to ensure that stacking yields improved model performance while being practical to implement. By paying attention to these factors, practitioners can maximize the benefits of stacking and create robust predictive models.

11.2 EXPLORING DATASETS IN OTHER FIELDS

In this book, we used datasets from packages like MASS, dslabs, and mlbench. These packages also contain several other datasets that you may wish to explore. Similarly, many datasets are available in other packages and in repositories like the UC Irvine's Machine Learning Repository.[1] The UCI Machine Learning Repository, a treasure trove of meticulously curated datasets by the University of California, Irvine, stands as a prominent repository in the field of machine learning research and education. It houses an extensive collection of diverse datasets spanning various domains, such as classification, regression, clustering, and recommendation systems. These datasets serve as benchmark standards for evaluating novel algorithms and techniques, ensuring the robustness and reliability of machine learning research endeavors.

You might also consider exploring the application of machine learning in your own field by reading books and published papers, and by implementing machine learning on datasets within your domain.

11.3 EXPLORING FRONTIERS OF MACHINE LEARNING

Machine learning has experienced rapid growth in recent years, fueled by advancements in computing power, algorithms, and the abundance of data. This evolution

[1] https://archive.ics.uci.edu

has opened up new avenues for machine learning systems, enabling them to excel in a wide range of applications. We'll delve into some of these exciting developments in the following sections.

11.3.1 DEEP LEARNING AND NEURAL ARCHITECTURES

Deep learning has revolutionized various fields within machine learning, particularly in computer vision, natural language processing (NLP), and speech recognition. The emergence of deep neural networks, comprising multiple layers of interconnected neurons, has empowered these models to autonomously learn hierarchical representations of data. **Convolutional neural networks** (**CNNs**) have achieved remarkable success in image and video analysis by capturing spatial dependencies, while transformers have excelled in sequence modeling tasks such as language translation and text generation. The frontier of deep learning continues to advance through ongoing efforts aimed at scaling models, enhancing efficiency, and improving interpretability. Research initiatives are focused on overcoming challenges like data efficiency, model robustness, and the ethical implications of deploying increasingly complex models in real-world applications.

11.3.2 REINFORCEMENT LEARNING AND AUTONOMOUS AGENTS

Reinforcement learning (**RL**) is another frontier that has garnered substantial attention, particularly for training autonomous agents to make sequential decisions in intricate environments. RL algorithms learn through trial and error by interacting with an environment to maximize cumulative rewards. Applications of RL span robotics, autonomous vehicles, game playing, and financial trading, among others. Recent advancements in RL include deep RL, which merges deep neural networks with RL algorithms to tackle more challenging tasks with high-dimensional state and action spaces. Current research focuses on enhancing the sample efficiency of RL algorithms, improving their stability, and addressing safety concerns in autonomous systems, thereby paving the way for more reliable and scalable deployments in real-world scenarios.

11.3.3 FEDERATED LEARNING AND PRIVACY-PRESERVING TECHNIQUES

Federated learning, a frontier addressing privacy concerns in centralized data processing, enables model training locally on distributed data sources (e.g., mobile devices, edge devices) without sharing raw data. Instead, only model updates are exchanged between devices or servers. This approach safeguards data privacy and confidentiality while facilitating collaborative model training across decentralized networks. Federated learning finds particular relevance in domains with stringent data privacy regulations, such as healthcare and finance. Ongoing research focuses on optimizing federated learning protocols, enhancing model aggregation techniques, and expanding its applicability beyond traditional centralized training paradigms.

11.3.4 META-LEARNING AND FEW-SHOT LEARNING

Meta-learning delves into techniques that empower models to learn how to learn effectively across diverse tasks or domains. This frontier holds immense value in scenarios where data is scarce or rapidly evolving. Within meta-learning, **few-shot learning** emerges as a subset, aiming to generalize from a mere few examples or even a single example per class. Promising advancements in meta-learning hold the potential to enhance the adaptability and scalability of machine learning systems, enabling them to swiftly adapt to novel tasks and environments with limited data. Research efforts in this domain focus on developing robust meta-learning algorithms, exploring novel meta-feature representations, and expanding the practical applications of few-shot learning to real-world scenarios across diverse fields, including personalized medicine, industrial automation, and adaptive user interfaces.

11.3.5 QUANTUM MACHINE LEARNING

Quantum computing and machine learning intersect at a frontier that holds the potential to solve computationally demanding tasks beyond the reach of classical computing. **Quantum machine learning** delves into how quantum algorithms and hardware can accelerate machine learning tasks, such as optimization, pattern recognition, and simulation. Quantum computers harness quantum properties like superposition and entanglement to perform calculations that classical computers find challenging. Ongoing research in quantum machine learning seeks to develop quantum algorithms that surpass classical algorithms on specific tasks, paving the way for groundbreaking advancements in fields like drug discovery, material science, and cryptography. However, challenges remain, including the scalability of hardware, error correction, and integrating quantum algorithms with existing machine learning frameworks.

11.3.6 ETHICAL AI AND BIAS MITIGATION

As machine learning systems increasingly integrate into societal decision-making processes, ethical considerations and bias mitigation have become paramount challenges in the field. **Ethical AI** focuses on developing guidelines and frameworks to ensure fairness, accountability, transparency, and interpretability in AI systems. Bias mitigation techniques aim to identify and address biases present in training data or learned models, thereby promoting equitable outcomes for diverse populations. Research efforts are dedicated to developing robust methodologies for detecting and mitigating biases, designing fair algorithms, and integrating ethical considerations throughout the entire machine learning lifecycle—from data collection and model development to deployment and monitoring.

11.4 GAINING A KNOWLEDGE OF MATHS

In this book, we've covered the fundamental principles of various machine learning algorithms. However, for those seeking a deeper understanding of the mathematical foundations of these algorithms, exploring courses at local universities, enrolling in online classes on platforms like edX or Swayam, or studying specialized books on the mathematics of machine learning can be highly beneficial.

11.5 EXPLORING TIDYMODELS

In this book, we've covered the fundamental concepts of tidymodels required for a beginner-friendly introduction to machine learning. However, tidymodels is a comprehensive package that provides a wide range of additional features and functionalities beyond what we've covered here. For a deeper exploration, readers may want to visit the CRAN page,[2] the package website,[3] or refer to resources such as the book *Tidy Modeling with R*.[4]

11.6 EXPLORING OTHER PACKAGES

Tidymodels is not the only machine learning package available for use with R. Other notable packages include caret (Kuhn, 2008), tensorflow (Allaire and Tang, 2024), and keras (Allaire and Chollet, 2025).

Depending on specific requirements, some of these packages may be more suitable than tidymodels. Therefore, readers are encouraged to explore these alternatives to find the most appropriate solution for their applications.

---

[2]https://cran.r-project.org/web/packages/tidymodels/index.html

[3]https://tidymodels.tidymodels.org

[4]https://www.tmwr.org

11.7 EXPLORING OTHER LANGUAGES

Machine learning is not limited to R; it can also be performed in languages such as Python[5] that has an extensive support base, and also in languages like C++,[6] Java,[7] and Haskell.[8] Exploring alternative languages for machine learning can be particularly advantageous when certain packages are only available for specific languages. This approach can be both functional and beneficial for various applications.

11.8 SCRUTINIZING NUANCES LIKE REGULARIZATION

Throughout this book, we've been focused on achieving a good fit and, during hyperparameter tuning, selecting the best model based on its fit. However, there's always a risk of **overfitting**, where the model becomes overly sensitive to minor variations in the training data and performs poorly on the test data. **Regularization** is a crucial technique in machine learning and statistics that aims to enhance model generalization by preventing overfitting. It achieves this by adding a penalty term to the model's objective function, which encourages simpler models or limits the magnitude of their parameters.

11.8.1 TYPES OF REGULARIZATION TECHNIQUES

There are three common methods of regularization: ridge regression, lasso regression, and elastic net regularization.

Ridge regression, also known as **L2 regularization**, introduces a penalty that is proportional to the squared magnitude of coefficients into the loss function. Mathematically, it transforms the objective function to:

$$\text{Loss}_{\text{ridge}} = \text{Loss} + \lambda \sum_{j=1}^{p} \beta_j^2$$

where λ (lambda) is the regularization parameter that controls the strength of regularization. Ridge regression shrinks the coefficients toward zero but does not set them exactly to zero, making it effective for reducing **multicollinearity**[9] and stabilizing model estimates.

[5]https://www.python.org

[6]https://isocpp.org

[7]https://www.java.com

[8]https://www.haskell.org

[9]Multicollinearity in machine learning occurs when two or more predictor variables in a multiple regression model are highly correlated, implying that one variable can be linearly predicted from the others with a high degree of accuracy. This condition poses challenges because it reduces the statistical significance of an independent variable.

In simple terms, L2 regularization discourages the model from giving excessive importance to any single feature by reducing the size of its coefficients. It achieves this by incorporating a term into the loss function that grows with the magnitude of the coefficients. This term functions as a "**penalty**" for having large coefficients, guiding the model toward simpler solutions characterized by smaller and more evenly distributed coefficients.

This technique helps stabilize the model, reducing its sensitivity to minor variations in the training data. Consequently, it enhances the model's ability to generalize and perform effectively on novel, unforeseen data.

Lasso regression, also known as **L1 regularization**, introduces a penalty that is proportional to the absolute value of coefficients to the loss function. This modification of the objective function penalizes large coefficients. The objective function becomes:

$$\text{Loss}_{\text{lasso}} = \text{Loss} + \lambda \sum_{j=1}^{p} |\beta_j|$$

In simple terms, L1 regularization encourages the model to use fewer features by penalizing large coefficients. This is akin to saying, "If a feature isn't truly useful, simply discard it entirely." By doing so, L1 regularization aids in feature selection, simplifying the model and enhancing its interpretability while still striving for optimal performance. It helps the model concentrate on the most significant features and disregard the less relevant ones. L1 regularization is particularly beneficial in managing high-dimensional data with numerous irrelevant or redundant features.

Elastic net regularization, a hybrid of ridge and lasso regression, merges L1 and L2 penalties to capitalize on their respective advantages. The objective function for this approach is:

$$\text{Loss}_{\text{elastic net}} = \text{Loss} + \lambda_1 \sum_{j=1}^{p} \beta_j^2 + \lambda_2 \sum_{j=1}^{p} |\beta_j|$$

where λ_1 and λ_2 control the strengths of L2 and L1 penalties, respectively. Elastic net is particularly useful when there are many correlated features in the data. It helps in determining which features are important while also maintaining the model's stability and reducing the likelihood of overfitting.

11.8.2 APPLICATIONS OF REGULARIZATION

Regularization techniques are widely used across various machine learning algorithms and applications. In linear regression, ridge and lasso regression are commonly employed to prevent overfitting and enhance prediction accuracy. These methods ensure that the model generalizes well to new data.

In logistic regression, regularization improves the model's stability and aids in feature selection by penalizing less useful coefficients. This results in a more robust model that performs better on unseen data.

For neural networks, techniques such as **weight decay** (L2 regularization) and **dropout**[10] are employed to prevent overfitting and enhance the model's ability to generalize. These methods help ensure that the neural network performs well across different datasets.

In support vector machines, regularization parameters manage the trade-off between maximizing the margin between classes and minimizing classification errors on the training data. This balance is crucial for building a model that performs effectively on new, unseen data.

11.8.3 CONSIDERATIONS AND BEST PRACTICES

When using regularization, selecting the appropriate value for the regularization parameter is paramount. This parameter controls the intensity of the regularization effect, and choosing the right value can substantially influence the model's performance. Usually, cross-validation techniques are employed to determine the optimal value for this parameter by assessing the model's performance on a separate validation set. This approach ensures that the selected regularization strength results in a model that effectively generalizes to novel data.

Another crucial aspect to consider is the equilibrium between model interpretability and complexity. Regularization aids in simplifying models by penalizing overly intricate ones, which can enhance interpretability. Nevertheless, it's crucial to determine the optimal level of regularization that minimizes complexity without compromising the model's capacity to discern significant patterns in the data. Achieving this balance can render the model more comprehensible while maintaining its effectiveness.

Regularization influences the **bias-variance trade-off**, which is the equilibrium between a model's capacity to fit the training data (**variance**) and its ability to generalize to new data (**bias**). Increasing regularization usually reduces variance and prevents overfitting, but it can also introduce bias if the model becomes overly simplistic. It's crucial to adjust regularization judiciously to maintain this balance and develop a model that performs effectively on both the training set and unseen data.

Overall, applying regularization requires careful tuning and consideration of the model's complexity, its performance on validation data, and the trade-off between bias and variance.

---

[10]Dropout in neural networks is a regularization technique that involves randomly selecting and ignoring neurons during training. This helps prevent overfitting by promoting the development of a more robust model.

11.9 LEARNING INTERPRETATION OF RESULTS

Interpreting machine learning results is crucial for understanding how models make predictions and ensuring their transparency and reliability. As machine learning models become increasingly integrated into decision-making processes across diverse sectors, comprehending their operational mechanisms is essential for validating their effectiveness and aligning them with their intended applications.

Model interpretability involves several techniques designed to explain how different features influence predictions. **Feature importance methods** are among the most common approaches. These methods help identify which features have the most significant impact on the model's outcomes. For example, **permutation importance** measures how the model's performance decreases when a feature's values are shuffled, highlighting features that are crucial for the model's predictions. Similarly, **SHAP (SHapley Additive exPlanations)** values provide a unified measure of feature importance by quantifying each feature's contribution based on game theory principles. These techniques enable stakeholders to understand the relative significance of various inputs and how they affect the model's decisions.

Partial dependence plots are another valuable tool for interpreting machine learning models. These plots visualize the relationship between a specific feature and the predicted outcome, while simultaneously averaging out the effects of other features. By demonstrating how alterations in a particular feature impact predictions, partial dependence plots enable users to comprehend the model's behavior in relation to individual features.

Local interpretability techniques offer insights at the level of individual predictions. For instance, **Individual Conditional Expectation (ICE)** plots illustrate how predictions vary for a single instance as a feature changes, providing a detailed view of the model's behavior in specific cases. **LIME (Local Interpretable Model-agnostic Explanations)** is another local interpretability method that approximates a model's decision boundary around a particular instance to offer understandable explanations for its predictions.

Global interpretability techniques offer a broader understanding of model behavior. **Decision trees**, with their straightforward decision rules and easily comprehensible paths from the root to the leaves, are inherently interpretable. **Rule extraction methods** further simplify complex models into human-readable rules, enhancing transparency, though sometimes with a little loss in predictive accuracy.

Some models incorporate **built-in interpretability features**. For instance, linear models directly reveal the strength and direction of the relationship between features and predictions through their **coefficients**. In contrast, tree-based models, like random forests or gradient-boosted trees, offer insights through **feature importance plots** and **decision paths**. These plots illustrate how various features contribute to the final predictions.

Evaluating interpretability requires ensuring that the interpretations align with domain knowledge, accurately represent the model's decision-making process, and remain resilient across diverse data and model configurations. Integrating **domain expertise** is essential for contextualizing interpretations and validating their practical relevance in real-world applications.

Sharing model results is crucial for establishing trust among all stakeholders. Ensuring that the results are comprehensible to all parties, regardless of their technical expertise, facilitates informed decision-making. Furthermore, it is paramount to maintain transparency and fairness in explaining the results, particularly when they hold significant implications or impact individuals' lives.

11.10 WRITING YOUR OWN MACHINE LEARNING CODE(S)

The algorithms used in machine learning were developed by individuals similar to yourself. Consequently, you can also create machine learning codes and packages that others can utilize. Writing code from the ground up offers an excellent opportunity to comprehend the underlying workings of an algorithm. You might even consider writing some, simply for the sake of it!

For instance, to implement k-Nearest Neighbors (kNN) classification on `Iris` dataset from scratch in R without utilizing the `knn()` function, we can follow these steps:

1. Calculate Euclidean distance: Implement a function to compute the Euclidean distance between two data points. The Euclidean distance between two points in Euclidean space is the length of the line segment connecting them. It can be calculated from their Cartesian coordinates using the Pythagorean theorem.
2. Find nearest neighbors: Implement a function to find the k nearest neighbors of a given test data point based on Euclidean distance.
3. Majority voting: Implement a function to perform majority voting among the nearest neighbors to determine the class label for the test data point.
4. Load and prepare data: Load a dataset, split it into training and test sets, and preprocess it as necessary.

Here's a code to achieve this:

```
# Function to calculate Euclidean
# distance between two points
euclidean_distance <- function(x1, x2) {
  x1 <- as.numeric(x1)
  x2 <- as.numeric(x2)
```

```
  sqrt(sum((x1 - x2)^2))
}

# Function to find knn for a test data point
find_neighbors <- function(
    train_data, test_point, k
  ) {
  distances <- apply(
    train_data[, -ncol(train_data)],
    1,
    function(train_point) euclidean_distance(
      train_point, test_point
    )
  )
  neighbors <- order(distances)[1:k]
  return(neighbors)
}

# Function for majority voting to determine the class
majority_vote <- function(train_data, neighbors) {
  majority_class <- names(
    sort(
      table(
        train_data[neighbors, "Species"]
      ), decreasing = TRUE
    )
  )[1]
  return(majority_class)
}

# Load example dataset (Iris dataset)
data(iris)

# Split data into training and testing
# sets (80% train, 20% test)
set.seed(2025)
train_indices <- sample(
  1:nrow(iris),
  0.8 * nrow(iris)
)
train_data <- iris[train_indices, ]
test_data <- iris[-train_indices, ]
```

```
# Predict using kNN from scratch
k <- 1  # Number of neighbors

predicted_classes <- sapply(
  1:nrow(test_data), function(i) {
  neighbors <- find_neighbors(
  train_data,
  as.numeric(test_data[i, -5]),
  k
  )
  majority_vote(train_data, neighbors)
  }
)

# Evaluate accuracy
accuracy <- mean(
  predicted_classes == test_data[, "Species"]
)
cat("Accuracy on test set:", accuracy, "\n")
```

```
## Accuracy on test set: 0.9333333
```

This basic kNN classification framework in R demonstrates distance computation, neighbor selection, and class prediction. Adjustments can be made for different datasets, k-values, or preprocessing steps, if required.

11.11 COMBINING STRATEGIES

Just like stacking, you can add, mix, and combine the aforesaid and other techniques to put an edge on your machine learning abilities. The only sure thing is that whatever you do, keep using and enjoying Machine Learning!

11.12 CHAPTER SUMMARY

- Stacking, a technique in machine learning, combines multiple diverse models to improve predictive performance. It involves training base models, using their predictions as meta-features for a meta-learner, and leveraging their strengths.
- When implementing stacking, consider model selection, meta-learner choice, data management, hyperparameter tuning, and performance evaluation.

- Machine learning has revolutionized various fields, particularly in computer vision, NLP, and speech recognition. Emerging frontiers include RL, federated learning, meta-learning, and quantum machine learning. Ethical considerations, bias mitigation, and mathematical foundations are crucial.
- Regularization techniques prevent overfitting and enhance interpretability by penalizing large coefficients.
- Interpreting machine learning results involves feature importance, partial dependence plots, and local/global interpretability techniques.

References

Allaire, J. and Chollet, F. (2025). *keras: R Interface to 'Keras'*. R package version 2.16.0.

Allaire, J. and Tang, Y. (2024). *tensorflow: R Interface to 'TensorFlow'*. R package version 2.16.0.

Blake, C. L. and Merz, C. J. (1998). *UCI Repository of Machine Learning Databases*. University of California, Irvine, Department of Information and Computer Sciences, Irvine, CA. Formerly available from http://www.ics.uci.edu/~mlearn/MLRepository.html.

Chang, W., Cheng, J., Allaire, J., Sievert, C., Schloerke, B., Xie, Y., Allen, J., McPherson, J., Dipert, A., and Borges, B. (2024). *shiny: Web Application Framework for R*. R package version 1.10.0.

Couch, S. P., Bray, A. P., Ismay, C., Chasnovski, E., Baumer, B. S., and Çetinkaya-Rundel, M. (2021). infer: An R package for tidyverse-friendly statistical inference. *Journal of Open Source Software*, 6(65):3661.

Crevier, D. (1993). *AI: The Tumultuous History of the Search for Artificial Intelligence*. Basic Books, Inc.

Eddelbuettel, D. and Balamuta, J. J. (2018). Extending R with C++: A Brief Introduction to Rcpp. *The American Statistician*, 72(1):28–36.

Falbel, D. and Luraschi, J. (2025). *torch: Tensors and Neural Networks with 'GPU' Acceleration*. R package version 0.14.2.

Firke, S. (2024). *janitor: Simple Tools for Examining and Cleaning Dirty Data*. R package version 2.2.1.

Frick, H., Chow, F., Kuhn, M., Mahoney, M., Silge, J., and Wickham, H. (2025). *rsample: General Resampling Infrastructure*. R package version 1.3.0.

Friendly, M. (2025). *HistData: Data Sets from the History of Statistics and Data Visualization*. R package version 0.9-3.

Grolemund, G. and Wickham, H. (2011). Dates and times made easy with lubridate. *Journal of Statistical Software*, 40(3):1–25.

hong Chan, C., Leeper, T. J., Becker, J., and Schoch, D. (2023). *rio: A Swiss-army Knife for Data File I/O*.

Horst, A. M., Hill, A. P., and Gorman, K. B. (2020). *palmerpenguins: Palmer Archipelago (Antarctica) penguin data*. R package version 0.1.0.

Hvitfeldt, E. (2025). *themis: Extra Recipes Steps for Dealing with Unbalanced Data*. R package version 1.0.3.

Irizarry, R. A. and Gill, A. (2024). *dslabs: Data Science Labs*. R package version 0.8.0.

Karatzoglou, A., Smola, A., Hornik, K., and Zeileis, A. (2004). kernlab – an S4 package for kernel methods in R. *Journal of Statistical Software*, 11(9):1–20.

Krizhevsky, A., Sutskever, I., and Hinton, G. E. (2012). Imagenet classification with deep convolutional neural networks. *Advances in Neural Information Processing Systems*, 25.

Kuhn, M. (2008). Building predictive models in r using the caret package. *Journal of Statistical Software*, 28(5):1–26.

Kuhn, M. (2025). *tune: Tidy Tuning Tools*. R package version 1.3.0.

Kuhn, M. and Falbel, D. (2025). *brulee: High-Level Modeling Functions with 'torch'*. R package version 0.5.0.

Kuhn, M. and Vaughan, D. (2025). *parsnip: A Common API to Modeling and Analysis Functions*. R package version 1.3.1.

Kuhn, M., Vaughan, D., and Hvitfeldt, E. (2025a). *yardstick: Tidy Characterizations of Model Performance*. R package version 1.3.2.

Kuhn, M. and Wickham, H. (2020). *Tidymodels: A Collection of Packages for Modeling and Machine Learning using tidyverse Principles*.

Kuhn, M., Wickham, H., and Hvitfeldt, E. (2025b). *recipes: Preprocessing and Feature Engineering Steps for Modeling*. R package version 1.3.0.

McCarthy, J., Minsky, M. L., Rochester, N., and Shannon, C. (1956). The dartmouth summer research project on artificial intelligence. *Artificial Intelligence: Past, Present, and Future*.

Milborrow, S. (2024). *rpart.plot: Plot 'rpart' Models: An Enhanced Version of 'plot.rpart'*. R package version 3.1.2.

Müller, K. and Wickham, H. (2023). *tibble: Simple Data Frames*. R package version 3.2.1.

Nilsson, N. J. (2009). *The Quest for Artificial Intelligence*. Cambridge University Press.

Ooms, J. (2025). *writexl: Export Data Frames to Excel 'xlsx' Format*. R package version 1.5.4.

R Core Team (2022). *R: A Language and Environment for Statistical Computing*. R Foundation for Statistical Computing, Vienna, Austria.

Rosenblatt, F. (1958). The perceptron: a probabilistic model for information storage and organization in the brain. *Psychological Review*, 65(6):386.

Samuel, A. L. (1959). Some studies in machine learning using the game of checkers. *IBM Journal of Research and Development*, 3(3):210–229.

Schliep, K. and Hechenbichler, K. (2025). *kknn: Weighted k-Nearest Neighbors*. R package version 1.4.0.

Schloerke, B., Cook, D., Larmarange, J., Briatte, F., Marbach, M., Thoen, E., Elberg, A., and Crowley, J. (2024). *GGally: Extension to 'ggplot2'*. R package version 2.2.1.

Skinner, B. (1938). The behavior of organisms: an experimental analysis.

Turing, A. M. (1950). Computing machinery and intelligence. *Mind*, LIX(236):433–460.

Vapnik, V. N. (1995). *The Nature of Statistical Learning Theory*. Springer, New York.

Vaughan, D. and Couch, S. (2025). *workflows: Modeling Workflows*. R package version 1.2.0.

Venables, W. N. and Ripley, B. D. (2002). *Modern Applied Statistics with S*. Springer, New York, fourth edition. ISBN 0-387-95457-0.

Wickham, H. (2016). *ggplot2: Elegant Graphics for Data Analysis*. Springer-Verlag New York.

Wickham, H. (2023a). *forcats: Tools for Working with Categorical Variables (Factors)*. R package version 1.0.0.

Wickham, H. (2023b). *stringr: Simple, Consistent Wrappers for Common String Operations*. R package version 1.5.1.

Wickham, H., Averick, M., Bryan, J., Chang, W., McGowan, L. D., François, R., Grolemund, G., Hayes, A., Henry, L., Hester, J., Kuhn, M., Pedersen, T. L., Miller, E., Bache, S. M., Müller, K., Ooms, J., Robinson, D., Seidel, D. P., Spinu, V., Takahashi, K., Vaughan, D., Wilke, C., Woo, K., and Yutani, H. (2019). Welcome to the tidyverse. *Journal of Open Source Software*, 4(43):1686.

Wickham, H. and Bryan, J. (2025). *readxl: Read Excel Files*. R package version 1.4.5.

Wickham, H., François, R., Henry, L., Müller, K., and Vaughan, D. (2023). *dplyr: A Grammar of Data Manipulation*. R package version 1.1.4.

Wickham, H. and Henry, L. (2025). *purrr: Functional Programming Tools*. R package version 1.0.4.

Wickham, H., Hester, J., and Bryan, J. (2024a). *readr: Read Rectangular Text Data*. R package version 2.1.5.

Wickham, H., Vaughan, D., and Girlich, M. (2024b). *tidyr: Tidy Messy Data*. R package version 1.3.1.

Xie, Y. (2016). *bookdown: Authoring Books and Technical Documents with R Markdown*. Chapman and Hall/CRC, Boca Raton, Florida.

Xie, Y., Dervieux, C., and Riederer, E. (2020). *R Markdown Cookbook*. Chapman and Hall/CRC, Boca Raton, Florida.

Index

abstraction, 42
accountability, 18, 25–28
accuracy, 48, 54, 56
accuracy metrics, 103, 126, 265
action potential, 297
activation function, 306
  leaky ReLU, 306
  ReLU, 306
  sigmoid, 306
  softmax, 307
AdaBoost, *see* adaptive boosting
adaptive boosting, 279
adversarial attack, 25
AI, *see* artificial intelligence
AI winter, 8, 302
Alan Turing, 8
Alex Krizhevsky, 303
Alexa, 299
AlexNet, 9, 303
algorithmic trading, 23, 299
ANN, *see* artificial neural network
anomaly detection, 15
anti-aliasing, 202
Arthur Samuel, 8
artificial intelligence, 5
artificial neural network, 298
assignment, 40
atomic vector, 43
augmented reality, 299
autonomous vehicle, 9, 15, 18, 23, 299, 303
axon, 296
axon terminal, 296

backpropagation, 298, 300, 302, 304
backward pass, 304
bagging, *see* bootstrap aggregating
balancing dataset, 157
  adjusting class weights, 158
  algorithmic approaches, 158
  bagging, 158
  boosting, 158
  ensemble method, 158
  oversampling, 158
  resampling, 158
  SMOTE, 158, 161, 163
  undersampling, 158
bayesian network, 6
BERT, 303
bias, 13, 20, 22, 24–27, 49, 53, 302, 338
bias-variance trade-off, 338
big data, 9
binary classification, 140
binary cross-entropy, 304
Bioconductor, 35, 36
black box, 24, 26, 301
body mass index, 149
boosted trees, 279
boosting, 158, 268
bootstrap aggregating, 267
bootstrap sampling, 64
bootstrapping, 273
box plot, 183, 228

C++, 33
camel case, 76
CART, *see* classification and regression tree
CatBoost, *see* categorical boosting
categorical boosting, 280
cell body, 296
chain rule, 305
chaining, 31
Charles Darwin, 72
class imbalance, 143
classification, 10–12, 48, 54
classification and regression tree, 243, 256, 269
climate modeling, 23
clustering, 13, 48
CNN, 298, *see* convolutional neural network
Cohen's Kappa, *see* kappa statistic
computer neural network, 295

input layer, 296
intermediate layer, 296
output layer, 296
computer vision, 9, 19, 22, 53
confusion matrix, 55, 154, 158, 190, 236
conjugate gradient method, 91
connection, 41
consent, 27
consistency, 87
control system, 6
convolution, 298
convolutional neural network, 9, 333
Cook's distance, 95
copy-on-modify semantics, 43
copyright, 27
correlation, 73
correlation coefficient, 73, 181
cost function, 90
CRAN, 32, 35, 36
cross-validation, 24, 52, 331
k-fold, 52
leave-one-out, 52
CSV, 31, 39
curse of dimensionality, 176, 218

Dartmouth Conference, 8
data, 9, 53
biometric data, 53
cleaning, 53
formats, 53
pre-processing, 78
quality, 53
semi-structured data, 53
sensor data, 53
social media data, 53
sources, 53
structured data, 53
textual data, 53
transactional data, 53
unstructured data, 53
data analysis, 3
data collection, 9
data frame, 39
data preparation, 10
data preprocessing, 63
data security, 302
dataset, 36
Boston Housing, 108, 308
Galton Families, 73
GaltonFamilies, 74
iris, 177, 179
MNIST, 202, 283
penguins, 223
Pima Indian Diabetes, 142
David Rumelhart, 302
decision boundary, 49, 221
decision forest, *see* random forest
decision path, 339
decision rule, 49
decision surface, 221
decision tree, 49, 63, 241, 243, 339
branch, 241
depth, 50
instability, 266
leaf, 241
root, 241
deep learning, 6, 9, 13, 24, 26, 299, 300, 333
challenges, 300
ethics, 302
hidden layer, 300
input layer, 300
output layer, 300
structure, 300
dendrite, 296
dependent variable, 47, 54, 70, 111
descriptive statistics, 32
digital value, 284
dimensionality reduction, 13, 14, 218
disaster forecasting, 23
discriminator, 299
distance
Euclidean distance, 176
Manhattan distance, 176
DNN, 298
domain expertise, 25, 340
domain specificity, 25
downsampling, *see* undersampling
dropout, 316, 318, 338

dummy variable, 99
dying ReLU, 306

EDA, *see* exploratory data analysis
efficiency, 87
elastic net regression, 91
elastic net regularization, 337
ensemble method, 158, 267
entropy, 244
environmental monitoring, 23
epoch, 318
equity, 302
ethical AI, 334
Euclidean distance, 176
eugenics, 73
evaluation, 10
expert system, 5, 6
exploitation, 17, 51
exploration, 17, 51
exploratory data analysis, 182, 227
extreme gradient boosting, 280

F-measure, 59
F1-score, *see* F-measure
factor, 44, 75
fairness, 25, 27, 28, 53, 302
false negative, 55
false positive, 55
feature, 47
feature engineering, 7, 53, 63
feature importance analysis, 274
feature importance method, 339
feature importance plot, 339
feature randomization, 273
feature scaling, 143
federated learning, 302, 333
feedback, 16
few-shot learning, 334
fine-tuning, 21
fitted value, 94, 118
forward pass, 304
Francis Galton, Sir, 72
Frank Rosenblatt, 8, 298, 302
fraud detection, 299
function, 36
svm_linear(), 234
accuracy(), 156
aes(), 79, 111
as_factor(), 208, 286
as_tibble(), 111, 123, 145, 205, 311
augment(), 102, 125, 134, 153, 166, 189, 199, 235, 264, 314, 323
bake(), 160
cbind(), 207, 286
clean_names(), 76
coefficients(), 115
colnames(), 207
conf_mat(), 154, 190, 236, 265
confint(), 97, 121
count(), 159
crossing(), 194
data(), 144, 250
data.frame(), 129
dim(), 180
enframe(), 94, 115
expand.grid(), 88, 318
export(), 31, 41
factor(), 77
file(), 41
filter(), 145, 259
finalize_workflow(), 134, 199, 323
fit(), 102, 125, 152, 189, 235, 313
geom_abline(), 86
geom_boxplot(), 183, 229
geom_point(), 79, 112
ggpairs(), 181, 227
ggplot(), 79, 111
glimpse(), 74, 110, 144, 250, 310
glm(), 148
gzfile(), 41
head(), 102, 126, 134, 179, 207, 314, 323
image(), 203, 284
import(), 31, 41
initial_split(), 100, 124, 151, 187, 233, 255, 312
install.packages(), 36
juice(), 160
labs(), 113

levels(), 146
library(), 36
lm(), 93, 114
load(), 40
logistic_reg(), 152, 165
mean(), 37
metric_set(), 131, 157, 191, 201, 237
metrics(), 103, 126, 135, 315, 324
mutate(), 77, 232
na.omit(), 225
nearest_neighbor(), 188
optim(), 92
plot(), 95, 119
poly(), 115
predict(), 150
prep(), 160
rand_forest(), 276
read.csv(), 39
read.delim(), 39
read_csv(), 40
read_excel(), 39
read_mnist(), 203, 283
read_tsv(), 40
readLines(), 41
recipe(), 152, 188, 234
relevel(), 147
rename(), 76
require(), 36
rotate(), 285
rpart.plot(), 257, 270
sample_n(), 205
save(), 40
select(), 111, 153, 166
select_best(), 133, 198, 322
set.seed(), 101, 124, 151, 188, 234, 312
show_best(), 132, 197, 321
step_downsample(), 160
step_dummy(), 101
step_normalize(), 152, 188, 234
step_poly(), 124
step_smote(), 163
step_upsample(), 162
step_zv(), 209
str(), 203, 283
str_to_title(), 77
summary(), 94, 119, 148, 180, 226
tail(), 208
testing(), 101, 124, 151, 187, 233, 255, 312
theme_classic(), 80, 113
training(), 101, 124, 151, 187, 233, 255, 312
tune(), 127, 192, 316
tune_grid(), 127, 131, 192, 196, 316, 320
update.packages(), 36
vfold_cv(), 130, 195, 319
workflow(), 102, 125, 152, 189, 235, 313
write.csv(), 40
write.table(), 40
write_csv(), 40
write_tsv(), 40
write_xlsx(), 40
writeLines(), 41
functional programming, 43
fuzzy logic, 6
Fβ-score, 60

Galton, 72
game playing, 15
GAN, *see* generative adversarial network
gaussian process, 51
GBM, *see* gradient boosting machine
GDPR, 25
generalization, 330
generalized linear model, 32, 105
generative adversarial network, 299
generator, 299
Geoffrey Hinton, 302, 303
Gini impurity, 245
Git, 34
GitHub, 36
global interpretability technique, 339
GOFAI, 5
GPT-3, 303
GPU, 24, 301, 303
gradient boosting, 268
gradient boosting machine, 280

gradient descent, 49, 51, 54, 90, 280, 305
 adaptive moment estimation (adam), 307
 batch gradient descent, 91
 mini-batch gradient descent, 91
 momentum, 307
 stochastic gradient descent, 91, 307
grid search, 87

Harrison, David Jr., 109, 309
heteroscedasticity, 104
hidden unit, 317
HIPAA, 25
homoscedasticity, 72
hyperparameter, 50, 127, 306, 315
hyperparameter tuning, 50, 63, 127, 176, 192, 315, 331
 bayesian optimization, 51
 constraints, 52
 ensemble method, 51
 gradient-based optimization, 51
 grid search, 51
 hyperparameter space, 51
 random search, 51
hypothesis testing, 32

Ian Goodfellow, 303
ICE, *see* individual conditional expectation
IDE, 33
Ilya Sutskever, 303
image classification, 20
image recognition, 15
ImageNet, 9, 22, 303
imputation, 63, 143
inclusivity, 28
independent variable, 47, 54, 69, 111
individual conditional expectation, 339
inference, 282
inferential statistics, 32
information gain, 244
information leakage, 331
instrumental conditioning, 15
integrated development environment, 31
interpretability, 24, 26
interquartile range, 183, 228
IQR, *see* interquartile range

Jürgen Schmidhuber, 303

k-fold cross-validation, 130, 195, 319
K-Nearest Neighbors
 unweighted, 189
 weighted, 189
kappa statistic, 60
Karl Pearson, 73
kernel parameter, 49
kernel trick, 223
knowledge, 9

L1 regularization, 337
L2 regularization, 336
labeled data, 49
lasso regression, 91, 337
LDA, *see* linear discriminant analysis
learning rate, 50, 90, 306
least squares method, 71
level, 75
leverage, 95
library, 31
light gradient boosting machine, 280
LightGBM, *see* light gradient boosting machine
LIME, *see* local interpretable model-agnostic explanations
linear discriminant analysis, 178
linear regression, 63, 69
 homoscedasticity, 72
 independence, 72
 linearity, 72
 multiple linear regression, 71
 normality, 72
 simple linear regression, 70
linearly separable, 221
local interpretability technique, 339
local interpretable model-agnostic explanations, 339
log-odds, *see* logit
logistic function, 141

logistic regression, 63, 139
logit, 140
long data, 44
Long Short-Term Memory, 303
loss function, 11, 304
LSTM, *see* Long Short-Term Memory

machine learning, 3
  applications, 22
  approaches, 10
  ecosystem disruption, 28
  environment, 24, 28
  ethics, 25, 26
  history, 8
  legal concerns, 25
  limitations, 24
  practice, 47
  social impact, 27
majority voting, 274
Manhattan distance, 176
margin, 223
Marvin Minsky, 302
Matthews Correlation Coefficient, 62
maximum likelihood estimation, 141
mean squared error, 54, 90, 304
meta ensembling, *see* stacking
meta feature, 329
meta learner, 329
meta learning, 334
meta model, 329
Minkowski distance, 176
ML, *see* machine learning
MLE, *see* maximum likelihood estimation
model, 48
model deployment, 10
model fitting, 10, 49
model parameter, 49
model selection, 10
MSE, *see* mean squared error
multicollinearity, 182, 227, 336
multivariate analysis, 32
myelin sheath, 296

natural language processing, 9, 15, 19, 22, 23, 53, 62, 298, 299, 303, 308
negative punishment, 16
neural architecture search, 302
neural network, 7, 8, 24, 26, 49, 63, 295
  artificial, 298
  backward pass, 304
  biases, 304
  biological, 295
  challenges, 299
  computer, 295
  convolutional, 298
  deep, 298
  feedforward, 298
  forward pass, 304
  generative adversarial, 299, 303
  quantum, 300
  recurrent, 298
  spiking, 299
  weights, 304
neuron, 295
  interneuron, 296
  motor, 296
  sensory, 296
neuroplasticity, 298
neurotransmitter, 296, 297
NLP, *see* natural language processing
noise, 24
normal distribution, 73
normal equation, 91
normal Q-Q plot, 95
normalization, 142, 143
NoSQL, 53

object detection, 20
odds ratio, 140
open-source, 32
operant conditioning, 15
optimization, 49
optimizer, 90, 307
ordered boosting, 280
Ordinary Least Squares, 84
out-of-fold predictions, 331
outcome variable, 47
outlier, 183, 229

overfitting, 13, 24, 49, 52, 54, 108, 242, 248, 267, 280, 330, 336
oversampling, 158

p-value, 94, 118
package, 31, 35
- palmerpenguins, 223
- bookdown, 34
- broom, 102, 125, 134, 153, 166, 189, 199, 235, 264, 314, 323
- brulee, 317
- datasets, 179
- dplyr, 36, 44, 74, 76, 77, 111, 145, 159
- dslabs, 203, 283
- forcats, 44
- GGally, 181, 227
- ggplot2, 32, 36, 44, 79, 111
- HistData, 74
- infer, 64
- janitor, 76
- kernlab, 224
- lubridate, 43, 44
- MASS, 109, 310
- mlbench, 144
- parsnip, 63, 188, 234, 276
- purrr, 44
- Rcpp, 33
- readr, 44
- readxl, 39
- recipes, 63, 160, 162, 163
- rio, 31, 41
- rpart.plot, 257, 270
- rsample, 64, 100, 124, 151, 187, 233, 255, 312
- Shiny, 33, 35
- stringr, 36, 43, 44
- themis, 144, 160, 162, 163
- tibble, 44, 111, 145
- tidyr, 43, 44
- tidyverse, 31, 40, 42
- torch, 317
- tune, 63, 132, 197, 321
- workflows, 64
- writexl, 40
- yardstick, 63, 156, 191, 265

pairs plot, *see* scatterplot
pairwise plot, *see* scatterplot
Palmer Penguin, 223
paperclip maximizer, 27
parallel processing, 332
partial dependence plot, 339
PCA, *see* principal component analysis
penalty, 318, 337
perceptron, 8, 298, 302
Perceptrons (book), 302
performance metrics, 54
permutation importance, 339
pipe, 31, 43
policy, 17
polynomial function, 107
polynomial kernel, 223
polynomial regression, 107
positive reinforcement, 16
post-pruning, 248
pre-pruning, 248
pre-training, 21
precision, 48, 54, 58
prediction, 47
predictor variable, 47
pretext task, 19
principal component analysis, 218
privacy, 22, 25–27, 53, 302
proactive interference, 22
pruning, *see* tree pruning
punishment, 16

quantum machine learning, 334

R, 31, 32
R Markdown, 34
R notebook, 35
R project, 35
R script, 35
R-squared value, 94, 118
radial basis function kernel, 223
random forest, 63, 272
RBF, *see* radial basis function kernel
recall, *see* sensitivity
receiver operating characteristic, 169
rectified linear unit, 306

recurrent neural network (RNN), 303
recursive partitioning, 243
regression, 10–12, 48, 49, 54
regression analysis, 32
regression coefficient, 94, 118
regularization, 24, 54, 91, 281, 316, 336
regularization parameter, 50, 91
reinforcement learning, 15, 333
relational database, 53
ReLU, *see* rectified linear unit
resampling, 158
reshaping data, 44
residual, 84, 94, 118
residuals vs fitted plot, 95
residuals vs leverage plot, 95
retraining, 13
reward, 16
ridge regression, 91, 336
RL, *see* reinforcement learning
RNN, 298
Robert Gentleman, 32
robustness, 18, 25
ROC, *see* receiver operating characteristic
Ronald A. Fisher, 177
Ronald Williams, 302
Ross Ihaka, 32
RStudio, 31, 33
Rubinfeld, Daniel L., 109, 309
rule extraction method, 339
rule-based system, 5, 8

S3, 43
scale-location plot, 95
scatterplot, 181, 227
script, 31
security, 25, 53
self-attention, 303
semi-supervised learning, 15
sensitivity, 57
sentiment analysis, 22, 23, 299
Sepp Hochreiter, 303
Seymour Papert, 302
SHAP, *see* SHapley Additive exPlanations
SHapley Additive exPlanations, 339
Siri, 299
SMOTE, 144, 158, 161, 163
snake case, 76
soma, 296
specificity, 58
speech recognition, 22
stacked generalization, *see* stacking
stacking, 268, 329
standardized residual, 95
stopping condition, 247
Subversion, 34
sum of the squared residuals, 84
supervised learning, 11
support vector, 49, 223
support vector machine, 9, 49, 63
SVM, *see* support vector machine
symbolic AI, 5
symbolic method, 8
synapse, 296
synaptic bouton, 296

target, 47
test set, 54
testing, 12, 54
tibble, 40, 123, 145, 311
tidy data, 44
tidymodels, 62, 99, 122, 150, 186, 232, 311
    infer, 64
    parsnip, 63
    recipes, 63
    rsample, 64
    tune, 63
    workflow, 64
    workflows, 64
    yardstick, 63
tidyverse, 42, 111
time series analysis, 32
token, 16
TPU, 24, 301
training, 10, 48, 54
training set, 52
transfer learning, 20, 301
transformer, 303
transmutation, 76
transparency, 18, 24–26, 28, 301

tree depth, 256, 269
tree pruning, 267
trolley problem, 27
true negative, 55
true negative rate, *see* specificity
true positive, 55
true positive rate, *see* sensitivity
TSV, 31, 39
tuning, 10

unbiased, 87
underfitting, 52
undersampling, 158, 159
unlabeled data, 49
unsupervised learning, 13
update rule, 90
upsampling, 161
utilitarianism, 27

V-fold cross-validation, 130, 195, 319
validation, 50, 52
  hold-out, 52
  stratified, 52
validation set, 52
vanishing gradient problem, 303
variance, 338
version control, 34
vignette, 37
virtual assistant, 299
virtual reality, 299
Vladimir Vapnik, 9
voting, *see* majority voting

Walter Pitts, 300, 302
Warren McCulloch, 300, 302
weight, 49
weight decay, 338
weighted K-Nearest Neighbors, 189
weighted least square, 99
wide data, 44
WKNN, *see* weighted K-Nearest Neighbors

XGBoost, *see* extreme gradient boosting
XOR problem, 302

For Product Safety Concerns and Information please contact our EU representative GPSR@taylorandfrancis.com Taylor & Francis Verlag GmbH, Kaufingerstraße 24, 80331 München, Germany

Batch number: 10733633

Printed by Printforce, the Netherlands